AERODYNAMICS OF THE
HELICOPTER

ALFRED GESSOW
GARRY C. MYERS, JR.

FREDERICK UNGAR PUBLISHING CO.
NEW YORK

DEDICATED TO MORRIS AND EMMA GESSOW

Third printing

Republished 1967 by arrangement with Alfred Gessow
and the estate of Garry C. Myers, Jr.

Printed in the United States of America

Library of Congress Catalog Card Number: 67-26126

PREFACE TO THE THIRD PRINTING

Since the first printing of *Aerodynamics of the Helicopter* in 1951, much has happened to the helicopter to justify the faith of the early enthusiasts who predicted a great future for the ungainly, noisy, vibrating aircraft which could barely lift their own weight on a hot day. The helicopter has fulfilled amply those expectations by proving its worth in a multitude of commercial and military tasks. It is especially gratifying that even in military operations, the helicopter has served primarily in a constructive and lifesaving capacity.

The impressive list of accomplishments achieved by the modern helicopter is obviously the result of marked improvement over the early models produced during the period when the text for this book was written. Why, then, republish the book in its original edition? To put the question in another way, why was it not fully revised?

Actually, the new printing was produced in response to numerous requests from engineers, professors, and students who were not able to obtain copies of the earlier printings. Obviously then, the basic treatment in the text of the various facets of helicopter aerodynamics is fundamental and just as valid for today's helicopters as it was for the earliest versions. The text was not revised for this reason and for fear of tampering with what experience has shown to be a successful format. Although much new material could have been added to make the book more complete, it might have been at the expense of simplicity of treatment. With a knowledge of the existing material, the student is well prepared to understand the current technical periodical literature which truly represents the forefront of knowledge in any technical field.

The new printing should prove useful also to the many engineers who are concerned with the design, development, and testing of the score of different VTOL configurations which have arisen during the past fifteen or so years. In spite of the great advances in aeronautical technology, particularly in power plants, which have taken place during that time, it is safe to say that a VTOL aircraft must incorporate a low or moderate disk loading rotor in the low speed end of its flight range in order to justify its dual-role complexity. In short, helicopter characteristics are required for successful VTOL operation, and helicopter principles are the basis of a good VTOL design.

I deeply regret that the untimely death of Garry C. Myers, Jr., prevented him from taking part in the many exciting helicopter developments that have taken place and are yet to come. He made the most of the years that he had, and the helicopter community, as well as his family and friends, have benefited from them.

Alfred Gessow

Washington, D. C.
May, 1967

PREFACE

Aerodynamics of rotating-wing aircraft, as the subject stands today, is the result of more than twenty years' work by many distinguished investigators such as Glauert, Lock, and Wheatley. While technical knowledge in many aspects of helicopter engineering is limited, aerodynamic theory has been reasonably well established over the years by tunnel and flight tests and has proved useful in the design and development of present-day helicopters.

Because the solutions of many problems connected with the design of helicopters—in areas of performance, vibration, stability, and stress—demand a clear understanding of fundamental aerodynamic principles, the authors felt that a significant contribution to the field could be made by presenting clearly and logically the aerodynamics of the helicopter as developed to date.

This book was written as a text for senior and graduate engineering students and engineers in the helicopter industry who are interested in obtaining a more thorough understanding of the rudiments of helicopter aerodynamics.

The greater part of the authors' training and experience in helicopters has been gained in the Flight Research Division of the Langley Aeronautical Laboratory of the National Advisory Committee for Aeronautics. The vast background of experimental and theoretical rotor work (comprising over seventy published papers) done by the NACA during the past fifteen years, served as a sound basis for the aerodynamic material developed in the book.

In presenting the material the authors have constantly endeavored to give physical concepts for the phenomena associated with rotating wings. Theory is developed in its most elemental form, refinements being added after students become familiar with the rudimentary material. Lengthy mathematical formulas have been avoided except where they are of fundamental significance. The aim has been to develop basic expressions to an extent where they are generally applicable and may be readily modified by the student to apply to specific design problems.

A word of explanation may be in order concerning the arrangement of material. In order that the aerodynamic theory would have greater significance to students who are unfamiliar with practical aspects of the helicopter, a chapter devoted to its mechanism and its general characteristics precedes the aerodynamic treatment. Readers are then introduced to aerodynamic theory through an analysis of hovering flight. In hovering, the underlying principles may be mastered without encountering the added complications that are present in forward flight analyses—complications arising from the variation in velocity around the disk. Similarly, the phenomenon of autorotation is first presented for the vertical flight condition.

To provide an understanding of the phenomena associated with forward flight, a thorough discussion of the physical concepts of flapping and feathering is presented. Basic force and moment equations which are then developed for the rotor in forward flight lead into a simple and logical method of performance prediction. Though extremely easy to use, this method is believed to be the most refined yet published and is shown to predict helicopter performance accurately. The theory is also shown to apply equally to autorotation and to powered level flight, the former being considered simply as the special case of zero shaft power. In all cases, theory is substantiated by sound experimental evidence.

The effects on performance of the various design parameters, such as disk loading, rotor solidity, and blade twist, are discussed in order to provide the student with an understanding of the aerodynamic side of design compromises. A thorough discussion is also given of the limiting effects of blade stall and compressibility on high-speed flight. Means are shown by which designers can postpone these limits and achieve high forward speeds. Finally, the physical aspects of helicopter stability and vibration are covered in detail in separate chapters to give students an understanding of the parameters that influence flying qualities of the helicopter, and of periodic forces and moments that excite vibrations in rotating-wing aircraft. Numerous figures and diagrams are given to illustrate the text.

A complete list of all NACA publications relating to rotating-wing aircraft is given in Appendix IIA. These papers, when used as references in the text, are referred to by Roman numerals followed by a number (e.g. reference IV–3). A representative list of other than NACA helicopter papers is contained in Appendix IIB. These papers are referred to in the text by a number alone (e.g. reference 24).

The authors wish to express their admiration and indebtedness to Mr. F. B. Gustafson, head of helicopter flight research at the Langley Laboratory, who, as supervisor, freely and patiently shared his understanding of engineering principles and extensive knowledge of rotating-wing aircraft. Their thanks are due also to Mr. T. E. McCorkle for his patience and efforts in illustrating the text.

Alfred Gessow

Garry C. Myers, Jr.

Hampton, Virginia

CONTENTS

1

THE DEVELOPMENT OF ROTATING-WING AIRCRAFT

The story of the helicopter is the story of the life works of men throughout the last five centuries. Early experimenters were seeking in the helicopter a means by which man might achieve flight. At the beginning of the twentieth century, however, before any successful rotating-wing aircraft was devised, man achieved flight in fixed wing, or "conventional" aircraft. Engineering effort concentrated on developing the fixed-wing aircraft until today the airplane has been developed to the point where it is one of the world's most important means of transportation.

Notwithstanding the development of the conventional airplane, men have been aware that they had still to achieve complete mastery of the air; namely, the ability to stay aloft without maintaining forward speed and to ascend and land vertically in restricted areas. Development of the helicopter continued to this end.

Three fundamental problems plagued experimenters: (1) keeping structural weight and engine weight down to the point where the machine could lift itself and some useful load; (2) counteracting rotor torque; and (3) controlling the machine in flight. The structural and engine-weight problems were responsible for the slow progress in early experiments. In recent years the torque and control problems predominated, resulting in a number of configurations of rotors and infinite variations of basic types.

About 1926 a Spaniard, Juan de la Cierva, produced a successful rotating-wing machine which employed a propeller for forward motion, as in an airplane, and a freely rotating rotor for lift. This aircraft he called the *autogyro*. While the autogyro was still not a direct-lift machine it required only small forward speeds to maintain its lift and could take off and land in extremely short distances. Autogyro development continued in Europe and in America until by 1936 the art had reached a state of considerable advancement. The economic depression of that period, however, together with the overpublicity of the machine in its early stages of development, brought progress almost to a standstill.

Paralleling autogyro development, progress was being made toward a successful helicopter. By 1937 Focke in Germany demonstrated a successful side-by-side, two-rotor machine, and in 1941 Sikorsky introduced the VS–300, a single lifting rotor machine with a vertical tail rotor for torque counteraction. Sikorsky's success gave tremendous impetus to helicopter development in this country and, in the few years that followed, at least forty serious and independent developments sprang up. It is interesting to note that for the most part the helicopter was developed independently from the autogyro, and many phases of the rotating-wing art which were quite familiar to autogyro engineers were completely relearned in helicopter development. Indeed, autogyros still compete with helicopters in regard to rotating-wing aircraft speed records, and it is only recently that helicopters have been built which surpass the autogyros of the nineteen thirties in their ratio of useful load to total weight.

Chronological Development of the Helicopter

In order to present a comprehensive picture of helicopter development, the experiments of some of the pioneers in the field are listed below. The list is by no means complete, for it is intended only to point out general trends in development.

B.C. It is known that before the days of the Roman empire the Chinese constructed "Chinese Tops." The top consisted of a propeller on a stick which was spun between the hands (Fig. 1–1). The "Chinese Top" probably represented the world's first helicopter.

15th century. *Leonardo da Vinci* considered the possibility of flight using a screw-type propeller to attain vertical lift (Fig. 1–2).

1796. *Sir George Cayley, England,* constructed several successful models driven by elastic substances such as whalebone and clock springs (Fig. 1–3). One rose to a height of 90 feet.

1842. *W. H. Phillips, England,* constructed a steam-driven model helicopter weighing 20 pounds. He proposed a full-sized machine to be fitted with three propellers—one to lift and two to steer the aircraft. The machine was never built.

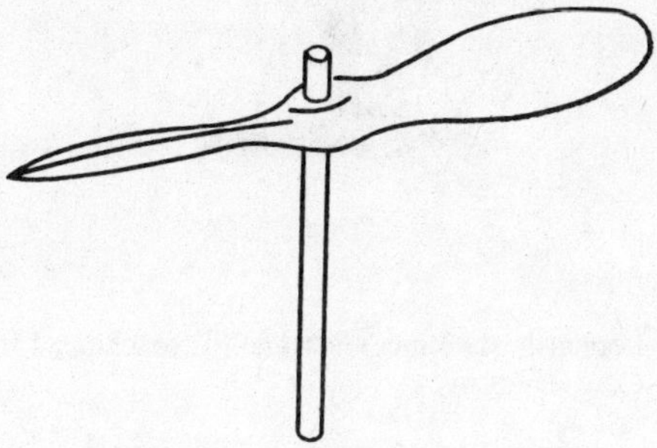

Fig. 1–1 The "Chinese Top."

1878. *Enrico Forlanini,* Italian professor of civil engineering, built a flying steam-driven model weighing only 7.7 pounds.

1880. *Thomas Edison, United States,* experimented with models. He built a test stand and tested several propellers, driven by an electric motor. He realized that what he needed was a lightweight engine which would produce a large amount of power. He experimented with an engine using guncotton for fuel but abandoned the project after a serious explosion in his laboratory.

1907. *Paul Cornu, France,* constructed a machine which carried a pilot aloft. The airframe consisted of a single beam with a rotor shaft at either end. Power was supplied from a 24-horsepower motor by belts to the two rotors, which rotated in opposite directions. The rotors had two blades of light, fabric-covered construction which were weighted at their two-thirds radius point to help centrifugal forces balance lift forces. Control was to be achieved by the reaction of rotor

Fig. 1–2 Leonardo da Vinci's vertical-lift machine, 15th century. Courtesy NACA.

Fig. 1–3 Sir George Cayley's helicopter and airplane, 1796. Courtesy NACA.

downwash on airfoils suspended below the rotor. The machine never flew untethered.

1908–1929. *Emile and Henry Berliner, United States.* Father and son spent most of their lives on helicopter development. In 1909 they

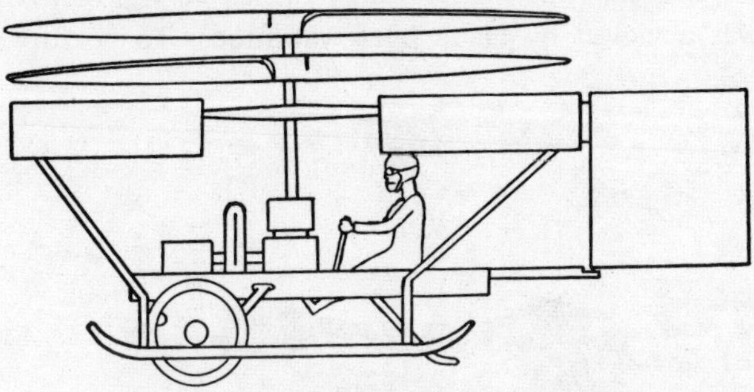

Fig. 1–4 Berliner coaxial helicopter, 1909.

built a two-engine craft with counterrotating rotors which lifted a pilot untethered (Fig. 1–4). Later they built side-by-side rotors over wings (Fig. 1–5). The rotors were rigid wooden propellers. Control was

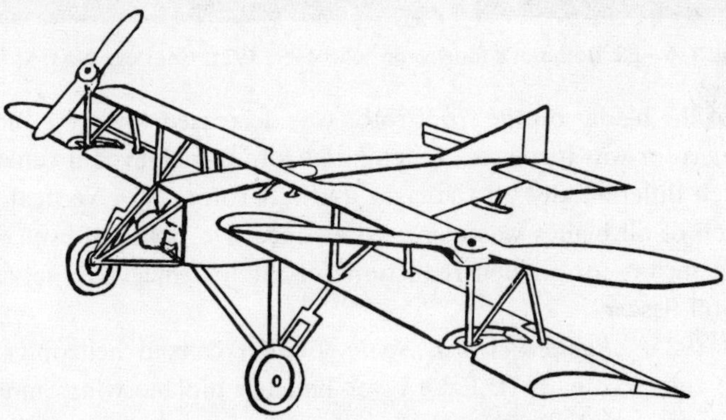

Fig. 1–5 Berliner side-by-side helicopter.

achieved by tilting the entire rotor with respect to the fuselage. The machines achieved limited success in hovering and flying slowly forward.

1921. *De Bothezat, United States,* built one of the largest helicopters of the time (Fig. 1–6). The machine had four rotors carried at the ends of intersecting beams. It flew at very limited altitudes at over 4000 pounds gross weight and carried three passengers. Power was produced by a 180-horsepower engine located at the intersection of the beams. Each rotor had six wide blades which turned very slowly. Complete control was achieved by blade pitch variations. To go forward, the

Fig. 1–6 De Bothezat's four-rotor helicopter, 1921. Courtesy NACA.

pitch of the blades on the front rotor was decreased while the pitch of the rear rotor was increased. Lateral control was achieved by changing the pitch differentially on the right and left rotors. For vertical flight the pitch of all blades was increased at once. De Bothezat even incorporated means for sudden reductions of pitch to negative values for power-off descent.

1919–1925. *Raoul Pescara, Spain*, built a coaxial helicopter with biplane rotors (Fig. 1–7). Each rotor had ten biplane wings mounted rigidly to the shaft. Pescara also employed a free wheeling device and negative pitch settings for power-off flight. Control was achieved through cyclic-pitch change obtained by warping the blades periodically as they rotated.

1924–1929. *Von Baumhauer*, a Dutch scientist, built the first single-rotor helicopter with a vertical tail rotor for torque counteraction.

(Such a design was patented in the United States by Emil Berliner in 1923 but was never built by him.) The fuselage consisted essentially of a tubular truss on one end of which was mounted a 160-horsepower engine. The other end carried a smaller 80-horsepower engine mounted at right angles to the truss and which turned a conventional propeller. The main rotor had two 25-foot blades with about 10 degrees of twist.

Fig. 1–7 Pescara coaxial helicopter, 1925. Courtesy NACA.

Blades were free to flap but were restrained by cables in such a way that when one blade flapped upward the other moved downward. Control was achieved by a swash-plate, cyclic-pitch mechanism, much like the modern single-rotor helicopter. The engine which drove the tail rotor was in no way connected with the main rotor, which caused difficulties in achieving precise directional control. The machine made numerous flights at a gross weight of about 2000 pounds but never rose more than a foot or two above the ground. The project was abandoned after a bad crash in 1929.

1930. *Dr. d'Ascanio, Italy*, built a coaxial helicopter which was very successful for its time and which held helicopter records for several years. The machine had two superimposed, two-bladed, counter-rotating rotors. The blades were pivoted at the root, free to flap and change pitch. Control was achieved by servo-tabs on the blades which were deflected periodically by a system of cables and pulleys. The tabs cyclically changed the pitch of the entire blade. For vertical flight, the tabs moved together so as to increase or decrease the pitch of all blades. The machine flew 3000 feet in five minutes, remained aloft for almost nine minutes, and achieved an altitude of 54 feet.

1930. *M. B. Bleeker, United States*, solved the torque problem by delivering power to propellers mounted on each blade. Power was supplied through an intricate system of gearing from an engine mounted in the center of the machine. The aircraft (Fig. 1–8) was controlled by

Fig. 1–8 Curtiss-Bleeker helicopter, 1930. Courtesy NACA.

auxiliary surfaces fastened to each blade as well as by a surface on the tail of the machine.

1930–1936. *Rene Breguet, France*, made notable advances in coaxial rotor development. He built a machine with two 54-foot diameter rotors (Fig. 1–9). Each rotor had two blades which were mounted with flap and lag hinges and were controlled in cyclic pitch. Directional control was achieved by increasing the torque on one rotor with respect to the other, which resulted in turning the fuselage. The rotor linkage was arranged so that the pitch of the blades decreased as the blades flapped upward, thus minimizing the flapping motion and helping avoid the possibility of the blades of one rotor hitting the other. The blades were highly tapered in plan form and thickness.

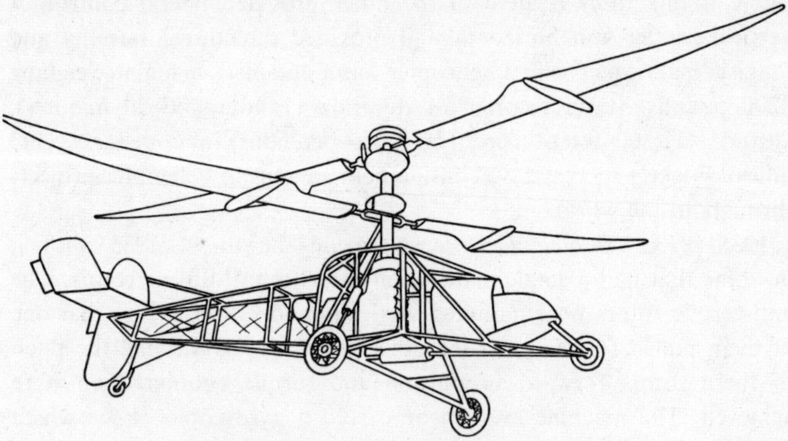

Fig. 1–9 Breguet coaxial helicopter, 1936.

1937. *Dr. Heinrich Focke, Germany,* built a successful machine using two rotors, side by side, rotating in opposite directions (Fig. 1–10). The rotors were inclined slightly inward to provide dihedral stability, just as is done with the wings of fixed-wing aircraft. The blades were

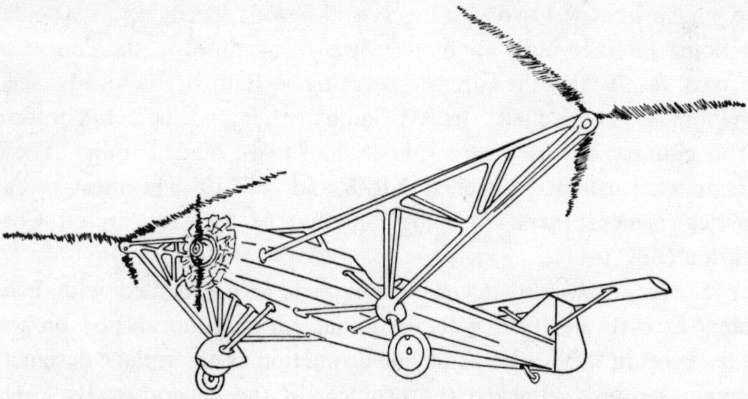

Fig. 1–10 Focke-Achgelis side-by-side helicopter, 1937.

tapered and attached at the root by both flapping and lagging hinges. Longitudinal control was achieved by tilting the rotors forward by means of a swash plate, cyclic-pitch mechanism, while directional control was gained by tilting the rotors differentially. Increasing the pitch of one rotor relative to the other provided lateral control. A vertical rudder and horizontal tail provided directional stability and longitudinal trim. The first helicopter was a one-place machine weighing 2200 pounds. It set records for duration (1 hour and 20 minutes), altitude (11,200 feet), speed (75 miles per hour), and distance (143 miles). Focke's success gave considerable impetus to helicopter projects throughout the world.

1935–1943. *Antoine Flettner, Germany,* began in 1935 with a machine that had a single main rotor and two anti-torque rotors. The anti-torque rotors were mounted vertically and positioned so that one of them pulled forward and the other rearward. By varying the pitch of these rotors forward propulsion and torque counteraction were achieved. The machine even incorporated a gyroscopic device which relieved the pilot of directional control coordination with changes in engine power. In 1937 Flettner abandoned the idea in favor of a side-by-side configuration with extreme intermeshing, which later became known as the *synchropter* type. The rotor shafts formed an angle of 24 degrees, with their center lines intersecting at a point below the fuselage. Each rotor had two blades and turned in opposite directions. The machine carried two passengers and flew at speeds up to 90 miles per hour. Flettner built about twenty-two machines in the course of the next few years, the aircraft reaching a state of quite advanced development. The Kellett Aircraft Company adapted the configuration in this country but used three-, instead of two-, bladed rotors. These aircraft were called the Kellett XR–8 and XR–10. The most recent American synchropters are being built by the Kaman Aircraft Corporation (Fig. 1–11).

1941. *Igor Sikorsky, United States,* had experimented with helicopters as early as 1907, with his initial efforts concentrated on the coaxial type. In 1938, after achieving distinction as an airplane designer, he again seriously attacked the problem of the helicopter. By 1941, Sikorsky had produced the VS–300, which embodied one main rotor and three auxiliary rotors, one vertical and two horizontal, at the tail of the machine. By the middle of 1941 Sikorsky had broken Focke's endurance record by remaining aloft one hour and thirty-two minutes. From its initial configuration the VS–300 encountered many changes and finally emerged as the well-known main-and-tail rotor configuration of today. Sikorsky had successfully built a relatively simple, completely controllable helicopter. He had found the practical solution to the single-rotor problem which Baumhauer had wrestled with ten years before.

Fig. 1–11 Kaman K–190 utility helicopter, 1949. Courtesy Kaman Aircraft Corporation.

Longitudinal and lateral control was obtained in the initial VS–300 configuration by means of two horizontal tail rotors. These controls were incorporated into the main lifting rotor of later configurations, in which only the vertical tail rotor was retained out of the original three auxiliary rotors for anti-torque and directional control purposes. In the later configuration, longitudinal and lateral control was achieved by tilting the main rotor by means of cyclic-pitch control, while directional control was achieved by pitch variations of the tail rotor. The tail rotor was driven by shafting from the main rotor transmission, so that in case of power failure the main rotor continued to turn the tail rotor so as to maintain directional control.

The pilot's controls consisted of a pitch stick at his side which he raised up and down for vertical flight, a control stick in front of him which he pushed in the direction he wished to go, and rudder pedals which he operated with his feet for directional control. The pitch stick

Fig. 1–12 Sikorsky S–52–1 helicopter, 1949. Courtesy Sikorsky Aircraft Division, United Aircraft Corporation.

changed the pitch of all three blades for climb and descent. The handgrip on the pitch stick was the throttle control, rotation of which increased or decreased engine power. Motion of the pedals changed the pitch of the tail rotor, swinging the machine to the right or left. Motion of the control stick tilted the rotor swash plate, producing a cyclic-pitch change such as to tilt the rotor in the desired direction. These controls are relatively standard on all present-day helicopters.

The VS–300 grew rapidly into the XR–4 and the YR–4 Army helicopters, and finally into the R–4 and other single-rotor production helicopters of which Sikorsky made many hundreds for the Army and Navy during the Second World War. A modern, small, single-rotor Sikorsky helicopter is shown in Fig. 1–12.

The above outline is but a brief sampling of projects which were significant in helicopter development and no attempt has been made to discuss the details of the many present-day helicopters. In reviewing the history of the art one is impressed by the large number of completely separate approaches to the problem. It would appear that progress was made more in proportion to technological developments of the time than to the results of experimenters who built on one another's work. Almost every configuration of rotors imaginable, and many which are almost unimaginable, have been experimented with.

Autogyro Development

One phase of the history of the rotating-wing not dealt with in the preceding outline is the development of the autogyro. While it does not have all the properties of the helicopter, the autogyro involves basically the same problems of rotor design as the helicopter. Autogyro development, which began about 1920 and which reached considerable advancement by 1935, had a great deal to do with the advent of the successful helicopter.

The story of the development of the autogyro is primarily the story of Juan de la Cierva. Cierva was particularly interested in making a flying machine which could land and ascend without high forward speed and which could not stall and drop to earth if the pilot reduced speed excessively. To Cierva, stalling was a tremendous limitation to the airplane, and rather than improve the stalling characteristics of airplanes he chose to devise an inherently different type of lifting surface. With his own funds and with a grant from the Spanish Government he ran wind-tunnel experiments on model rotors and established many basic facts of rotor behavior. He found that a rotor tilted slightly back in a wind could produce a sizable lift even at a very low speed. He further discovered that best results were obtained when the blades of his "windmills" were set at low positive angles.

Cierva flew his first autogyro in 1923. The rotor was mounted above an airplane fuselage and acted simply as a wing, rotating in the wind and supporting the machine. An engine and propeller pulled the machine through the air as in a normal airplane. Control was achieved by conventional airplane surfaces which tilted the machine and changed the direction of rotor lift. Cierva built three machines before he achieved success. His third machine incorporated freely hinged blades which Cierva invented as a means of equalizing the lift on the two sides of his rotor in forward flight. Although the principle of flapping blades had actually been suggested by Renard, a Frenchman, in 1904, Cierva rediscovered and first applied the principle.

Fig. 1–13 Kellett YG–1B direct control autogyro. Courtesy NACA.

Early autogyros started their rotors turning by taxiing around the airport. Later, a geared connection with the engine was provided to bring the rotor up to speed.

Considerable credit is to be given to the Pitcairn and the Kellett Aircraft Companies in this country who, as licensees of the Cierva Company, coordinated with Cierva in further autogyro developments. By 1932, autogyros were developed in which control was achieved by tilting the rotor with respect to the fuselage and thus conventional aircraft surfaces became unnecessary. This step was a considerable advantage inasmuch as the conventional control surfaces were not very effective in changing the attitude of the machine at low speeds. In the so-called *direct control* machine (Fig. 1–13), the pilot actually tilted the axis of the rotor with the control stick. Directional control was still achieved by a rudder, as in an airplane.

Raoul Hafner in England advanced the art further by introducing cyclic-pitch control as a means of effectively tilting the rotor without encountering the undesirable forces which were transmitted through the rotor hinges in the direct control machine. Another interesting version of the autorotating rotor was introduced in the United States by E. Burke Wilford, who built a four-bladed gyroplane whose rotor differed basically from other rotors in that it was nonarticulated; that is, bending stresses in the spars were not relieved by hinges. Balance and control were achieved by feathering the blades substantially about their span axis.

The final phase of autogyro development was the introduction of the "jump take-off." This involved the overspeeding of the rotor on the ground with the blades set at zero pitch and the subsequent use of this stored energy to lift the machine into the air by a sudden increase in pitch. With jump take-off the autogyro closely rivaled the helicopter in flight characteristics. It was still unable, however, to approach a landing spot and back away if the spot appeared unsuitable.

2

AN INTRODUCTION TO THE HELICOPTER

Later chapters of this book will deal primarily with the behavior of the helicopter rotor in various conditions of flight. The fact that the rotor has a fuselage, source of power, and means of control will be taken for granted, and very little attention will be given to the details of mechanical design. The purpose of the present chapter is to give the reader a picture of the helicopter as a whole—its geometrical configurations, its means of control, its general design features, its performance characteristics, and its flying qualities. This will, it is hoped, provide a background and permit a clearer understanding of the following chapters.

Helicopter Configurations

Helicopter configurations may be classified into five main types and several subclasses. Each type has its unique characteristics, advantages, and disadvantages. These are discussed below.

THE SINGLE ROTOR. In terms of the number of machines in operation today, the *single-rotor machine* with tail rotor (Fig. 2–1) is by far the most common type. It has the advantage of being relatively simple—one rotor, one set of controls, one main transmission. While the tail rotor uses about 8 to 10 per cent of the engine power in hovering and 3 to 4 per cent in forward flight, the simplicity of the configuration and the saving in weight as compared with other means of torque counteraction probably compensate for this loss. One disadvantage is the danger of the vertical tail rotor to ground personnel, the whirling blades being behind the pilot and thus not under his precise control. The *gyrodyne*, a type of helicopter in which the torque counteracting rotor points forward, has the advantage of using the anti-torque rotor instead of the main rotor to pull the machine through the air. This results in more efficient operation of the main rotor in forward flight

Fig. 2–1 Bell H13–B single-rotor helicopter. Courtesy NACA.

since it avoids the tilting forward of the rotor and the accompanying radial dissymmetry in blade angle of attack. On the other hand, the gyrodyne torque rotor must be mounted on a relatively short arm in order to avoid excessive parasite drag, and the engine power required to counteract torque at the shorter moment arm is accordingly higher.

The *jet rotor* (Fig. 2–2) provides the simplest solution to the torque problem. The rotor torque is supplied by units at the blade tips rather than by shaft torque so that the fuselage may be simply supported on a bearing, the only torque transmitted to the fuselage being the bearing friction. Fuselage directional control may then be achieved by a vane or rudder which utilizes the rotor downwash in hovering and the air stream in forward flight. Jet thrust may be provided by tip units, as in the ram jet rotor, or by an engine-driven blower from which air is ducted to rearward-pointing nozzles at the blade tips. The jet rotor has the advantage of simplicity and small storage space and the disadvantage of high specific fuel consumption as compared with a conventional machine. Development will depend primarily on jet engine development. Ultimately, the jet helicopter may very well prove to be the most practical configuration.

Fig. 2–2 McDonnell "Little Henry" ram jet helicopter. Courtesy McDonnell Aircraft Corporation.

COAXIAL ROTORS. In the coaxial machine (Fig. 2–3), fuselage torque is eliminated by utilizing two superimposed rotors, rotating in opposite directions. These rotors may or may not have the same diameter or turn at the same speed. The only requirement is that they both absorb the same torque. The coaxial design has the advantage of having its over-all dimensions defined only by the rotor diameter and of a saving of power over the single rotor-tail rotor design. On the other hand, the rotor hubs and controls become more complex and rotor weights tend to increase.

SIDE-BY-SIDE ROTORS. The basic advantage of the side-by-side configuration (Fig. 2–4) is that the laterally displaced rotors effect a reduction in power required to produce lift in forward flight, similar to the aspect ratio effect on an airplane wing. This advantage becomes important in large multi-engine helicopters where standards require that level flight be possible with one engine dead, since the reduction in power necessary to maintain level flight in the side-by-side ship permits bigger loads to be carried. The configuration has the disadvantage of

Fig. 2–3 Bendix experimental coaxial helicopter. Courtesy NACA.

having either high fuselage parasite drag or high structural weight, for as the supporting pylons become thin and aerodynamically clean they become heavy. The supporting pylons, however, may act as lifting surfaces and unload the rotors in forward flight, effecting a sizable gain in efficiency at high speed. As compared with the single-rotor machine, the side-by-side configuration has the disadvantage of requiring relatively complex gearing and shafting. Its over-all dimensions are greater than the single-rotor machine, this depending, of course, on the degree of overlap. The *synchropter* (Fig. 1–11), in which rotors are intermeshed to the point of approaching the single rotor, sacrifices some lifting efficiency gains for compactness and transmission simplifications.

TANDEM ROTORS. The main advantage of the tandem configuration (Fig. 2–5) lies in its clean fuselage possibilities, together with a large available center-of-gravity range. The useful load may be distributed between the two rotors in varying proportions. Disadvantages in transmission and shafting weights are similar to the side-by-side configuration. One main disadvantage lies in the loss in lifting efficiency in forward flight, for just as the side-by-side configuration is more efficient

Fig. 2–4 McDonnell XHJD–1 side-by-side helicopter. Courtesy McDonnell Aircraft Corporation.

than a single rotor in this flight condition, the tandem configuration is less efficient than the single rotor because one rotor is working in the wake of the other. The loss in lifting efficiency in forward flight may be minimized by stagger, i.e., by placing the rear rotor above the front rotor.

Tandem designs also include variations in the relative size of the front and rear rotors. These dimensions are important from the point of view of forward flight stability and handling qualities.

MULTI-ROTORS. Helicopters with many rotors have been proposed for special uses and generally for large machines. Three or more rotors offer simplifications in control system design inasmuch as control in all directions may be achieved by simply increasing the thrust of one rotor relative to the others. For large machines, use of multi-rotors

Fig. 2–5 Piasecki HJP–1 tandem helicopter. Courtesy Piasecki Helicopter Corporation.

offers the further advantage of influencing a large mass of air without having blades of unwieldy dimensions.

Helicopter Control Methods

Having established the geometrical shapes of helicopters, it is well to gain an appreciation of the manner by which each type of machine is controlled in flight. The purpose of the following paragraphs is to discuss control methods, first from the over-all point of view of the forces and moments applied to the helicopter, and second, from the point of view of the levers which the pilot moves.

CONTROL REQUIREMENTS. To control completely the position and attitude of a body in space requires control of the forces and moments about all three axes. This involves six independent controls (Fig. 2–6). Thus, if the body drifts to the side, a force may be exerted to return it to its original position. If it rolls over, a moment may be exerted to right it again. It would be exceedingly difficult, however, for a man to coordinate the controls of any machine having six independent control systems. Fortunately, it is possible to reduce this number by coupling

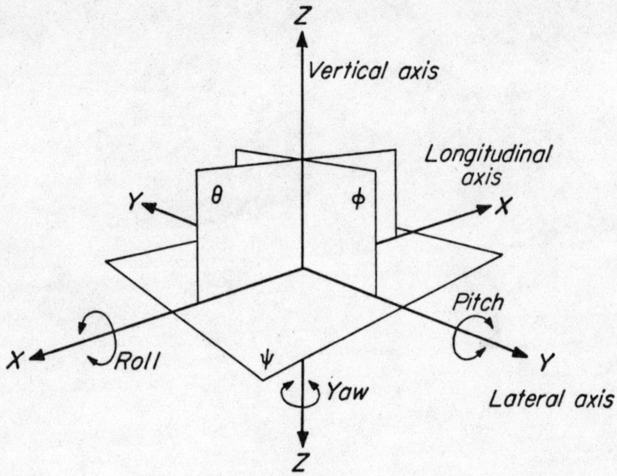

Fig. 2–6 System of control axes.

together independent controls. Such couplings involve some sacrifice of complete freedom of control of position and attitude in space, but the sacrifice may actually be desirable.

The pilot of the helicopter, for example, does demand the ability to produce moments about all axes in order to right himself as when disturbed by a gust. He does not, however, demand that he be able to produce moments (a pitching moment, for example), without producing an accompanying force—in this case in the longitudinal direction. He therefore sacrifices the ability to maintain force equilibrium, as in hovering, and to rotate his fuselage in pitch at will so as to attain a desired attitude. By thus coupling pitching moments with longitudinal forces the necessity for one of the six independent controls is eliminated.

Actually, four independent controls are adequate for the helicopter. These are discussed and illustrated below.

(1) *Vertical control.* This is necessary to fix the position of the helicopter in the vertical direction. It is achieved by increasing or decreasing the pitch of the rotor so as to increase or decrease the thrust.

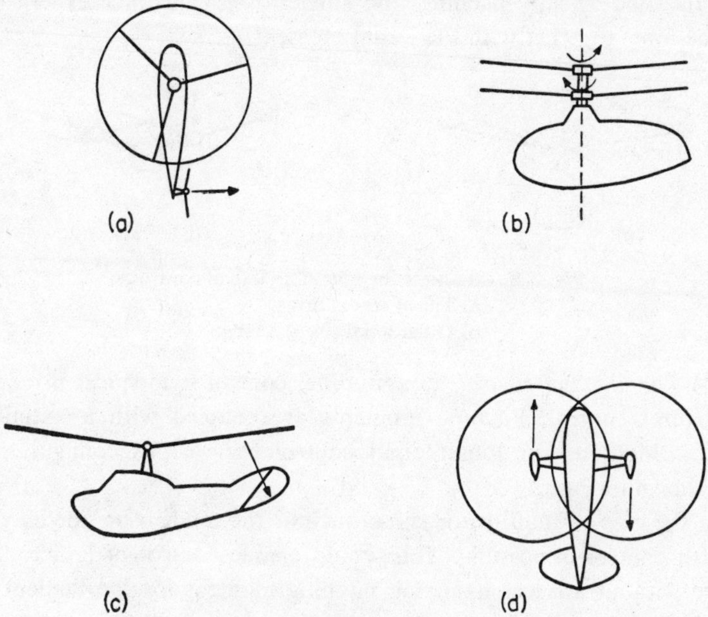

Fig. 2–7 Means for achieving directional control.
(a) Anti-torque rotor
(b) Differential torque between two rotors
(c) Vane in rotor slipstream
(d) Differential tilt of rotor thrusts

(2) *Directional control.* Directional control fixes the attitude of the helicopter in rotation about the vertical axis, permitting the pilot to point the ship in any horizontal direction. Means for achieving directional control are shown in Fig. 2–7. (Note that moment control is not basically coupled with force control about the vertical axis.)

(3) *Lateral control.* Lateral control involves the application of both moments and forces. When the pilot applies lateral control a rolling moment is produced about the aircraft center of gravity which tilts the

helicopter. As a consequence of the tilt, a component of the rotor thrust vector acts in the direction of tilt. The application of lateral control has therefore resulted in a tilt and sideward motion of the helicopter. Methods for obtaining lateral control are shown in Fig. 2–8. Note that while the initial effect of lateral control is a pure moment for the side-by-side machine, the single-rotor helicopter experiences a side force together with the initial moment.

Fig. 2–8 Methods for obtaining lateral control.
(a) Tilt of rotor thrust
(b) Differential thrust change

(4) *Longitudinal control.* Longitudinal control is identical in nature to lateral control. Pitching moments are coupled with longitudinal forces. Methods for longitudinal control for various configurations are shown in Fig. 2–9.

In the case of multi-rotor ships, such as the tandem or side-by-side, a fifth control is possible. This would enable control of longitudinal force without an accompanying pitching moment for the tandem, or control of the side force without accompanying rolling moments for the side-by-side. Rather than introduce this fifth control, force control is usually coupled directly with moment control as described above. A fifth control for longitudinal trim (moment which is independent of horizontal force) may be available in the tandem by differentially adjusting the pitch of the two rotors just as the horizontal tail is trimmed in the single-rotor or side-by-side machine.

Cross effects are, in general, undesirable. For example, in the single-rotor machine an increase in vertical force necessitates an increase in rotor torque so that a correction is required in directional control to maintain the fuselage direction. Such cross effects necessitate considerable coordination on the part of the pilot and result in longer periods of training in order to control the machine.

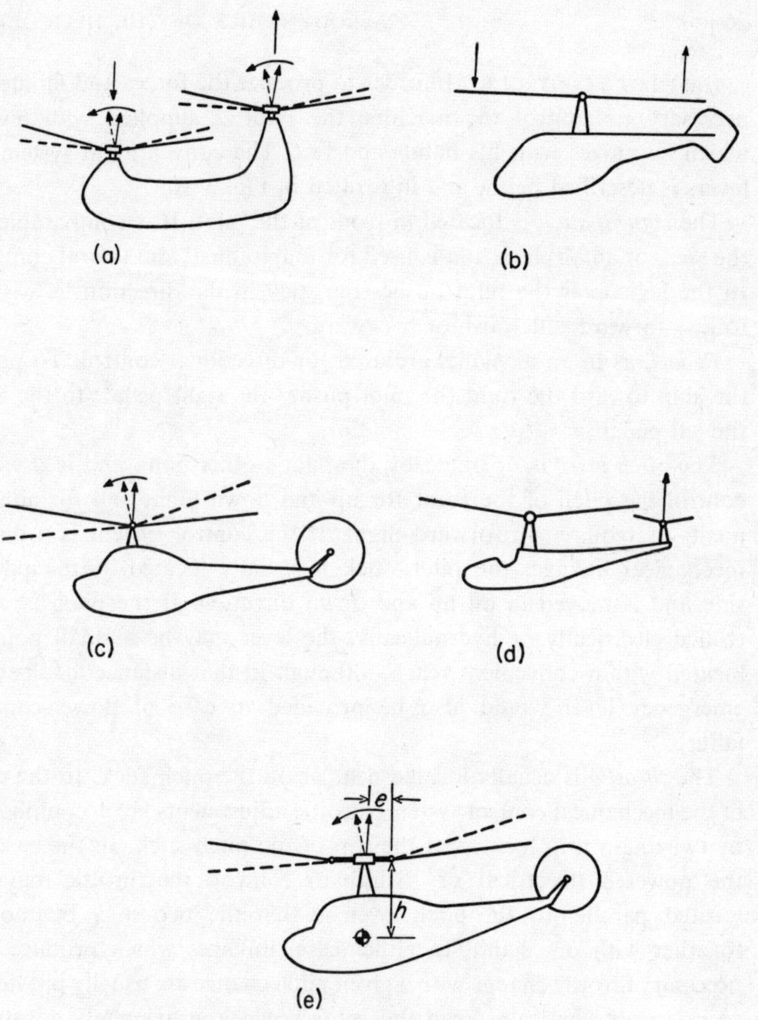

Fig. 2–9 Methods for obtaining longitudinal control.
(a) Differential thrust change
(b) Thrust variation around azimuth
(c) Tilt of rotor thrust
(d) Horizontal tail rotor
(e) Tilt of rotor thrust coupled with offset-flapping hinges

THE PILOT'S CONTROLS. In order to produce the forces and moments necessary to control the machine the pilot is supplied with levers which he moves with his hands and feet. The conventional system of levers is described below and illustrated in Fig. 2–10.

The *control stick* is located in front of the pilot. It is comparable to the stick of an airplane and is used for longitudinal and lateral control. In the helicopter the pilot pushes the stick in the direction he wishes to go—forward, sideward, or backward.

Pedals, as in an airplane, are used for directional control. To point the ship toward the right the pilot pushes the right pedal; to the left, the left pedal.

The *pitch lever* is operated by the pilot's other hand and is used to control the pitch of the rotor for up and down flight and for adjustments as required in forward flight. If the control system is a direct mechanical linkage, the pitch stick is usually located at the pilot's side and is moved in an up and down direction. If the pitch is controlled electrically or hydraulically, the lever may be a small pointer located within convenient reach, although in that instance a full-sized emergency lever would also be provided in case of power-control failure.

The *throttle* is usually located near or on the pitch stick. In the case of the mechanical control system, throttle adjustments are accomplished by twisting a grip located at the top of the pitch stick. In the case of the powered (electrical or hydraulic) control, the throttle may be located parallel to the pitch lever so that the two may be moved together with one hand. In either case, linkages which produce the necessary throttle change with a given pitch change are usually provided, so as to automatically keep the rotor speed approximately constant. In fact, a constant speed governor may be provided which relieves the pilot of either the pitch or throttle control. In the first case, the governor adjusts rotor pitch to absorb the engine power and to maintain a constant rotational speed. In the second case, the governor adjusts the throttle so as to supply the proper power for a given pitch setting, again so as to maintain constant rotor speed.

Devices are sometimes employed which automatically decrease the rotor pitch so as to maintain a certain minimum rotor speed in order

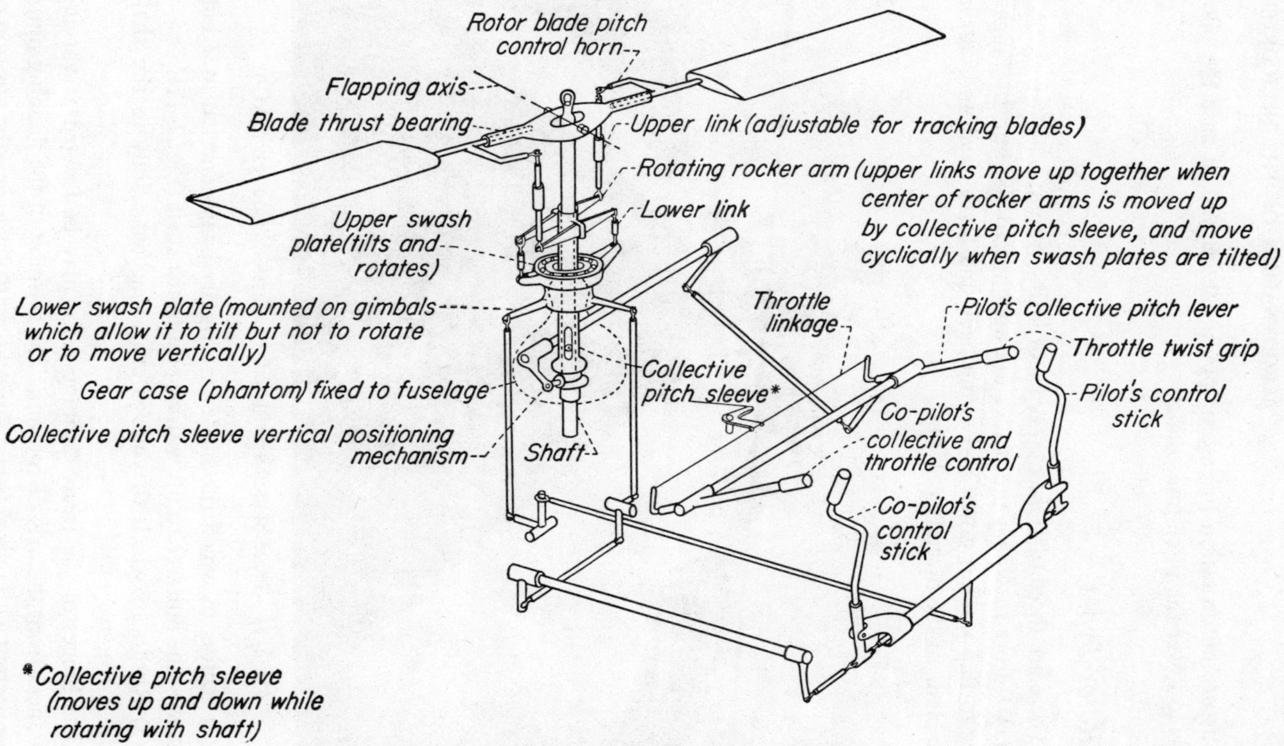

Fig. 2–10 Control system of conventional helicopter.

to assure autorotation in case of power failure and in case the pilot fails to lower the pitch immediately.

Rotor Types

There are three fundamental types of lifting rotors:

(1) Rotors in which the blades are attached to the hub by hinges, free to flap up and down and swing back and forth (lead and lag) in the plane of the disk (Fig. 2–11).

Fig. 2–11 Three-bladed articulated rotor system. Courtesy NACA.

(2) Rotors in which the blades are rigidly interconnected to a hub but with the hub free to tilt with respect to the shaft (Fig. 2–12).

(3) Rotors in which the blades are connected rigidly to the shaft (Fig. 2–13).

The hinges of the freely flapping rotor may be located at varying distances from the axis of rotation. The position of the flapping hinge is important with regard to stability and control, whereas the position of

Fig. 2–12 Two-bladed "see-saw" rotor system. Courtesy Bell Aircraft Corporation.

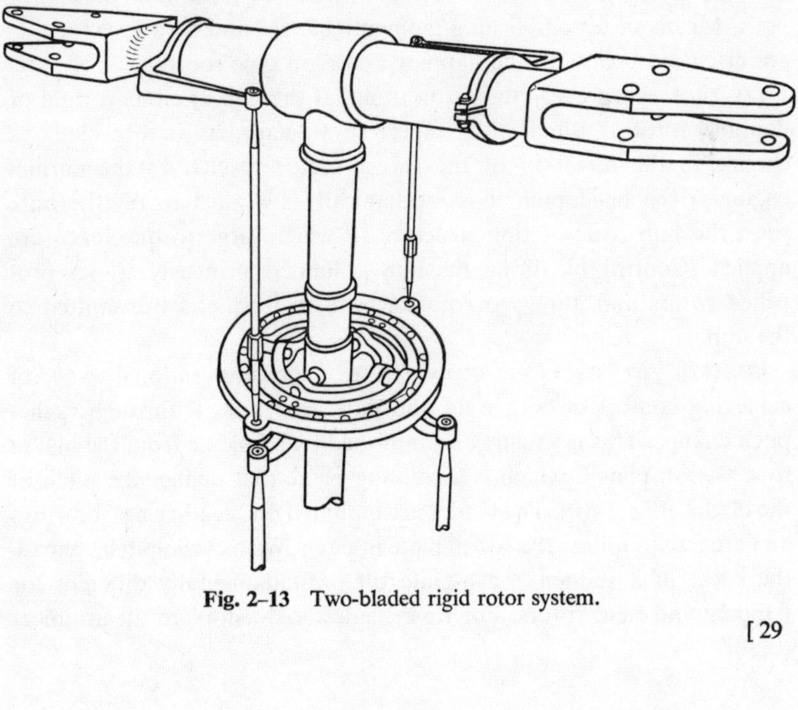

Fig. 2–13 Two-bladed rigid rotor system.

the lag hinge is important primarily in regard to vibration. Hinged rotors usually have dampers which prevent excessive motion about the lag hinge. Rotor types (1) and (2) differ chiefly in regard to the lag motions that are permitted in case (1) but which are restrained in case (2). In the discussions of rotor control which will follow, the flapping motions of rotor type (2) are equivalent to a rotor of type (1) in which the flapping hinges are located on the axis of rotation.

Rotors may have one, two, three, four, or more blades, the choice depending on such factors as vibration characteristics, rotor weight, mechanical complexity, and storage space required. In general, increasing the number of blades decreases vibration problems and increases rotor weight and, usually, mechanical complexity.

Mechanics of Rotor Control

As pointed out in the preceding section on helicopter control methods, the helicopter is controlled by (1) producing moments about the rotor hub, (2) tilting the resultant rotor lift vector, or (3) a combination of both. Means of accomplishing moment changes and thrust vector tilts are discussed below for the flapping and rigid type rotors.

CONTROL BY TILTING THE ROTOR HUB. If the hub of either a rigid or flapping rotor is tilted with respect to fuselage, as in Fig. 2–8a, a change in the direction of the thrust vector results. In the normal engine-driven helicopter, it is mechanically awkward to tilt the hub, since the hub is a rotating structure to which large torque loads are applied. Control by tilting the hub is limited primarily to jet propelled rotors and autogyro rotors where no torque is transmitted to the hub.

CONTROL BY CYCLIC-PITCH CHANGE. The conventional way of achieving control in both rigid and flapping rotors is through cyclic-pitch change. This is usually accomplished by a linkage from the blades to a "swash plate," which is a rotating plane that defines the pitch of the blades (Fig. 2–10). The blades are mounted on "feathering" bearings and are free to follow the swash plate in pitch. With cyclic-pitch control, the effect of a sudden swash-plate tilt is fundamentally different for flapping and rigid rotors. For rigid blades, a swash-plate tilt produces a moment about the rotor hub in the direction of the swash-plate tilt, owing to the difference in lift on the feathered blades (Fig. 2–14). For flapping blades with hinges on the axis of rotation, a swash-plate tilt results in a tilt of the rotor vector. Because the blades are freely

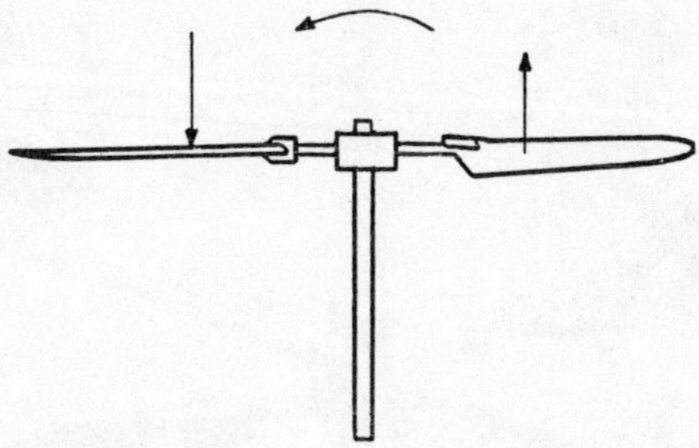

Fig. 2–14 Moment produced by thrust vector offset.

hinged, no moments may be transmitted, and the swash-plate tilt has the same effect as a corresponding shaft tilt (Fig. 2–9c). When the flapping hinges are moved outboard, the tilt of the rotor caused by a swash-plate tilt results in a moment about the hub as well as a thrust vector tilt. This moment is caused by the blade mass forces acting on the hub.

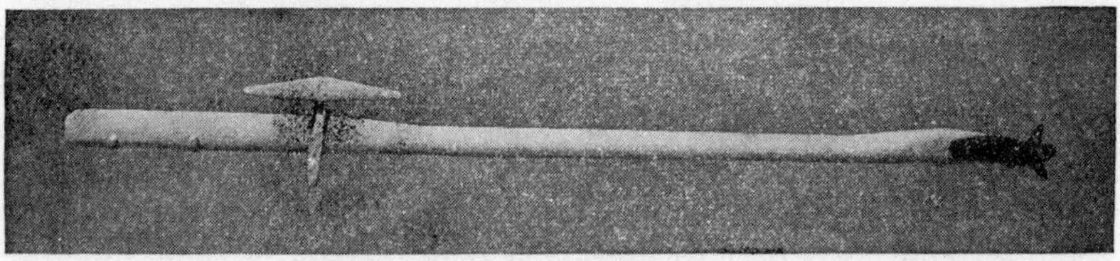

Fig. 2–15 Kaman rotor blade with servo-tab. Courtesy NACA.

ALTERNATIVE MEANS OF ACCOMPLISHING CYCLIC-PITCH CHANGE. In addition to the direct swash-plate linkage discussed above, blade pitch change may be accomplished by connecting the swash plate to a *servo-tab* on each blade, as in Fig. 2–15, or connecting the swash plate to a *servo-rotor* which in turn acts as the swash plate for the main rotor (Fig. 2–16). The advantages of such systems are that they prevent the feedback of forces from the rotor into the control system and that they may be arranged so as to produce favorable effects on the stability of the machine in flight.

Fig. 2–16 Hiller servo-rotor control system. Courtesy United Helicopters, Inc.

Conventional Helicopter Design Features

ROTOR BLADES. The blades of conventional helicopter rotors are about fifteen to twenty times as long as they are wide. Airfoils are used which have low pitching moment coefficients, usually the NACA 00 series (0012, 0015, etc.) or the NACA 230 series (23012, 23015, etc.). Airfoil thickness ratios vary between 9 per cent and 20 per cent, thicker sections being used only on the inner portions of the blade.

Blades vary both in plan form and amount of twist. It will be shown later that the best blade from an aerodynamic standpoint incorporates both twist and taper. However, gains resulting from twist and taper are oftentimes relatively small (depending on the type of helicopter and the task it is primarily designed for), and oftentimes factors such as cost of production win out and blades of simple rectangular plan form without twist are used. Typical rotor blade shapes are shown in Fig. 2–17.

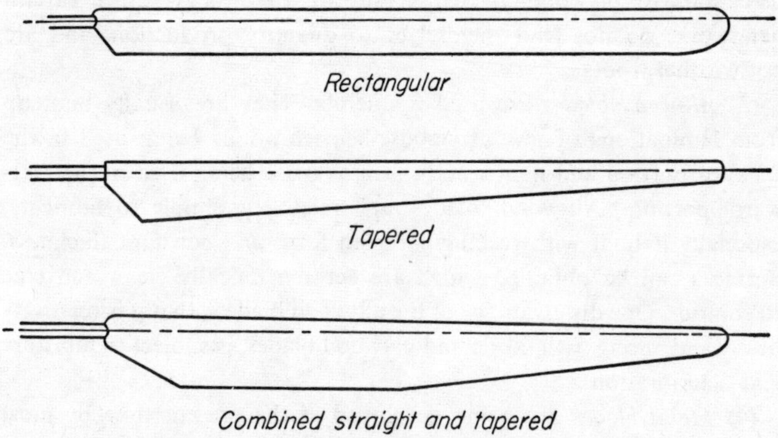

Fig. 2–17 Representative rotor blade plan forms.

Several methods of blade construction are outlined below:

(1) *Steel spar, fabric covering.* Most early rotor blades employed this type of construction. The blades are reasonably simple to fabricate but have very definite disadvantages in that it is difficult to avoid surface irregularities and fabric distortions in flight. The primary structural member of the typical fabric-covered blade consists of a steel spar which is usually step-tapered. Spars are drawn as one continuous tube with no discontinuities in the structure of the metal occurring at the steps. The ribs are usually cut from plywood and are fastened to the steel spar by metal collars. The collars are riveted to the rib and are spot-welded or glued (cycle-welded) to the spars. The leading edge is built up of solid wood—spruce or mahogany—often with a metal strip to help to keep the blade center of gravity forward. The forward portion of the blade is covered with plywood back about to the spar line. The entire blade is covered with fabric which is sewed to each rib. Blades are vented by small holes, usually on the under surface, in order

to relieve the internal pressure created by the centrifugal pumping action of the blade.

(2) *Plywood-covered blades.* Most of the objectionable features of the fabric-covered blade can be overcome by using the same basic structure and covering the entire blade with thin plywood. Some of the objections to plywood-covered blades are that they require careful handwork, do not lend themselves to quantity production, and are not weatherproof.

(3) *All-wood blades* are used frequently. They are usually built up from laminations of several woods, heavier woods being used in the forward portion and light woods such as balsa being used in the rearward portion. All-wood blades are relatively simple to fabricate, especially if built with rectangular plan form and constant thickness. Surfaces can be obtained which are aerodynamically clean and true to contour. One disadvantage of the all-wood blade is that it is relatively heavy and, along with fabric and plywood blades, is subject to moisture and deterioration.

(4) *Metal blades* are being developed at the present time by most manufacturers. Blades are either built up from pieces of sheet stock or utilize extrusions together with sheet metal. Probably the simplest blade yet fabricated involves an extruded D-spar which forms the leading edge and a V-shaped sheet metal trailing edge joined to the D-spar by flush rivets. Entire blade sections have been extruded successfully. Extrusions lend themselves well to quantity production. It is probably safe to say that all-metal blades will eventually become standard for helicopter rotors.

ROTOR HUBS. The hub is a main structural member of the rotor and is usually forged from steel or dural. Designs differ according to the hinge offsets and number of blades employed. Usually the forging houses needle-bearing hinges on which the blades flap.

ROTOR CONTROL LINKAGES. The rotor control mechanism usually consists of a swash plate, connecting links, and blades which rotate in their sockets (free to feather). The swash plate consists of a central nonrotating disk and an outer ring which rotates with the rotor (Fig. 2–10). These parts are connected by thin-race ball or roller bearings. The inner portion of the swash plate is universally mounted and connected to a linkage which allows it to move up and down and tilt in any direction. Blades are connected to the swash plate by links so that the pitch of the blade is determined by the plane of the swash plate. Care is usually taken to proportion the linkage so that in its normal operating attitude the blade will not change pitch as it flaps or lags. Furthermore, linkages are arranged to minimize these couplings of pitch with flap or lag in all positions of the blade. Changes of pitch with flapping, if moderate, have some desirable effects and are often purposely incorporated. Large changes of pitch with lag angle, however, are undesirable and are usually avoided as much as possible.

THE CONTROL SYSTEM. A typical direct-linkage control system for a single-rotor helicopter may be understood by again referring to Fig. 2–10. It is seen that the control stick is connected so as to tilt the swash plate in the direction in which the stick is moved. The pitch stick raises and lowers the pitch sleeve while retaining any tilt imposed by the control stick. The mechanical advantage between control stick and blade is usually such that 1 inch of stick motion results in 1 to 2 degrees of cyclic-pitch change.

In multi-rotor configurations, control systems are necessarily modified. In the side-by-side machine the swash plate may be free to tilt only in a fore and aft direction. In this case, lateral control is achieved by raising one swash plate and lowering the other, thus tilting the ship and producing sideward motion. Lateral control may also involve a swash-plate tilt which is coupled with the collective pitch change so as to tilt the thrust vector laterally as well as roll the machine. The same remarks apply to tandem machines in regard to longitudinal control.

FUSELAGE DESIGN. Several factors which influence fuselage design are listed below:

(1) A streamlined shape for low parasite-drag and moment coefficients.

(2) Good visibility for the pilot.

(3) Area for disposable load located as nearly under the rotor as possible to avoid center of gravity shifts.

(4) Easy accessibility to engines and transmissions.

(5) Accommodation of a tail rotor and/or stabilizing surfaces at a reasonable moment arm.

In order to increase the range of center of gravity travel for a given control tilt and in order to improve the control of the machine, it is desirable to keep the center of gravity as far below the rotor center as possible.

LANDING GEAR. Landing gears of both the three-wheel and four-wheel types are used. Landing gear design is comparable to normal airplane design except that the stroke available in the shock absorber of the helicopter is usually considerably longer to provide softer action in landings and provide damping for "ground resonance." Alternate gear arrangements, which permit operation from all possible types of terrain, are sometimes supplied by the manufacturer. Thus flotation gear which is suitable for water, land, and marsh operation is available, as well as skid gear for high forward-speed landings on rough, plowed ground as well as on improved surfaces, and ski gear for soft snow and rough ice. In all cases, it is important that the landing angle, as determined by tail wheel or tail skid position, be sufficient to permit high pitch-up attitudes of the fuselage for flare-outs in autorotation landings.

TRANSMISSION SYSTEMS. Transmission systems usually involve gear ratios between engine and rotor of the order of 10:1. Planetary gear trains are most efficient from a weight point of view but are expensive and often noisy. Bevel gears, along with a single-stage-planetary-gear train, are frequently used. The drive system is also supplied with a clutch which is engaged either manually by the pilot or centrifugally when a certain engine speed is reached. In addition to the clutch, a free-wheeling unit or overriding clutch is incorporated so that the engine may drive the rotor, but the rotor cannot drive the engine in case of power failure. In a single-rotor machine, the tail rotor is geared directly to the main rotor so that in case of engine failure, the main rotor turns the tail rotor.

Flight Characteristics of the Helicopter

An appreciation of the flight characteristics of the helicopter involves an understanding of its performance characteristics, its vibration characteristics, and its stability and control characteristics. The following paragraphs deal with these topics in a qualitative manner.

PERFORMANCE CHARACTERISTICS. Power must be supplied to the rotor of the hovering helicopter for two reasons:

(1) Power is required to produce lift. This is referred to as *induced* power.

(2) Power is required to drag the blades through the air. This is called *profile-drag* power.

The helicopter rotor produces thrust to support the helicopter in air by imparting momentum to a mass of air. The rotor imparts a downward velocity to a large mass of air and, in so doing, realizes an upward thrust. It is clear that power must be expended to produce this jet of air. The power is, in fact, proportional to the downwash velocity for a given weight of helicopter. The downwash velocity, in turn, depends upon the amount of air to which velocity is imparted in producing the rotor thrust. A large diameter rotor, then, can lift a weight with much less induced loss than a small diameter rotor.

Profile-drag power arises entirely from the fact that air is a viscous fluid and that when a body is pulled through this fluid frictional forces are exerted on the body.

For the normal helicopter in hovering, induced losses account for about 60 per cent to 70 per cent of the total rotor power required; profile-drag losses account for about 30 per cent to 40 per cent. The engine must supply sufficient power for the rotor and, in the case of a single-rotor machine, for the tail rotor in order that the helicopter may hover. If more power is applied to the rotor than is required to overcome the induced and profile losses, then the helicopter will climb.

In forward flight power must be supplied to drag the fuselage through the air as well as to overcome the induced and profile-drag losses. The power required to drag the fuselage through the air increases as the cube of the forward speed and becomes large at higher speeds. In one of the early production helicopters, for example, one-half of the available engine power was used in overcoming fuselage drag at 80 miles per hour.

While parasite-drag power increases rapidly with airspeed, the power required to produce lift—the induced power—decreases with increasing speed. As the rotor moves forward, it encounters a larger mass of air per second. To produce its thrust it therefore needs to impart less velocity to each mass of air and the energy imparted to the air is thereby reduced.

The profile-drag power increases slightly as forward speed is increased, the increase becoming very rapid at high forward speeds. The trends of induced, profile, and parasite power with airspeed are shown in Fig. 2–18. The sum of these three components at any forward

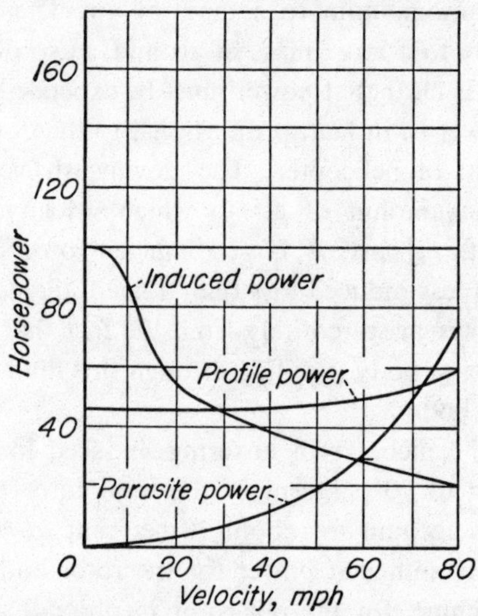

Fig. 2–18 Breakdown of helicopter power losses.

speed gives the total power required for level flight at that speed. The resultant power-required curve is shown in Fig. 2–19. The numbers given are typical for a small, two-place helicopter. It is seen that the power required to hover is relatively high, the power decreasing rapidly in the low speed range (because of decreased induced losses) and increasing again at high speeds due to fuselage drag.

Minimum power is required in level flight at about 40 miles per hour in the example shown. It is characteristic of almost all helicopters that minimum power falls somewhere in the 40- to 60-mile per hour range. Also shown in Fig. 2–19 is a horizontal line which represents the power available at the helicopter rotor. The power available is the rated engine power minus the tail rotor power (if an auxiliary rotor is used), as well as the frictional losses in the transmission, and losses from powering a blower to cool the engine.

It is clear that the performance capabilities of a helicopter are determined by the level of the power-available curve with respect to the power-required curve. If, for example, the power available is just equal to the power required to hover, as in curve (a) of Fig. 2–19, the performance of the machine is marginal. It is only barely able to hover

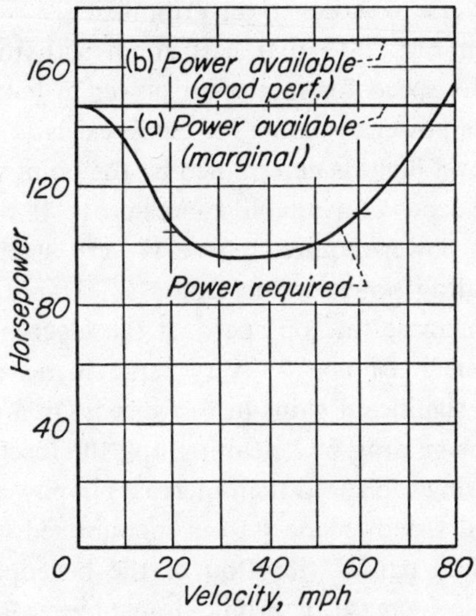

Fig. 2–19 Power-available and power-required curves.

and unable to climb vertically. A slight overload would increase the induced power required and the machine would be unable to hover.

Actually, a helicopter is able to hover very near the ground even when it has insufficient power to hover away from the ground. This is because of a phenomenon known as *ground effect*. The ground stops the rotor downwash, or induced velocity, thus decreasing the induced power required.

It will also be noted that because of the reduction of power required with forward speed, an overloaded helicopter may take off in a wind or by making a run on the ground to attain a small forward speed. While the machine can fly forward in this overloaded condition, it cannot hover. The marginal hovering performance of many present-day helicopters has resulted in the loss of several machines in the hands of inexperienced pilots. When flying close to the ground there is a tendency to fly by ground speed rather than according to the airspeed indicator. If winds are involved, a "downwind turn" may result in zero airspeed, so that the machine will settle to the ground. Again, the helicopter may be hovering in a wind above some obstacle, such as a row of trees. When the helicopter drops below the trees, where the wind is decreased, it is unable to hover and settles to the ground.

It is clear from Fig. 2–19 that best climb with the helicopter will occur at about the speed for minimum power in level flight, for here the greatest excess power available for climb exists.

Top speed in level flight is determined by the point where the power-required curve and power-available curves cross. It will be noted that the slope of the power-required curve is very steep at high speed, because fuselage-drag power is increasing at a rapid rate. Thus, it is very difficult to increase the top speed of the machine appreciably by increase in power, inasmuch as very large increases in power are required to make significant gains in top speed. On a percentage basis, reductions in fuselage drag by "cleaning up" the fuselage are far more effective in increasing top speed than increases in power.

At high forward speeds blade stall is encountered over a portion of the rotor disk. Stall causes vibration of the helicopter and controls and a considerable increase in profile-drag power. Blade stall is due to the difference in velocity encountered by the advancing and retreating blades in forward flight. As a lifting rotor moves forward, the advancing blades encounter progressively higher velocities and the retreating blades progressively lower velocities. In order to maintain its lift, the retreating blade must operate at progressively higher angles of attack as forward speed increases. It follows that at some forward speed the angles of attack on the retreating side will reach the stall.

While the dissymmetry can be reduced by turning the rotor faster, thus permitting a higher forward speed for a given amount of stall, this method soon leads to excessively high velocities over the advancing blade and accompanying power losses and roughness due to compressibility effects. The fundamental limits to the high speed of the helicopter are therefore blade stall and blade Mach number. It will always be difficult to build helicopters which can reach speeds very much greater than about 200 miles per hour.

In case of power failure, the helicopter is able to glide, its rotors continuing to whirl in autorotation as does the rotor of the autogyro. In vertical descent the rotor is about as effective as a parachute of the same diameter in allowing the machine to descend slowly. At its best gliding speed the rotor lets the helicopter down at about one-half the vertical autorotative rate of descent, or about 15 to 20 feet per second. As the helicopter approaches the ground the pilot may pull back on his stick and "flare out," trading his energy of forward motion for additional lifting power. In this manner, he is able to settle slowly to the ground with very little forward speed. He may also take advantage of the energy in the rotor and increase the blade pitch, producing additional thrust while decelerating the rotor.

CONTROL FORCES. Stick forces in the helicopter are quite important in regard to the pilot's impressions of the machine. Pilots tend to fly aircraft by the "force feel" of the stick rather than by stick displacements. Without accurate reference points it is extremely difficult to judge the number of inches which a stick has been displaced. Most pilots like a moderate force gradient always resisting a motion of the stick. For steady flight, desirable stick force characteristics require that the pilot push with moderate but increasing force to move the stick forward and pull with increasing force to move the stick aft. When released, the stick should return to a neutral position.

In maneuvers, the forces which feed back into the pilot's stick have considerable influence on his impressions of the stability of the machine. If in a maneuver forces are created which tend to move the pilot's hand in a direction to aggravate the maneuver, the pilot experiences difficulty in properly controlling the machine.

While stick forces are quite important in regard to flying qualities, they are difficult to control in the helicopter. Stick forces in the helicopter do not arise from straightforward sources as in an airplane. In an airplane a motion of the control stick deflects a hinged control surface. Because of the deflection of the hinged surface a moment is created which is transmitted to the pilot's stick. In the simple cyclic-pitch control rotor, on the other hand, a motion of the stick changes the pitch of the blades as they rotate. In the helicopter, all stick forces must therefore arise from pitching moments on the blades themselves. When airfoil sections are chosen and mounted so as to have no pitching moments at any pitch angle, then there should be no stick forces for the pilot to overcome.

Actually, pitching moments exist on rotor blades because of several secondary effects such as airfoil imperfections, blade-bending distortions, chordwise mass balance, etc. Stick forces are caused from these secondary effects rather than from straightforward moments as in the airplane.

Control forces consist of both oscillating forces and steady forces. Oscillating forces may occur at frequencies of 1/rev. and even multiples of the number of blades. Oscillating forces with a frequency of 1/rev. are entirely chargeable to differences in pitching moments between the rotor blades. For example, if only one blade of a three-bladed rotor experiences a pitching moment, this one blade exerts a steady force on the swash plate as it rotates. This rotating force is transmitted to the control stick. Because the pilot's hand is not a rigid support, the control stick yields under the rotating force and describes a small circle. Helicopters are often characterized by a small 1/rev. stick shake.

Higher frequency oscillations in the control stick can arise at integral multiples of the number of rotor blades. A three-bladed rotor may, therefore, experience a 3/rev., 6/rev., 9/rev., etc., oscillation in the controls. These higher frequency oscillations arise from periodic changes in the pitching moment of each blade. The periodic changes, in turn, arise from periodic air force changes in a rotor in forward flight or from periodic blade deflections. For example, a three-bladed rotor with equal pitching moments on all three blades will produce a 3/rev. motion of the control stick in forward flight because of the change in velocity on the advancing and retreating blades. As the blade comes forward and experiences higher velocities, its pitching moment is increased, and since this happens three times for each rotor revolution, a 3/rev. shake of the control stick results.

Oscillating forces in the control system can be prevented from annoying the pilot by the use of irreversible controls or by making the effective mass of the pilot's stick large enough to absorb the oscillating force.

One convenient means of accomplishing the latter is by the use of an inertia damper which may consist of a weight on a high-pitch screw, the weight being forced to rotate when axial force is applied to the screw. The inertia damper is simply a means of producing a large, effective mass without heavy weights.

Steady forces on the control stick come entirely from 1/rev. variations in blade-pitching moments. These again may arise from periodic air forces, periodic blade deflections, or both.

Sometimes tabs are used on blades intentionally to produce pitching moments. In forward flight these pitching moments then vary periodically as the velocity over the blade varies, becoming large on the advancing blade and small on the retreating blade. A steady stick force results.

Usually stick forces are relatively small in small rotors (20 to 30 feet in diameter). As diameters increase, however, stick forces increase rapidly. It is, therefore, very difficult to use directly connected controls in large rotors since extreme care must be taken in blade construction to avoid the small secondary effects which produce the annoying control forces. Servo controls, which relieve the pilot from these feedback forces, seem to be necessary for very large helicopters.

VIBRATION CHARACTERISTICS. As in the case of control oscillations, the helicopter itself is subject to vibrating forces from the rotor. If it is assumed that the force input from each rotor blade is identical, then the only frequencies at which oscillations occur are again even multiples of the number of rotor blades. Vibrating forces may be in a vertical, fore and aft, or sideward direction. The reduction of the input forces from the rotor blade is primarily a question of blade design, involving the proper distribution of masses and air forces along the span. The effect of the input forces on the vibration characteristics of the machine depends considerably upon the natural frequencies of the fuselage structure. If the fuselage structure has any mode of vibration, such as vertical bending, fuselage torsion, etc., in resonance with an exciting frequency, vibrations may become quite disagreeable. One of the main problems in minimizing vibrations in the helicopter lies in avoiding natural exciting frequencies. This is difficult since exciting frequencies cover a wide range (3/rev., 6/rev., 9/rev., etc.).

GROUND RESONANCE. The helicopter is a very complex dynamic system, involving masses and springs in several degrees of freedom. Sometimes couplings occur between blade lag motion in the rotating system and sideward or fore and aft motion of the shaft. If insufficient damping is present in the system, a self-excited vibration may result, the ship shaking back and forth with increasing amplitude and the blades moving back and forth in the plane of rotation, out of pattern. This phenomenon is commonly known as *ground resonance* because the most frequent place of occurrence of the vibration is on the ground where the machine is supported on its relatively soft tires, giving a low natural frequency of the machine in sideward motion. Ground resonance, if it begins, often results in complete destruction of the machine within a few seconds. Several helicopters and autogyros have been destroyed by these self-excited vibrations.

STABILITY CHARACTERISTICS. Hovering in the conventional helicopter requires considerable practice on the part of the pilot, although most of the problems tend to disappear with experience. The novice pilot must learn to put up with high control sensitivity in roll, or, in other words, the high rate of roll per inch of stick displacement. Helicopters with conventional control systems are subject to high control sensitivity which, according to reference VI–7 (Appendix IIA), can lead to overcontrolling, and in turn, to a short-period, pilot-induced lateral oscillation. High control sensitivity is apparently due to low rotor damping, which for the helicopter is a fraction of that for airplanes (reference VI–7, Appendix IIA). This problem is therefore minimized with large diameter rotors, which have greater damping than small rotors.

The normal helicopter in hovering is somewhat sluggish in response to a sudden control deflection. This sluggishness is not due to a lag in the response of the rotor to an applied control motion, for the rotor follows almost instantaneously a motion of the stick, but rather to the fact that velocity changes or displacement of the helicopter in space do not follow the inclination of the thrust vector immediately because of the mass of the helicopter. During training the student pilot overcomes his first impressions of lag by learning to control the helicopter's accelerations.

If disturbed from hovering equilibrium with control stick fixed, the helicopter will describe a slow, translational oscillation and move back and forth with slowly increasing amplitude. The conventional helicopter is thus dynamically unstable in hovering. The instability is easily controllable, however, and is not considered a serious handicap in the machine inasmuch as the period of the oscillation is long enough to allow for the pilot's reaction time in perceiving and correcting the motion. There are several means of stabilizing the helicopter in hovering, all of which utilize gyroscopic forces.

In forward flight, the normal helicopter may exhibit some undesirable tendencies in the longitudinal control. Helicopters have a tendency to "zoom" at low forward speeds following take-off, pitching upward abruptly, and sometimes requiring full forward stick motion to regain control of the machine. At normal cruising speed the conventional helicopter is usually dynamically unstable in pitch, but only mildly so, with a long enough period to allow the pilot to recognize the disturbance and correct it. This instability is primarily due to low rotor damping and to the fact that the conventional helicopter rotor is statically unstable with angle of attack, the instability becoming greater as the forward speed is increased. Because of this increase of angle-of-attack instability with forward speed, the attentiveness required of the pilot to correct for a disturbance increases rapidly as the helicopter's top speed is reached, in that at those speeds the disturbance builds up so rapidly that corrective control must be applied in a few seconds. The elimination of this pitching instability is considered to be of primary importance in achieving satisfactory flying qualities. Many experimental helicopters, and several production machines, already incorporate devices which either increase the damping in pitch or add positive static stability with angle of attack.

Most helicopters on the market today have definite instabilities in some regimes of flight and by airplane standards have very poor flying qualities. The stability and control characteristics of the helicopter represent a major present-day development problem, but because they are receiving considerable attention from manufacturers and government research agencies, it may be stated with confidence that a rapid improvement of helicopter flying qualities will not be long forthcoming

3

AN INTRODUCTION TO HOVERING THEORY

It is best to start the study of rotor theory with hovering and vertical flight, inasmuch as there is no dissymmetry of velocities across the rotor disk in those flight conditions. In other words, contrary to forward flight, the same conditions exist at a blade element operating at a given radial distance, irrespective of its azimuth position.

The hovering theory presented in the following chapters is based on the combination of the momentum theory, originated by W. J. M. Rankine and developed by R. B. Froude, and the blade-element theory. Identical equations may be derived by means of the vortex theory, but it is believed that the combination of the momentum and blade-element theories has greater physical significance and can be more easily grasped.

Momentum Considerations

DEVELOPMENT OF THRUST. The momentum theory stems from Newton's second law of motion, $F = ma$, and is developed on the basis that the axial velocity u of the fluid through the airscrew disk is generally higher than the speed V with which the airscrew is advancing through the air. The increase in velocity of the air from its initial value V to its value at the airscrew disk, which arises from the production of thrust, is called the *induced* or *downwash* velocity, and is denoted by v. The thrust developed by the airscrew is then equal to the mass of air passing through the disk in unit time, multiplied by the total increase in velocity caused by the action of the airscrew.

In hovering or vertical climb, the action of the air influenced by a helicopter rotor is as shown in Fig. 3–1. Because of the increase in velocity of the air mass by the rotor, there is a gradual contraction of the slipstream as it approaches the rotor, the maximum contraction or

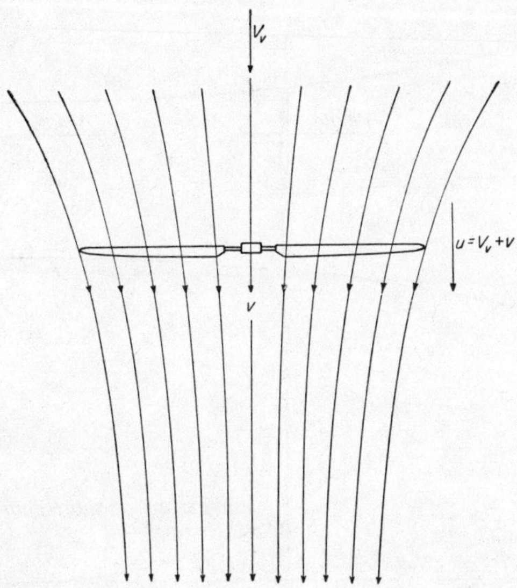

Fig. 3–1 Flow in vertical climb.

speedup of the air being accomplished within one rotor-diameter behind the disk.

ASSUMPTIONS OF THEORY. The simple momentum theory assumes the following:

(1) The rotor is composed of an infinite number of blades and may therefore be considered as an "actuator disk," uniformly accelerating the air through the disk with no loss of thrust at the blade tips.

(2) The power required to produce the thrust is represented only by the axial kinetic energy imparted to the air composing the slipstream. A frictionless fluid is assumed so that there is no blade friction or

profile-drag losses. Rotational energy imparted to the slipstream is ignored.[1]

(3) The disk is infinitely thin so that no discontinuities in velocity occur on the two sides of the disk.

INDUCED VELOCITY RELATIONSHIPS. Consider an airscrew (Fig. 3–2a) advancing to the left so that the velocity of the free stream relative to

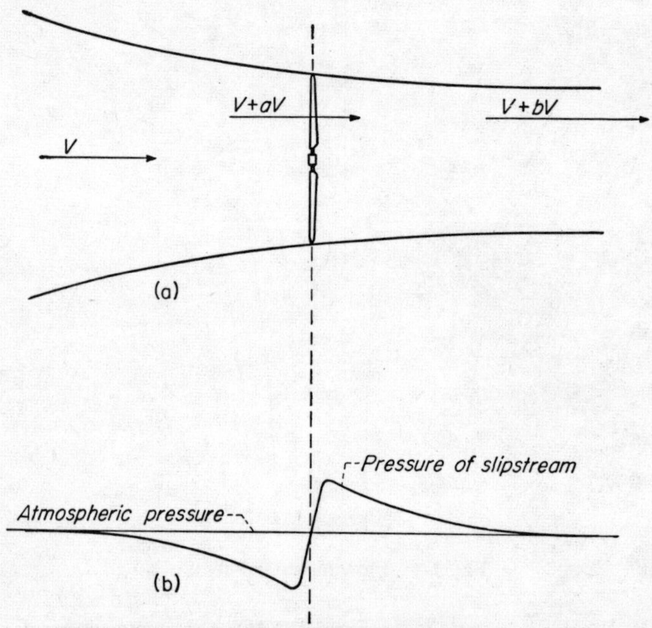

Fig. 3–2 Airscrew in forward flight.
(a) Airstream velocities
(b) Pressure distribution

it is V. Let aV be the increase in velocity at the disk and bV the increase in velocity over free air at an infinitely large distance downstream.

Under the stated assumptions, the work done in unit time by the thrust of the airscrew on the air must be equal to the increase of the

[1] Slipstream rotation, caused by the reaction of the torque on the rotor, is considered in the general momentum theory developed by A. Betz. The theory is completely described by H. Glauert in reference 18 (Appendix IIB).

kinetic energy of the slipstream in unit time. The relationship may be expressed as follows:

$$T(V + aV) = \Delta \text{ K.E.} \tag{1}$$

If A is the area of the disk and ρ the mass density of air in slugs per cubic foot, the thrust developed by the airscrew can be expressed as the change of the axial momentum of air in unit time.

$$T = [\rho A(V + aV)]bV \tag{2}$$

The increase of the kinetic energy of the air effected by the airscrew is equal to one-half of the mass of air passing through the disk multiplied by the difference in the squares of the velocities infinitely in front of and behind the disk.

$$\Delta \text{ K.E.} = \tfrac{1}{2}\rho A(V + aV)[(V + bV)^2 - V^2] \tag{3}$$

Substituting the expressions for T and Δ K.E. into equation (1)

$$[\rho AV(1 + a)bV][V(1 + a)] = \tfrac{1}{2}\rho AV^3(1 + a)(b^2 + 2b)$$

and solving for b in terms of a

$$b = 2a \tag{4}$$

Equation (4) states that the induced velocity at the airscrew disk is one-half of the total increase in velocity imparted to the air column. It might be mentioned that the static pressure of the slipstream at distances infinitely in front of and back of the airscrew disk is equal to the free stream static pressure. As shown in Fig. 3–2b, there is a falling pressure gradient all along the slipstream, except at the airscrew disk at which place energy is added to the air.

INDUCED VELOCITY IN HOVERING. The relationship expressed by equation (4) may be used to obtain the velocity induced by a rotor in the hovering state. For this condition, $V = 0$ and the increase in velocity at the rotor disk aV is equal to the total velocity through the disk, or the induced velocity v. Then, from equation (4), the increase in velocity downstream from the disk, bV, is equal to $2v$. If R is the rotor radius, the thrust T of the hovering rotor may be written from equation (2) as

$$T = (\rho\pi R^2 v)2v \tag{5}$$

Solving for v

$$v = \sqrt{\frac{T}{2\rho\pi R^2}} \tag{6}$$

Inasmuch as the major power loss in hovering is the induced loss, represented by Tv, it follows that the disk loading $T/\pi R^2$ should be kept as low as possible for efficient hovering performance.

It is important to realize that the assumption of uniform inflow by the momentum theory results in a minimum induced power loss for a given thrust.[1] The mathematical proof of this concept involves setting up the integrals of the thrust and kinetic energy loss produced by an annulus of the disk and then, by the calculus of variations, showing that if the thrust integral is kept constant, the energy integral will be a minimum when the induced velocity is constant over the disk. This conclusion can also be arrived at as follows. The thrust of a hovering rotor is produced by imparting momentum to the air, which depends on the first power of the induced velocity. The induced or kinetic energy losses depend upon the square of the velocity. For a given thrust the kinetic energy losses will be a minimum when the induced velocity distribution over the disk is equal at every point to the mean velocity, for only in this condition are velocity peaks, and thus large losses, avoided.

IDEAL AND ACTUAL LOSSES. The analysis made by the simple momentum theory was idealized, because it neglected profile-drag losses, nonuniformity of induced flow (including the energy losses incurred by the spilling of air about the blade tips, commonly called tip losses), and slipstream rotational losses. Thus, an actual rotor would require more power to hover with a given load than an "ideal" rotor (i.e., a rotor having zero profile drag and uniform inflow), and therefore would be less efficient.

The order of magnitude of the rotor losses not considered by the simple momentum theory, expressed as a percentage of the total power required, is as follows:

Profile-drag losses	30 per cent
Nonuniform inflow	6 per cent
Slipstream rotation	0.2 per cent
Tip losses	3 per cent

[1] This is true only when rotational and profile-drag losses are ignored. Rotational energy losses arise from the rotational velocity imparted to the slipstream by the production of rotor torque.

In actual practice, values of thrust as high as 83 per cent of the thrust produced by an ideal rotor have been measured on full-scale rotors.

Rotor Figure of Merit

In estimating the effectiveness of a lifting rotor in the hovering condition, it is not possible to apply the propeller criterion of efficiency, namely

$$\eta = \frac{\text{useful power}}{\text{total power}} = \frac{TV}{P}$$

because a standard is needed to judge the effectiveness of a rotor in producing thrust when its velocity of translation is zero. In the hovering condition, for example, power P is expended in producing thrust T, while the translational speed of the rotor V is zero, thus making $\eta = 0$. Obviously the lifting rotor needs some other standard of efficiency whereby its lifting ability may be judged.

A very reasonable way to estimate the efficiency of a lifting rotor is to compare the actual power required to produce a given thrust with the minimum possible power required to produce that thrust. The minimum possible power required to produce a given thrust is obtained with an ideal rotor, inasmuch as such a rotor has minimum induced power (constant inflow), zero profile-drag power, and zero rotational and tip losses.

Consequently the criterion of rotor hovering efficiency may be defined as

$$M = \frac{\text{minimum possible power required to hover}}{\text{actual power required to hover}} = \frac{Tv}{P} \qquad (7)$$

where M is called the rotor *figure of merit*. If equation (6) is substituted into equation (7), then the general expression for rotor figure of merit is

$$M = \frac{1}{\sqrt{2}} \frac{T}{P} \sqrt{\frac{T}{\rho \pi R^2}} \qquad (8)$$

Thus, the larger the figure of merit for a given rotor, the smaller the power required to produce a given thrust, or the larger the thrust per unit horsepower.

IDEAL FIGURE OF MERIT. Inasmuch as the ideal rotor was defined as a rotor producing thrust with the minimum amount of power, it follows that the figure of merit of an ideal rotor, or the *ideal* figure of merit M_i should equal unity.

Although the figure of merit of an ideal rotor is simply equal to unity, the maximum figure of merit of a rotor having zero profile drag but with non-uniform inflow distribution is not a unique number, in that it varies with thrust coefficient and cannot easily be obtained as an analytical expression. (As will be explained later, uniform inflow can only be obtained with blades having *ideal* twist; untwisted blades (i.e., blades having no spanwise variation in blade angle) result in a nonlinear inflow distribution.) Over most of the operating range of thrust coefficients, however, the maximum figure of merit of untwisted blades can be considered approximately equal to 0.94 if no profile-drag losses are present.

By means of the ideal figure of merit, the upper limit to the hovering performance of any helicopter at various altitudes can be quickly estimated, if the engine power and the rotor diameter are known. The use of the ideal figure of merit enables one to discount the claims of many overenthusiastic rotor designers by showing that the combination of thrust and power claimed for their rotor is more optimistic than could be realized with an ideal rotor and therefore never could be attained in practice.

Example: It is required to compute the maximum thrust that could be produced by a 40-foot diameter rotor to which is transmitted 170 horsepower, assuming sea-level operation. Also compute the disk and power loadings of such a rotor.

From equation (8)

$$1 = 0.707 \frac{T}{P} \sqrt{\frac{T}{\rho \pi R^2}}$$

or

$$1 = 0.707 \frac{T}{170 \times 550} \sqrt{\frac{T}{0.002378 \times \pi \times (20)^2}}$$

and

$$T = 3740 \text{ pounds}$$

also

$$\text{disk loading} = \frac{W}{\pi R^2} = \frac{3740}{\pi \times (20)^2} = 2.98 \text{ pounds per square foot}$$

$$\text{power loading} = \frac{W}{P} = \frac{3740}{170} = 22.0 \text{ pounds per horsepower}$$

The ideal figure of merit may be used to obtain a simple relation between disk loading and power loading that is useful for quickly finding the maximum possible performance of a given rotor design.

It was seen in equation (8) that

$$M = \frac{1}{\sqrt{2}} \frac{T}{P} \sqrt{\frac{T}{\rho \pi R^2}}$$

In terms of disk loading, D.L., and power loading, P.L.,

$$M = \frac{1}{\sqrt{2}} \text{ P.L.} \sqrt{\frac{\text{D.L.}}{\rho}} \tag{9}$$

Assuming sea-level conditions, the above equation reduces to

$$\text{P.L.} = 38M \frac{1}{\sqrt{\text{D.L.}}} \tag{10}$$

This relation is shown graphically in Fig. 3–3 for the ideal figure of merit, $M = 1$, along with curves for $M = 0.75$ and $M = 0.50$. The curve for $M = 1$ is by definition the upper limit for any rotor, since it represents a rotor with zero profile drag and with uniform induced flow. The curve for $M = 0.75$ may be considered typical of good rotors; the curve for $M = 0.50$ represents poor rotors. Figure 3–3 is useful in making preliminary estimates of helicopter hovering performance.

NONDIMENSIONAL FIGURE OF MERIT. As defined by equation (8), the figure of merit is not expressed in terms of nondimensional quantities. It can be written, however, in a completely nondimensional and more compact form. Before proceeding to do so, it is necessary to introduce nondimensional coefficients for thrust, torque, and power. These coefficients, which are similar to propeller coefficients and are based on rotor disk area, are defined by means of the following expressions:

$$\left.\begin{aligned} T &= C_T \pi R^2 \rho (\Omega R)^2 \\ Q &= C_Q \pi R^2 \rho (\Omega R)^2 R \\ P &= C_P \pi R^2 \rho (\Omega R)^3 \end{aligned}\right\} \tag{11}$$

It should be noted that $C_Q = C_P$ because

$$C_Q = \frac{Q}{\pi R^2 \rho (\Omega R)^2 R} \times \frac{\Omega}{\Omega} = \frac{P}{\pi R^2 \rho (\Omega R)^3} = C_P$$

The coefficients effect the simplification of performance equations and, once determined for a particular rotor for a certain operating

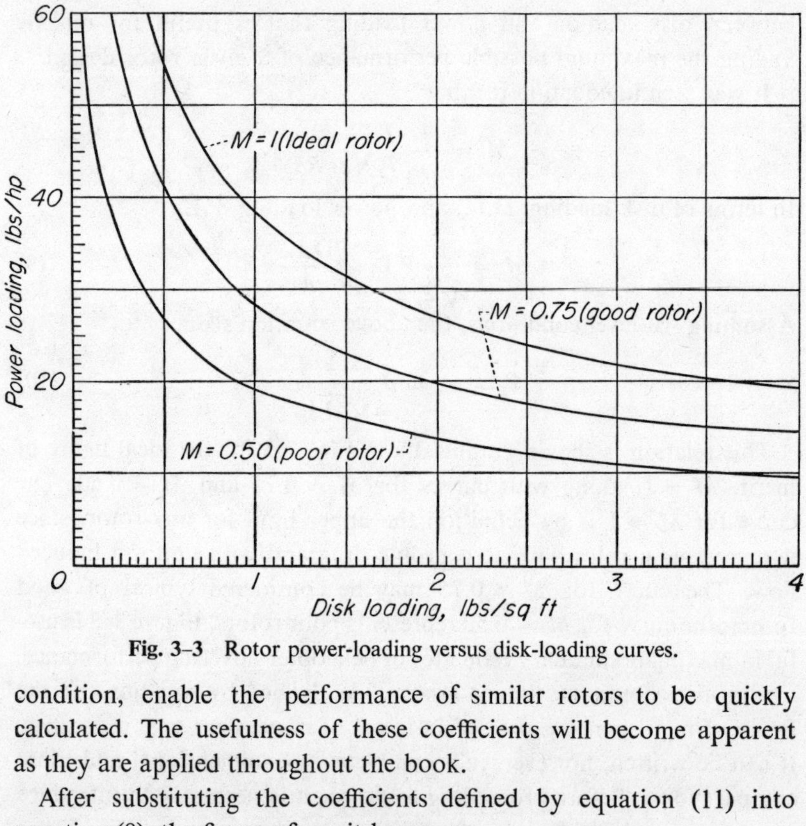

Fig. 3–3 Rotor power-loading versus disk-loading curves.

condition, enable the performance of similar rotors to be quickly calculated. The usefulness of these coefficients will become apparent as they are applied throughout the book.

After substituting the coefficients defined by equation (11) into equation (8), the figure of merit becomes

$$\begin{aligned} M &= \frac{1}{\sqrt{2}} \frac{C_T \pi R^2 \rho (\Omega R)^2}{C_P \pi R^2 \rho (\Omega R)^3} \sqrt{\frac{C_T \pi R^2 \rho (\Omega R)^2}{\pi R^2 \rho}} \\ &= \frac{1}{\sqrt{2}} \frac{C_T{}^{3/2}}{C_Q} \\ &= 0.707 \frac{C_T{}^{3/2}}{C_Q} \end{aligned} \tag{12}$$

Blade-Element Considerations

The primary limitation of the momentum theory is that it provides no information as to how the rotor blades should be designed so as to produce a given thrust. Also, profile-drag losses are ignored. The blade-element theory provides means for removing these limitations. The blade-element theory, which was first put in practical form by Drzewiecki, is based on the assumption that each element of a propeller or rotor can be considered as an airfoil segment that follows a helical

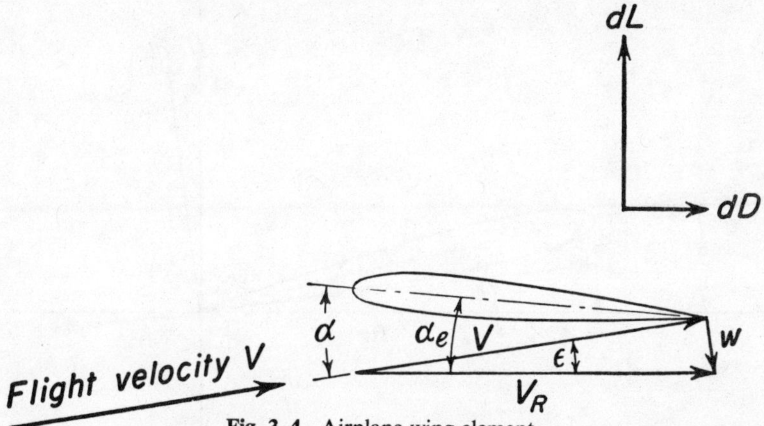

Fig. 3–4 Airplane wing element.

path. Lift and drag are then calculated from the resultant velocity acting on the airfoil, each element being considered independent of the adjoining elements. The thrust and torque of the propeller or rotor are obtained by integrating the individual contribution of each element along the radius.

In order to obtain the resultant velocity at a blade element, the total flow through the disk must be known. For a rotor in the climb condition, the flow through the disk is composed of the climb velocity, V_v, plus the induced velocity, v. The inclusion of the induced velocity enables two-dimensional flow characteristics to be assumed. This practice is similar to that of wing theory, in which lift and drag are determined from the effective angle of attack α_e, which is equal to the geometrical angle, α, minus an induced or downwash angle ϵ. Thus, referring to Fig. 3–4, $\alpha_e = \alpha - \epsilon$, where $\epsilon = w/V$. The differential lift and drag act perpendicular and parallel to the resultant velocity V_R.[1]

In a similar manner, velocities and the forces acting on a blade element of a rotor in the climb condition are shown in Fig. 3–5. The angle ϕ is the inflow angle of attack and contains the induced velocity v as well as the climb velocity V_v. The blade pitch angle θ is measured from the line of no lift of the airfoil section to the plane of the rotor

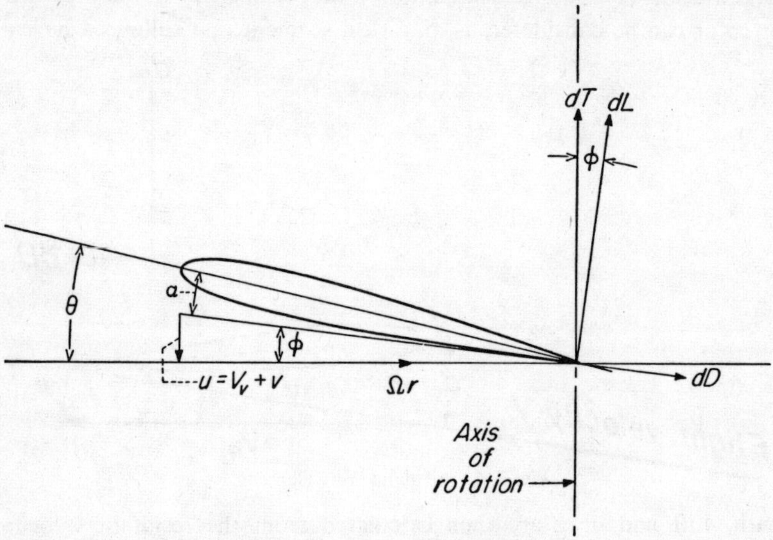

Fig. 3–5 Rotor blade element.

disk. The two-dimensional blade angle of attack is therefore seen to be the difference between the blade pitch and inflow angles.

The calculation of the effect of profile drag on the power required by a hovering rotor by means of the blade-element theory will illustrate the effectiveness of this powerful analytical tool.

Effect of Profile Drag on Figure of Merit

The effect of blade profile-drag losses on the efficiency of a hovering rotor can be estimated by developing an equation for the power (i.e., C_Q) term in the expression for figure of merit given by equation (12).

[1] Note that the definitions of the symbols used in Fig. 3–4 apply only to this paragraph.

EXPRESSION FOR THRUST. From simple blade-element theory, the expression for the differential lift dL on b rotor blade elements operating at a distance r from the axis of rotation and rotating with an angular velocity Ω may be written as

$$dL = c_l(bc\,dr)(\tfrac{1}{2}\rho V_R^2) \tag{13}$$

For purposes of simplification, advantage may be taken of the fact that the flow through the disk is small compared with the tangential velocity, so that thc following can be assumed:

$$\begin{aligned} \sin \phi &= \phi \\ \cos \phi &= 1 \\ V_R &= \Omega r \end{aligned} \tag{14}$$

The blade-element lift coefficient may be expressed as follows:

$$\begin{aligned} c_l &= a\alpha_r \\ &= a(\theta - \phi) \end{aligned} \tag{15}$$

where a is the slope of the curve of section lift coefficient against angle of attack.

With the aid of equations (14) and (15) the differential lift or thrust expression becomes:

$$dT = dL = b\tfrac{1}{2}\rho(\Omega r)^2 a(\theta - \phi)c\,dr \tag{16}$$

For simplicity of integration, it will be assumed that the pitch angle of a blade element will vary with its radial position r in the following manner

$$\theta = \theta_t \frac{R}{r} \tag{17}$$

where θ_t is the pitch angle at the blade tip. It will be shown in the following chapter that the pitch distribution given by equation (17) results in a uniform inflow distribution along the blade span. Such a distribution is therefore labeled *ideal twist*. It will also be shown that ideal twist results in the following variation of inflow angle ϕ with r

$$\phi = \phi_t \frac{R}{r} \tag{18}$$

where ϕ_t is the inflow angle at the blade tip.

After substituting equations (17) and (18), equation (16) becomes

$$dT = b\frac{1}{2}\rho(\Omega r)^2 a\frac{R}{r}(\theta_t - \phi_t)c\,dr \qquad (19)$$

Integrating equation (19) over the blade radius, assuming the blade chord c constant, the thrust of the rotor is

$$T = \frac{b}{2}\rho\Omega^2 a\frac{R^3}{2}(\theta_t - \phi_t)c \qquad (20)$$

Equating this equation to the expression for thrust as given by equation (11), an expression for the rotor thrust coefficient is obtained.

$$T = b\frac{1}{2}\rho\Omega^2 a\frac{R^3}{2}(\theta_t - \phi_t)c = C_T\pi R^2\rho(\Omega R)^2$$

$$C_T = \frac{a}{4}\frac{bc}{\pi R}(\theta_t - \phi_t) \qquad (21)$$

The term *solidity* will now be introduced. The solidity, σ, of a rotor having rectangular blades[1] may be defined as the ratio of the total blade area to the rotor disk area. Thus

$$\sigma = \frac{bcR}{\pi R^2} = \frac{bc}{\pi R} \qquad (22)$$

Substituting $\sigma = bc/\pi R$ into equation (21), the expression for C_T becomes

$$C_T = \frac{\sigma}{4}a(\theta_t - \phi_t) \qquad (23)$$

Equation (23) expresses the thrust of an ideally twisted, constant-chord blade.

EXPRESSION FOR TORQUE. By referring again to Fig. 3–5, it can be seen that the drag of the blade element is composed of two parts: (1) the profile drag and (2) the induced drag, which consists of the components of lift in the plane of rotation arising from the tilt of the lift vector caused by the inflow velocity.

Inasmuch as the torque about the axis of rotation resulting from the drag on the element is dD times r, then

$$dQ = b\tfrac{1}{2}\rho(\Omega r)^2 c(c_{d_o} + \phi c_l)r\,dr \qquad (24)$$

[1] The calculation of the solidity of a rotor having blades of any plan form will be considered in a later section of this chapter.

Before this equation can be integrated, the variation of c_{d_o} with r must be known, which in turn requires a knowledge of the variation of α_r with r. (The variation of ϕc_l is known because ideal twist was assumed.)

The expression for the blade-section angle of attack for an ideally twisted blade is

$$\alpha_r = \frac{R}{r}(\theta_t - \phi_t) = \frac{1}{x}(\theta_t - \phi_t) \tag{25}$$

Because the angle of attack of the element varies with the spanwise position of the element, the profile-drag coefficient will do likewise, unless the blade section characteristics are such that the variation of c_{d_o} with α_r is relatively small over the operating range of the blade. The assumption of a constant profile-drag coefficient is satisfactory in many cases, provided that its limitations are understood. For example, the drag coefficient increases rapidly with increasing angles of attack above the stall. If this fact is overlooked, the use of the constant drag coefficient will result in highly optimistic rotor performance for conditions in which blade stall is present.

For the present case, assume that $c_{d_o} = \delta$, where δ is an average blade profile-drag coefficient. Substituting the following relations

$$\begin{aligned} c_l &= a\frac{R}{r}(\theta_t - \phi_t) \\ \phi &= \phi_t \frac{R}{r} \\ c_{d_o} &= \delta \end{aligned} \tag{26}$$

into equation (24), the differential expression for torque becomes

$$dQ = b\frac{1}{2}\rho\Omega^2 r^3 c\left[\delta + \phi_t \frac{R^2}{r^2}(\theta_t - \phi_t)a\right] dr \tag{27}$$

Upon integrating over the blade, the torque equation is

$$Q = \frac{b}{4}\rho\Omega^2 R^4 c\left[\frac{\delta}{2} + a\phi_t(\theta_t - \phi_t)\right] \tag{28}$$

Equating the above to the expression involving the torque coefficient as given by equation (11), and substituting $\sigma = (bc)/(\pi R)$, equation (28) becomes

$$C_Q = \frac{\sigma}{4}\left[\frac{\delta}{2} + a\phi_t(\theta_t - \phi_t)\right] \tag{29}$$

Substituting equation (23) into (29), and rearranging terms

$$C_Q = \frac{\sigma\delta}{8} + \phi_t C_T \tag{30}$$

In order to use equation (30) easily, it is necessary to replace ϕ_t by parameters that are known or easily determined. This is done as follows: From equation (6)

$$v = \sqrt{\frac{T}{2\rho\pi R^2}} = \sqrt{\frac{C_T \pi R^2 \rho (\Omega R)^2}{2\rho\pi R^2}} = \Omega R \sqrt{\frac{C_T}{2}} \tag{31}$$

and, by definition

$$\phi_t = \frac{v}{\Omega R} \tag{32}$$

Combining equation (31) and (32)

$$\phi_t = \sqrt{\frac{C_T}{2}} \tag{33}$$

Substituting equation (33) into (30), the torque equation becomes

$$C_Q = \frac{C_T^{3/2}}{\sqrt{2}} + \frac{\sigma\delta}{8} \tag{34}$$

M AS A FUNCTION OF C_T AND δ. Equation (34) expresses the hovering performance of an ideally twisted constant-chord blade of solidity σ and average profile-drag coefficient δ. The first part of the equation represents the induced loss, while the second part expresses the profile-drag loss.

If equation (34) is substituted into the formula for rotor figure of merit, an expression is obtained that will help bring about an understanding of the factors to be considered in the design of a rotor for the hovering condition. This expression is

$$M = 0.707 \frac{C_T^{3/2}}{C_Q}$$

$$= 0.707 \frac{C_T^{3/2}}{\dfrac{C_T^{3/2}}{\sqrt{2}} + \dfrac{\sigma\delta}{8}} \tag{35}$$

If $\delta = 0$ (zero profile drag), equation (35) yields the same result that was previously found, namely that $M = 1$. In addition, it can be seen again that M is independent of rotor operating conditions such as disk loading or tip speed.

The variation of M with C_T for typical values of σ and δ is shown in Fig. 3–6. The shape of the curve may be explained as follows. If the rotor is operating at a small thrust coefficient it can be seen from equation (35) that M would be small, too, inasmuch as the profile-drag term in the denominator is fixed and large compared to the numerator.

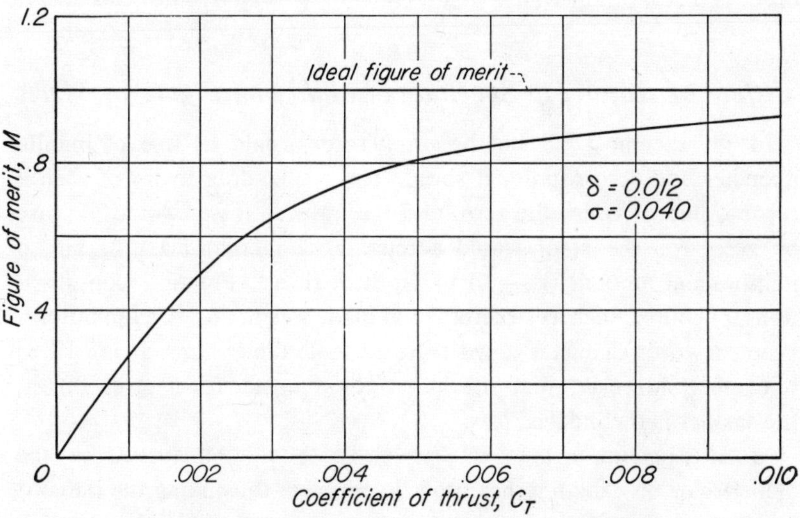

Fig. 3–6 Figure of merit versus thrust coefficient.

As C_T increases, however, the relative importance of the δ term decreases, thus allowing M to increase. This continues until at a large enough C_T, the profile-drag term is so small compared to the induced terms that any increase in C_T results in an extremely small increase in M.

In other words, if the rotor is producing little thrust, or if the rotor tip speed is very high, both factors resulting in a low C_T, the profile drag will be comparatively high for the thrust produced because the blade elements have drag even when their lift is zero. Thus the rotor is not operating very efficiently, for it is absorbing power and producing little thrust. As the thrust of the rotor increases from zero, however, its figure of merit increases, for the rate of increase of thrust is greater than the increase in power required to produce thrust. At very high thrust values, the rate of change of figure of merit with C_T is not as rapid, for the induced losses become a larger percentage of the total loss. If δ is fixed, equation (35) shows that M will asymptotically approach 1 as C_T is increased. For an actual design, however, large increases in C_T would result in stalling of the blade sections, which in turn would result in a large increase in profile drag and a falling off of thrust, both of which would contribute to a large decrease in M.

Effect of Rotor Tip Speed and Solidity on Figure of Merit

Ideally the most efficient hovering rotor would be one of infinite diameter and zero rotational speed. The profile-drag losses of such a rotor would of course be zero, and the induced losses would likewise be zero, for the rotor would accelerate an infinite mass of air an infinitesimal amount ($v \rightarrow 0$) to produce thrust. Practical considerations, of course, such as structural and blade weight and size limitations, keep the rotor diameter down to reasonable dimensions. It is well to remember, however, that the larger the diameter for a given thrust, the smaller is the induced loss.

OPTIMUM COMBINATIONS OF TIP SPEED AND SOLIDITY. Once the diameter or disk loading has been determined, thus fixing the induced losses, the next step is to fix the optimum combination (for the hovering condition) of solidity and rotor tip speed for minimum profile-drag losses. The choice of these two parameters is dependent upon two considerations that are somewhat interrelated, (1) the rotor should operate at the mean lift coefficient closest to the stalling angle of the blade section, and (2) the rotor should operate at the lowest feasible tip speed. The use of a low tip speed is equivalent to having the greatest possible solidity, or in this case, the greatest blade chord, inasmuch as the rotor diameter has already been fixed.

Both of the above considerations are based on the fact that rotor thrust varies as the tip speed squared, whereas profile power varies as the cube of the tip speed. For a given thrust, the cube law results in the smallest amount of profile-drag power loss at the smallest tip speed. It follows immediately, therefore, that since the thrust depends on the mean lift coefficient $\overline{C}_L$ at which the rotor operates as well as the blade area and the rotor tip speed, the thrust should be produced by a high mean lift coefficient and a low tip speed (and hence a high solidity).

EFFECT OF TIP SPEED. The effect of rotor tip speed on the hovering performance of a helicopter of fixed weight and rotor solidity is shown for a typical case in Fig. 3–7 (from reference 1–7, Appendix IIA).

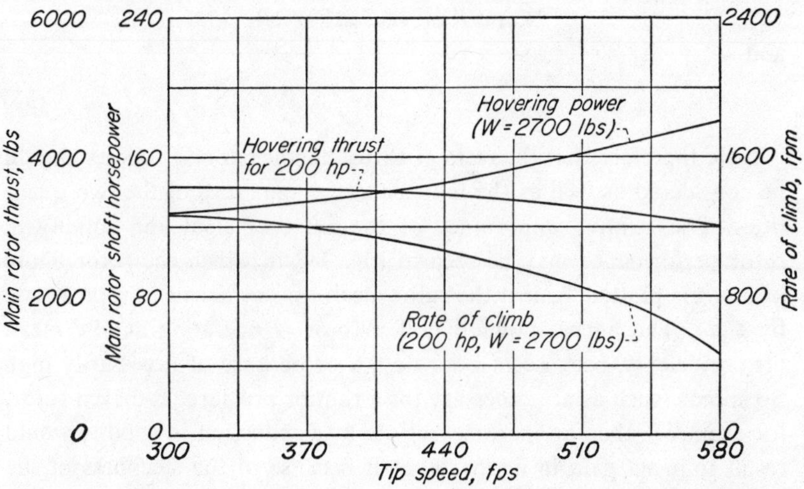

Fig. 3–7 Effect of rotor tip speed on hovering and vertical climb performance.

A reduction in tip speed from 600 feet per second to 400 feet per second results in a 50-horsepower (26 per cent) decrease in power required to produce a fixed thrust, or an 800-pound (25 per cent) increase of thrust for a constant power input. These changes are more significant when useful load is considered; for example, although an increase in take-off thrust of 800 pounds is a 25 per cent increase in total thrust, it represents approximately a 100 per cent increase in useful load for the helicopter considered. These gains are also shown in Fig. 3–7 in terms of increased vertical climb performance. Thus, for a fixed power and helicopter gross weight, a reduction in tip speed from 600 feet per second to 400 feet per second results in changing the rate at which the helicopter can climb vertically from 200 feet per minute to approximately 1150 feet per minute.

RELATIONSHIP BETWEEN $\overline{C}_L$, C_T, AND σ. The rotor solidity σ, mean lift coefficient $\overline{C}_L$, and tip speed ΩR are not independent of each other, but are related by a simple equation. In the hovering condition, the gross weight of the helicopter W is equal to the rotor thrust and rotor lift. Therefore,

$$W = \overline{C}_L \int_0^R b\tfrac{1}{2}\rho(\Omega r)^2 c\, dr = C_T \pi R^2 \rho(\Omega R)^2$$

$$\tfrac{1}{6}\overline{C}_L \rho\Omega^2 R^3 bc = C_T \pi R^2 \rho(\Omega R)^2$$

and

$$\overline{C}_L = 6\,\frac{C_T}{\sigma} \tag{36}$$

Note therefore that the ratio of thrust coefficient and solidity should be considered as well as the independent magnitudes of the two quantities. The relative importance of the lift coefficient and solidity in rotor performance may be seen in Fig. 3–8 in which the rotor figure of merit is plotted against the rotor mean lift coefficient as represented by C_T/σ. The figure shows that if a rotor is operating at low mean lift coefficients, because of poor design or because of necessarily high tip speeds (such as are necessary for a ram or pressure jet-driven rotor, for example), then an increase in C_T/σ by a reduction in solidity would result in a net gain in figure of merit because of the steepness of the figure of merit curves in that region. At high values of C_T/σ, however, a net gain in performance can only be obtained (at a fixed value of C_T), by an increase in solidity, because of the relative flatness of the curves in that region.

It might be remarked that the comparative hovering performance of various rotors may be shown to advantage on such a plot of M against C_T/σ, for by comparing the rotors at the same mean lift coefficient, the factor of possible differences in operating conditions is removed, and the rotors can be judged by their physical design alone. Note that although plotting against C_T/σ provides a comparision at equal mean lift coefficients, it does not altogether eliminate the primary effects of solidity, for the lower solidity rotor must operate at a higher tip speed to provide the same thrust at the same value of C_T/σ. This increased tip speed acts to increase the profile-drag power losses, and thus results in a lower figure of merit, as can be seen from Fig. 3–8.

Practical considerations, of course, place a lower limit on rotor speed. One of the most important is safety in case of sudden power failure, for the more kinetic energy in the blades, the greater is the time margin possessed by the pilot to lower his pitch to a value that

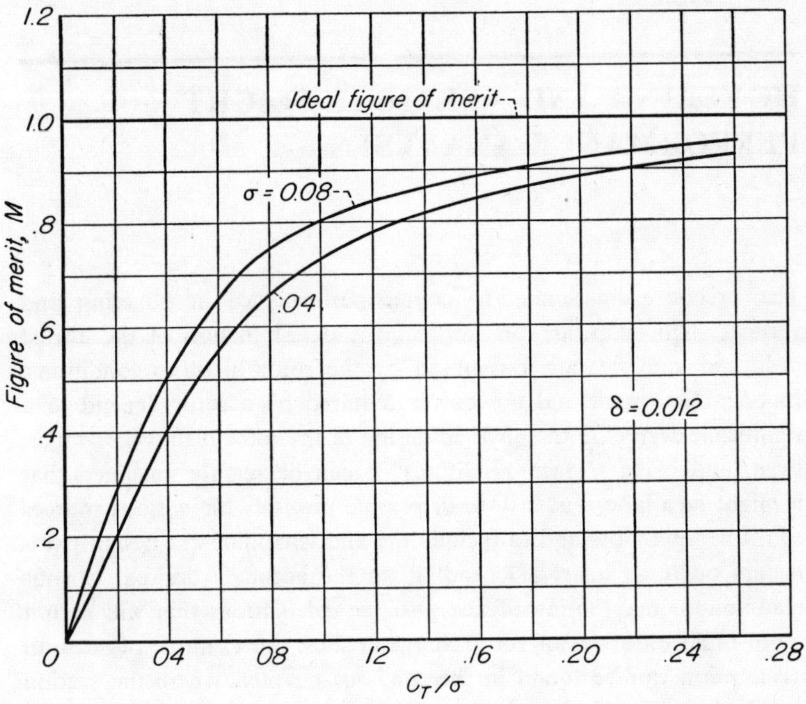

Fig. 3–8 Figure of merit versus thrust coefficient solidity ratio.

will permit autorotation. Large coning angles and the poor psychological effects of slowly revolving blades also tend to fix a lower limit on rotor speeds.

It should be emphasized that this section concerns the hovering rotor and does not constitute design criteria for rotors which must also operate in conditions of high forward speed. It will be shown in later chapters that requirements for efficient and smooth operation at high speeds conflict with hovering design requirements so that a design compromise must be reached.

4

HOVERING AND VERTICAL-FLIGHT PERFORMANCE ANALYSES

The precise estimation of helicopter performance in hovering and vertical flight depends on an accurate determination of thc thrust produced and the power required by the rotor in those conditions. In turn, the thrust and the power required by a rotor depend to a significant degree on the physical design of the rotor blades (i.e., taper, twist, and blade surface condition). It can be readily imagined that it might be a laborious task to determine precisely the various sources of rotor power loss and to include the effects of different design parameters on these losses. Depending on the accuracy desired, various assumptions may be introduced into the calculations that will permit more rapid estimates of rotor characteristics. An example of such an assumption can be found in the previous chapter, where the section profile-drag coefficient was assumed not to vary with the radial position of the section in calculating the gross effects of profile drag on figure of merit.

In this chapter a general method will be developed for accurately computing rotor performance which will involve a minimum of assumptions. The results of this analysis will be compared with experimental data on full-scale rotors in order to prove the validity of the theory. A method will then be developed whereby the performance of different rotors may be quickly estimated for preliminary design and evaluation purposes.

The value of the induced velocity at each spanwise blade station must be known before it is possible to calculate the angle of attack and the thrust and torque contributed by each blade element. The calculation of the induced velocity will thus be considered first.

General Equation for Induced Velocity

The momentum and blade-element theories may be combined to derive a general expression for the velocity induced at any point on a helicopter rotor that is hovering or climbing vertically.

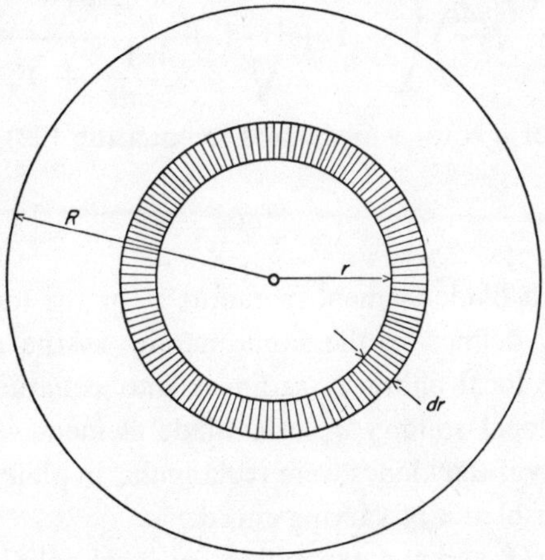

Fig. 4–1 Annulus of rotor disk.

From equation (16) of Chapter 3, the differential thrust produced by b blade elements, as given by simple blade-element theory, is

$$dT = b\tfrac{1}{2}\rho(\Omega r)^2 a(\theta - \phi)c\, dr \tag{1}$$

An expression for the differential thrust can also be obtained from momentum considerations. Consider an annular ring of an airscrew disk, of thickness dr and of radius r (Fig. 4–1). Thrust produced by this ring is equal to the mass of air passing through it per unit time, multiplied by the change in velocity of the mass. In equation form

$$\begin{aligned} dT &= (2\pi r\, dr\, \rho u)\, 2v \\ &= 4\pi\rho u v r\, dr \end{aligned} \tag{2}$$

The two thrust expressions can be equated as follows:

$$b\frac{1}{2}\rho\Omega^2r^2a\left(\theta - \frac{u}{\Omega r}\right)c\,dr = 4\pi\rho uvr\,dr \tag{3}$$

Substituting $u = V_v + v$ and expanding,

$$4\pi v^2 + \left(4\pi V_v + \frac{b}{2}\Omega ac\right)v - \frac{b}{2}c\Omega^2 ra\left(\theta - \frac{V_v}{\Omega r}\right) = 0 \tag{4}$$

Solving the above quadratic for v gives

$$v = \left(\frac{V_v}{2} + \frac{bca\Omega}{16\pi}\right)\left(-1 + \sqrt{1 + \frac{2\Omega r\left(\theta - \dfrac{V_v}{\Omega r}\right)}{\dfrac{4\pi V_v^2}{bca\Omega} + V_v + \dfrac{bca\Omega}{16\pi}}}\right) \tag{5}$$

The solidity of a rotor was defined in equation (22) of the preceding chapter as

$$\sigma = \frac{bc}{\pi R} \tag{6}$$

The solidity of a blade element at radius x, or the local solidity, may be conveniently defined in the same manner as the rotor solidity by substituting the local chord at radius x into equation (6). With this definition, the local solidity σ_x of a blade element would then equal the rotor solidity if the blades were rectangular in plan form, but would differ from it for blades of varying chord.

Putting $x = r/R$ and the expression for local solidity into equation (5), the general expression for the induced velocity at a blade element becomes

$$v = \left(\frac{V_v}{2} + \frac{\sigma_x a\Omega R}{16}\right)\left(-1 + \sqrt{1 + \frac{2(\theta x\Omega R - V_v)}{\dfrac{4V_v^2}{\sigma_x a\,\Omega R} + V_v + \dfrac{\sigma_x a\Omega R}{16}}}\right) \tag{7}$$

The identical expression for v may be derived by means of the vortex theory. Such a derivation can be found in reference I–15 (Appendix IIA).

Once the induced velocity is known, the inflow angle at a blade element can be found by means of the following expression (see Fig. 3–5):

$$\phi = \frac{v + V_v}{\Omega Rx} \tag{8}$$

It can be seen that equation (7) is completely general in regard to blade plan form and pitch distribution, for it enables the calculation of the inflow velocity or angle produced by a blade element operating at a radial distance x and having any chord c and pitch angle θ. The application of the equation to the calculation of the hovering and vertical-climb performance of a rotor of any plan form, twist, and airfoil section will be demonstrated later in the chapter.

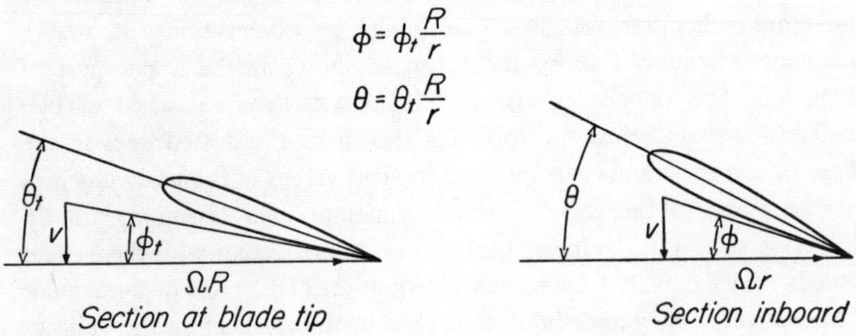

Fig. 4–2 Element diagrams illustrating ideal twist.

Referring to equation (7), it is seen that the induced velocity v can be made constant across the disk by allowing the pitch angle θ to vary inversely with x, so that

$$\theta = \theta_t \frac{R}{r} = \frac{\theta_t}{x} \tag{9}$$

where θ_t is the blade-tip pitch angle. In terms of the blade-tip pitch and inflow angle, the angle of attack becomes

$$\alpha_r = \frac{\theta_t}{x} - \frac{\phi_t}{x} = \frac{1}{x}(\theta_t - \phi_t) \tag{10}$$

The pitch-angle distribution given by equation (9) was defined in Chapter 3 as the *ideal* twist, because it yields the minimum induced loss for a given thrust. Such a twist appears to be logical, because the reduced tangential velocity near the inboard end of the blade increases the inflow angle of attack inversely with the radius so that the pitch angle must be likewise increased in order to maintain a positive angle of attack; this point is illustrated in Fig. 4–2. Strictly speaking, the assumption of uniform inflow distribution (ideal twist) demands a different twist distribution for each thrust coefficient or rate of climb because a higher C_T or V_v calls for a larger blade-tip angle because of the greater inflow. Inasmuch as the root angle is 90 degrees in both cases, the distribution of pitch angle along the blade must then vary.

An experimental check of the theoretical induced velocity equation in hovering was obtained by British flight tests of a full-scale helicopter (reference 13, Appendix IIB). The shape of the slipstream beneath the hovering helicopter was first determined by observations of smoke filaments introduced above the rotor, as shown in the upper part of Fig. 4–3. The velocity of the slipstream was then measured at two different heights below the rotor (as shown by the dotted lines in the figure), and the results extrapolated to yield values of induced velocities at the plane of the rotor. The experimental data (assuming no tip losses) is shown in the lower part of Fig. 4–3, together with theoretical values of v calculated by means of equation (7). Excellent agreement between the theory and the data is shown, thus indicating the validity of the combined blade-element-momentum theory.

General Performance Equations in Hovering

It is not always feasible to obtain explicit analytic expressions for the characteristics of rotors having arbitrary plan forms and airfoil sections in terms of a nonuniform inflow distribution. In such cases, use must be made of numerical integration methods for performance calculations similar to those used in standard propeller analyses. The general method for carrying out such calculations will now be discussed.

The thrust per unit of blade span may be expressed in coefficient form as

$$\frac{dC_T}{dx} = \sigma_x \frac{a}{2} \alpha_r x^2 \tag{11}$$

The rotor torque is composed of the induced and profile-drag contributions. The induced part, or the torque caused by the components of the lift vectors in the plane of rotation, is written as

$$\frac{dC_{Qi}}{dx} = \sigma_x \frac{a}{2} \phi \alpha_r x^3 \tag{12}$$

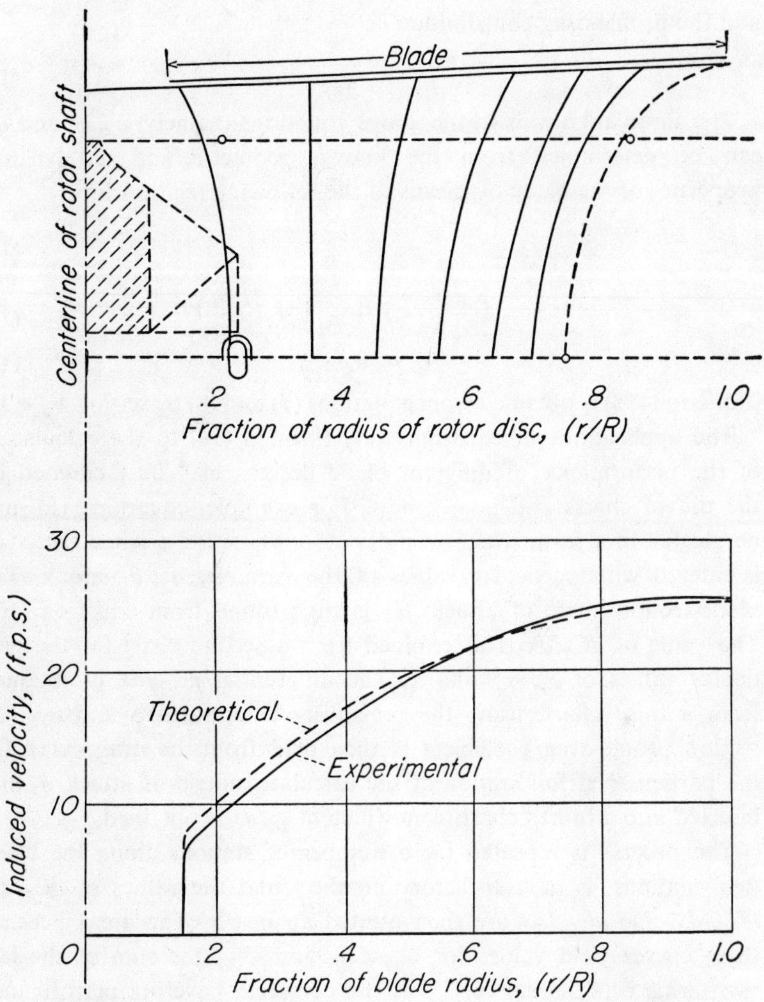

Fig. 4–3 Rotor inflow measurements in flight (from Appendix IIB, reference 13).

and the profile-drag contribution is

$$\frac{dC_{Q_o}}{dx} = \frac{\sigma_x}{2} c_{d_o} x^3 \tag{13}$$

The three unknowns in the above equations (namely, σ_x, ϕ, and α_r) can be determined from the known geometric and aerodynamic properties of the blade by means of the following relationships:

$$\sigma_x = \frac{bc}{\pi R} \tag{14}$$

$$\phi = \frac{\sigma_x a}{16x}\left(-1 + \sqrt{1 + \frac{32x\theta}{\sigma_x a}}\right) \tag{15}$$

$$\alpha_r = \theta - \phi \tag{16}$$

[Equation (15) is obtained from equations (7) and (8) by setting $V_v = 0$.]

The applications of equations (11) through (16) to the calculation of the performance of different blade designs may be facilitated by the use of charts and nomograms. For example, equation (15) may be plotted in a form which would yield a value for ϕ when the chart is entered with particular values of the variables σ_x, θ, and x. The blade-section angle of attack, α_r, is then found from equation (16). The value of dC_T/dx is determined from a second chart for the particular values of α_r, σ_x, and x. The differential dC_{Q_i}/dx is obtained from a third chart, using the previously determined ϕ and α_r. The section profile-drag coefficient is then read from the drag curve for the particular airfoil section at the calculated angle of attack α_r and inserted into a fourth chart from which dC_{Qo}/dx is obtained.

The process is repeated for a number of stations along the blade (ten stations is a satisfactory number) and the values of dC_T/dx, dC_{Q_i}/dx, and dC_{Qo}/dx are then plotted against x. The areas beneath these curves yield values for C_T, C_{Q_i}, and C_{Q_o}, the sum of the last two terms being equal to C_Q. If the complete hovering performance is desired for a rotor over a large range of thrust coefficients, the process must be repeated for a number of pitch angles θ.

Tip Losses

The procedure just described is particularly useful in obtaining comparative estimates of the performance of various rotors. (An example of the application of the procedure to rotors differing in blade twist and plan-form taper will be found in the following chapter.) When the analysis is used to calculate the "absolute" performance of a rotor, however, a factor should be included in the calculations to take into account the reduction in thrust near the blade tips resulting from the use of a finite number of blades. These thrust losses, or "tip losses," are caused by the "spilling" of air around the blade tip.

In order to produce lift there is of necessity a difference of pressure between the upper and lower surfaces of the blade. Air at the blade tip thus tends to flow from bottom to top, destroying the pressure difference, and thus the lift in the tip region. The extent of tip losses depends on the load per foot carried by the blade. The thrust coefficient and the number of blades are therefore important variables in determining tip loss, in that the thrust coefficient defines the total loading, and the number of blades defines the loading per blade. Tip losses increase as C_T increases and decrease as the number of blades increase. Inasmuch as the loss is dependent on the load per foot of blade it is essentially independent of the particular combination of blade chord (solidity) and angle of attack which make up this load. The tip-loss factor is therefore essentially independent of solidity.

An approximate solution to the problem of calculating tip losses was published in 1927 by Prandtl and Betz and is contained in reference 18 (see Appendix IIB). Their analysis is considered a good approximation for lightly loaded rotors and yields, when further approximated, a simple and physically understandable result; namely,

$$B = 1 - \frac{\sqrt{2C_T}}{b} \tag{17}$$

where B is tip-loss factor; blade elements outboard of radius BR are assumed to have profile drag but no lift

C_T is rotor thrust coefficient

b is number of blades per rotor

Thus it is seen that the tip loss varies as the square root of the thrust coefficient and inversely as the number of blades. Tip losses are zero for zero loading or for an infinite number of blades. Equation (17) is plotted in Fig. 4–4 for rotors of two, three, and four blades.

A more rigorous solution to the problem of three-dimensional flow around a propeller blade was published by S. Goldstein in 1929 (reference 17 of Appendix IIB), and applied to propeller data by Lock during the following year (reference 11 of Appendix IIB). The results of the Goldstein theory can be applied to helicopter rotors if greater accuracy in performance calculations is desired. (See reference 12 of

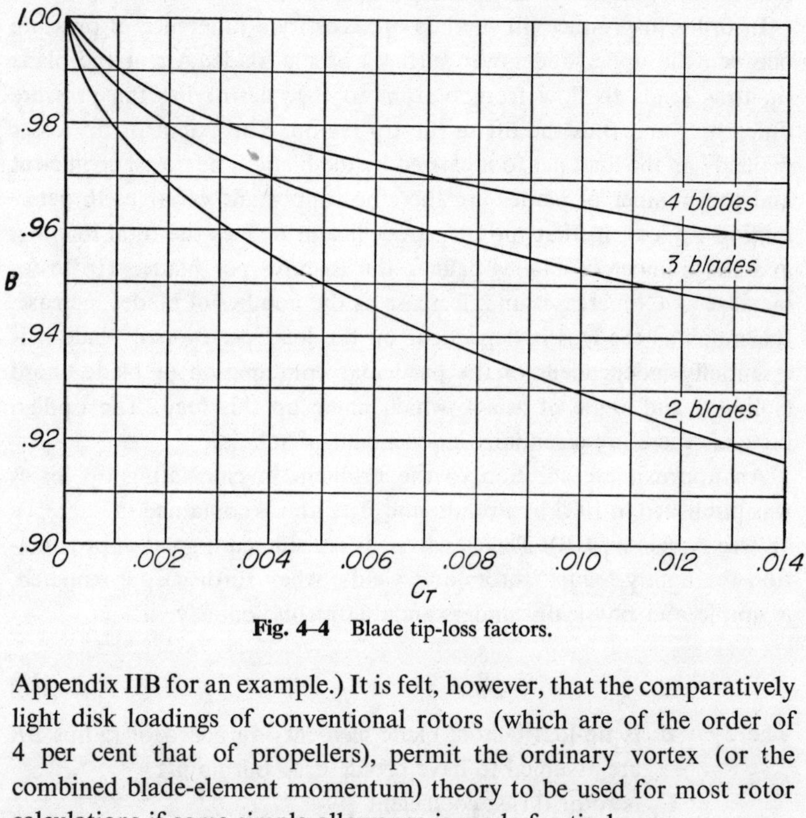

Fig. 4-4 Blade tip-loss factors.

Appendix IIB for an example.) It is felt, however, that the comparatively light disk loadings of conventional rotors (which are of the order of 4 per cent that of propellers), permit the ordinary vortex (or the combined blade-element momentum) theory to be used for most rotor calculations if some simple allowance is made for tip losses.

One such allowance which has been widely used and which yields satisfactory results in performance calculations is to assume that *a tip length equal to one-half the tip chord* develops no thrust [or that $B = 1 - (\text{tip chord}/2R)$]. For blades of conventional plan form and design, this assumption results in a tip-loss factor B equal to about

0.97. This figure was substantiated by the previously mentioned flight tests of reference 13 (Appendix IIB) which reported the fact that tip losses reduce the effective radius by approximately 3 to 4 per cent. The manner in which tip losses are taken into account will be shown subsequently in the chapter.

Experimental Data

For many years, reliable experimental data on the performance of helicopters have been extremely limited. Although a fair amount of wind-tunnel data on model rotors were available, the small scale at which these tests were conducted and the lack of knowledge of wind-tunnel wall corrections for rotors cast considerable doubt upon the results. In general, experience has indicated that even comparative performance tests on small-scale models should be avoided. The results of such tests are of value only in conjunction with careful theoretical analysis, and any significant conclusions would have to be checked by full-scale tests before they could be universally accepted. Also, the usual ranges of tolerance in the construction of small models, as regards surface roughness and distortion of contour, usually introduce significant errors in profile-drag and lift-slope values.

It should be realized that these comments do not necessarily cover all types of small-scale rotor testing. Stability tests on dynamic models, for example, are of interest with regard to the effects of large changes in the various parameters affecting the stability of the helicopter. Model tests are also valuable in vibration studies and the investigation of different control type arrangements and various rotor configurations. The effect of wind-tunnel wall constrictions and the presence of a ground should, however, be studied in all types of test work to see if any such effects are present.

An example of careful analysis of the results of small-scale static thrust tests on model rotors was the investigation made by Montgomery Knight and reported in reference I–15 of Appendix IIA. Knight investigated the rotor parameters that are functions of scale and concluded that the primary ones are the slope of the lift curve and the profile-drag curve. He then used his test results as a basis for deriving full-scale values for these parameters. It was found that up to the stall region a value of lift curve slope equal to 5.75 would closely represent the symmetrical airfoil sections used in conventional rotors.[1] Assuming a parabolic profile-drag curve of the form $c_{d_o} = \delta + \epsilon\alpha_r^2$, the median full-scale value of ϵ was found to be equal to 0.3 and it was shown that δ varied from 0.006 for large rotors with thin blade profiles, to 0.012 for small rotors with thick profiles, and that $\delta = 0.010$ represented the full-scale median.

Although Knight's results provided a rapid and fairly accurate method of estimating and checking rotor hovering performance, the theory could not be considered verified until full-scale flight data were obtained with precise instrumentation and with careful experimental techniques. Such data were obtained by flight tests reported in references I–4 and I–10 of Appendix IIA.

The tests were conducted with two sets of main rotor blades, each set differing from the other in solidity, surface condition, amount of twist, and airfoil section. The experimentally determined performance of the two rotors is shown in Fig. 4–5, together with theoretical performance curves computed by the graphical integration method (obtained from reference I–5 of Appendix IIA). The calculations assumed a tip-loss factor of $B = 0.97$. Profile-drag coefficients of the "rough" fabric-covered set of blades, in accordance with the results of wind-tunnel tests on sections of the actual blade, were increased over the "semi-smooth" blade values by a factor of 1.28. The other set of blades, which were of lower solidity, were of smooth plywood construction, and the "semi-smooth" polar with no roughness correction factor was assumed. It may be seen from the figure that the discrepancy between theory and experiment is only a few per cent and that the theory is well within engineering-design accuracy. It was thus shown that the over-all predictions of the theory may be used with confidence for engineering design and analysis.

In addition to providing a check on hovering theory, the test results of reference I–4 of Appendix IIA indicated the important effects of

[1] Most NACA work is based on a value of lift curve slope equal to 5.73 for convenience of calculation, for then the lift coefficient equals 0.1 of the angle of attack in degrees.

rotor solidity and blade-surface condition on hovering performance. The comparative performance of the two rotors tested is shown in Fig. 4–6 in terms of rotor figure of merit. It can be seen that the plywood covered rotor was about 15 per cent more efficient than the fabric-covered rotor. This difference corresponds to an increase in thrust at a fixed power of about 330 pounds for a 2700-pound machine.

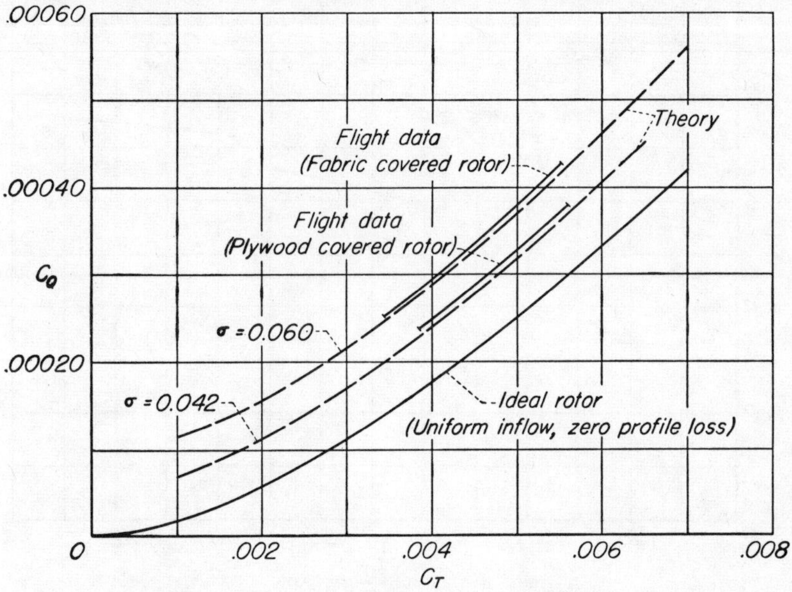

Fig. 4–5 Flight check of hovering theory.

Theory indicates that approximately one-half of this performance difference may be accounted for by the difference in solidities of the two rotors. A reduction in solidity is advantageous when the tip speed is fixed and the blade sections are operating at an angle of attack below the optimum (C_T/σ too low). Thus, in the present case, the 0.042 solidity rotor operated at mean blade angles of attack closer to the optimum than did the 0.060 solidity rotor with a consequent saving in profile-drag power.

Aside from the few per cent savings in induced power brought about by blade twist, most of the remaining performance gains may be attributed to the smoother, more rigid surfaces of the plywood-covered blades of the lower solidity rotor, as compared with the fabric-covered surfaces of the 0.060 solidity rotor blades. The importance of smooth, rigid-surfaced rotor blades was also demonstrated by flight tests in the power-on and power-off forward flight conditions.

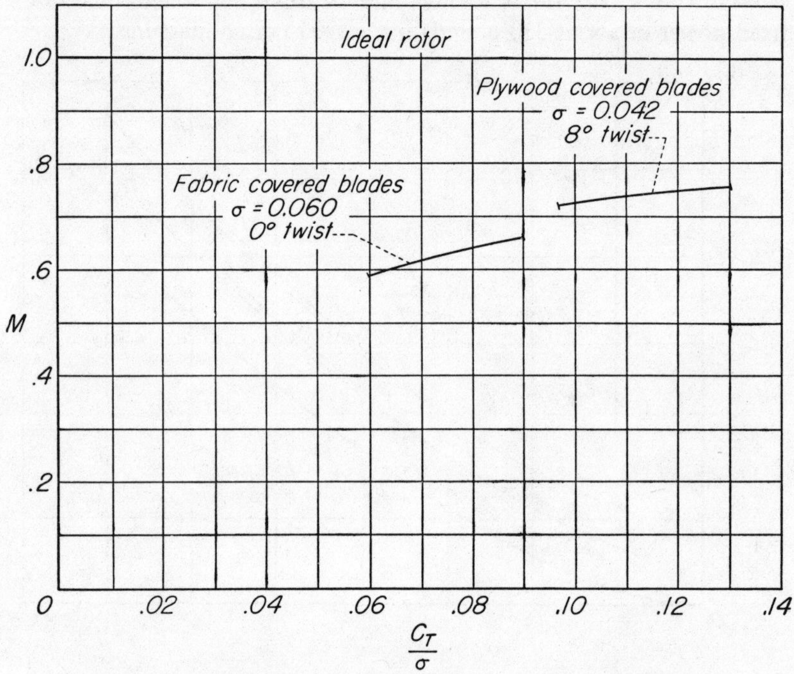

Fig. 4–6 Comparative hovering performance measurements (from Appendix IIA, reference I–4).

Performance of Ideally Twisted, Constant-Chord Blades

One of the few types of rotor-blade designs for which explicit thrust and torque expressions can be obtained is blades having ideal twist (i.e., uniform inflow) and rectangular plan form. Although present-day rotors are not constructed with nonuniform twist, the performance of such a rotor is very useful as an easily computed base from which to judge practical rotor designs.

Thrust and torque equations have already been developed in the previous chapter for a hovering rotor having ideally twisted, constant-chord blades. The profile-drag torque was calculated for that rotor with the assumption that each element of the blade operated at a constant drag coefficient regardless of the section angle of attack. Although this assumption is correct enough for certain calculations if the limitations arising from it are understood, the use of a constant drag coefficient can result in optimistic performance predictions, especially when the rotor is operating at angles of attack in the vicinity of the stall. In this section, the restriction of a constant profile-drag coefficient is removed by employing an analytical expression to represent the actual variation of the drag coefficient with blade-section angle of attack up to the stalling angle. The equations are also made more general by including vertical climb in addition to hovering and by taking tip losses into account.

EXPRESSION FOR THRUST. The differential thrust expression for a rotor having b blades is

$$dT = b\tfrac{1}{2}\rho\Omega^2 r^2 c_l c \, dr \tag{18}$$

For the case of ideal twist,

$$c_l = a(\theta - \phi) = a\frac{R}{r}(\theta_t - \phi_t) \tag{19}$$

Substituting equation (19) into equation (18) and replacing r by xR, the total expression for thrust becomes

$$T = \int_0^B b\tfrac{1}{2}\rho\Omega^2 R^3 a x(\theta_t - \phi_t)c \, dx \tag{20}$$

The upper limit of the integral is taken to be $x = B$ instead of $x = 1.0$ in order to take into account the loss of thrust at the rotor tip.

Upon integrating equation (20), the following is obtained:

$$\begin{aligned} T &= b\frac{1}{2}\rho\Omega^2 R^3 a(\theta_t - \phi_t)c\left.\frac{x^2}{2}\right]_0^B \\ &= b\frac{1}{2}\rho\Omega^2 R^3 a(\theta_t - \phi_t)c\frac{B^2}{2} \end{aligned} \tag{21}$$

Equating equation (21) to

$$T = C_T \pi R^2 \rho(\Omega R)^2$$

and letting $\sigma = bc/\pi R$, the generalized thrust expression becomes

$$\frac{C_T}{B^2} = \frac{\sigma a}{4}(\theta_t - \phi_t) \tag{22}$$

EXPRESSION FOR INDUCED AND CLIMB TORQUE. The total power expended by the rotor in the vertical-climb condition is used in producing thrust (represented by the induced loss), in overcoming the profile drag of the blades and in raising the helicopter at a fixed vertical velocity. (Inasmuch as the fuselage parasite drag is usually small in climb, it will be ignored.)

The power employed in causing the helicopter to climb is absorbed by the rotor in a manner similar to the action of induced torque (i.e.,

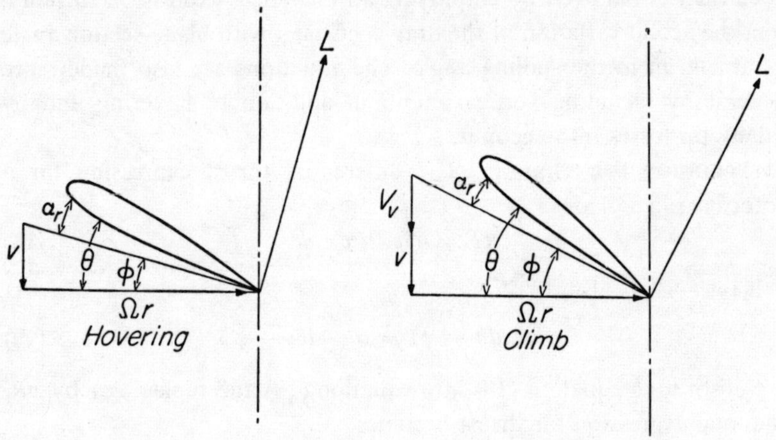

Fig. 4–7 Effect of climb on inclination of lift vector.

torque resulting from a component of the lift vector in the plane of rotation). Figure 4–7 illustrates how the lift vector of a blade element in climb is inclined further from the plane of rotation than it is in hovering, because of the climb velocity. The increased component of lift in the plane of rotation in climb results in an increase in rotor torque. Figure 4–7 also shows that a higher pitch angle is required in the climb condition in order to obtain the same angle of attack (and the same thrust) as in hovering.

Inasmuch as the induced and climb torques are produced in the same manner, a combined expression for both may be written as follows:

$$Q_i + Q_{\text{climb}} = \int_0^B b\,\frac{1}{2}\,\rho(\Omega r)^2 a\,\frac{R}{r}\,(\theta_t - \phi_t)\,\frac{R}{r}\,\phi_t rc\,dr \qquad (23)$$

Placing $x = r/R$, equation (23) becomes

$$Q_i + Q_{\text{climb}} = \int_0^B b\tfrac{1}{2}\rho\Omega^2 R^4 ax(\theta_t - \phi_t)\phi_t c\, dx \tag{24}$$

The upper limit to the above integral was taken as $x = B$ as was done in the expression for thrust, inasmuch as both Q_i and Q_{climb} are dependent upon the generation of thrust. Integrating equation (24),

$$Q_i + Q_{\text{climb}} = \frac{B^2}{4} b\rho\Omega^2 R^4 a(\theta_t - \phi_t)\phi_t c \tag{25}$$

Also

$$Q_i + Q_{\text{climb}} = (C_{Q_i} + C_{Q_{\text{climb}}})\pi R^2 \rho(\Omega R)^2 R \tag{26}$$

Equating expressions (25) and (26) and substituting the solidity term, the following nondimensional expression for the induced and climb torque is obtained:

$$C_{Q_i} + C_{Q_{\text{climb}}} = \frac{B^2}{4}\sigma a(\theta_t - \phi_t)\phi_t \tag{27}$$

EXPRESSION FOR PROFILE-DRAG TORQUE. The remaining source of energy loss (namely, that caused by profile drag) will now be considered. This loss will, of course, depend on the type of drag curve assumed for the airfoil section employed. As previously discussed, many treatments of hovering theory assume that the blade sections operate at an average drag coefficient, δ. Inasmuch as the section angle of attack varies inversely along the span for an ideally twisted blade, and because the drag coefficient for conventional airfoils varies with angle of attack, it is obviously more correct to try to use the actual drag coefficient corresponding to the calculated angle of attack.

Knight (reference I–15 of Appendix IIA) first expressed the relation between the drag coefficient and angle of attack by a parabolic expression of the form $c_{d_o} = c_{d_{o\min}} + \epsilon\alpha_r^2$. Sissingh (reference II–21 of Appendix IIA) then used an analytical expression for c_{d_o} involving three terms, of the form

$$c_{d_o} = \delta_0 + \delta_1\alpha_r + \delta_2\alpha_r^2 \tag{28}$$

In reference II–18 of Appendix IIA, Bailey made a study of the drag coefficients of conventional airfoils and derived a method by which the constants in equation (28) could be determined for a particular airfoil section. The following equation was presented in that report as representative of the drag characteristics of "good" practical-construction blades of conventional airfoil section:

$$c_{d_o} = 0.0087 - 0.0216\alpha_r + 0.400\alpha_r^2 \tag{29}$$

The corresponding profile-drag curve is shown in Fig. 4–8 together with experimental data obtained in the Langley two-dimensional

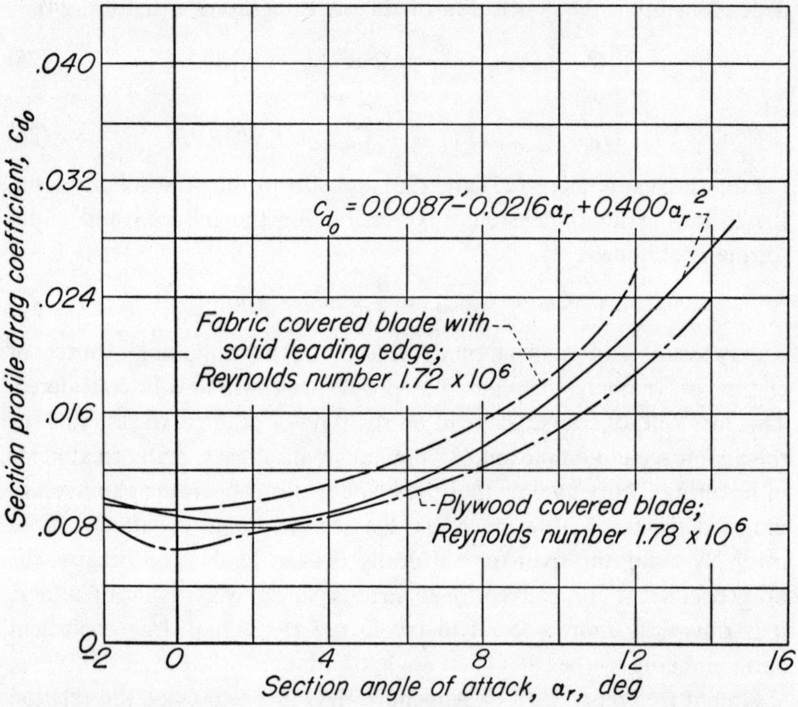

Fig. 4–8 Analytical and measured profile-drag curves (from Appendix IIA, reference II–16).

low-turbulence pressure tunnel on practical-construction, rotor-blade specimens having NACA 0012 airfoil section.

The profile-drag torque may be written as

$$Q_o = \int_0^1 b\tfrac{1}{2}\rho\Omega^2 R^4 x^3 c c_{d_o}\, dx \tag{30}$$

Because profile-drag forces are present in the region of unsteady flow at the blade tip, the expression is integrated along the entire length of the blade.

With the substitution of equation (28), equation (30) becomes

$$Q_o = \frac{b\rho}{2}\Omega^2R^4c\int_0^1 x^3\left[\delta_0 + \frac{\delta_1}{x}(\theta_t - \phi_t) + \frac{\delta_2}{x^2}(\theta_t - \phi_t)^2\right]dx \tag{31}$$

Integrating, Q_o is found to be

$$Q_o = \frac{b}{2}\rho\Omega^2R^4c\left[\frac{\delta_0}{4} + \frac{\delta_1}{3}(\theta_t - \phi_t) + \frac{\delta_2}{2}(\theta_t - \phi_t)^2\right] \tag{32}$$

Equating (32) to $Q_o = C_{Q_o}\pi R^2\rho(\Omega R)^2R$ and substituting for σ, the profile-drag torque coefficient is

$$C_{Q_o} = \frac{\sigma}{2}\left[\frac{\delta_0}{4} + \frac{\delta_1}{3}(\theta_t - \phi_t) + \frac{\delta_2}{2}(\theta_t - \phi_t)^2\right] \tag{33}$$

FINAL PERFORMANCE EQUATION. The total torque absorbed by the rotor is represented by the sum of equations (27) and (33).

$$C_Q = \frac{\sigma}{2}\left[\frac{B^2}{2}a(\theta_t - \phi_t)\phi_t + \frac{\delta_0}{4} + \frac{\delta_1}{3}(\theta_t - \phi_t) + \frac{\delta_2}{2}(\theta_t - \phi_t)^2\right] \tag{34}$$

In order for equation (34) to be directly applicable to performance calculations, θ_t and ϕ_t have to be replaced by C_T, so that C_Q will be in terms of known parameters. If equations (7), (8), and (22) are solved simultaneously, then

$$\left.\begin{aligned} \phi_t &= \frac{1}{2}\sqrt{\left(\frac{V_v}{\Omega R}\right)^2 + \frac{2C_T}{B^2}} + \frac{1}{2}\frac{V_v}{\Omega R} \\ \theta_t &= \frac{1}{2}\sqrt{\left(\frac{V_v}{\Omega R}\right)^2 + \frac{2C_T}{B^2}} + \frac{4}{\sigma a}\frac{C_T}{B^2} + \frac{1}{2}\frac{V_v}{\Omega R} \end{aligned}\right\} \tag{35}$$

When equations (35) are substituted into equation (34), then the total torque coefficient is expressed by

$$C_Q = \frac{1}{2}C_T\sqrt{\left(\frac{V_v}{\Omega R}\right)^2 + \frac{2C_T}{B^2}} + \frac{1}{2}\frac{V_v}{\Omega R}C_T + \frac{\sigma\delta_0}{8} + \frac{2}{3}\frac{\delta_1}{a}\frac{C_T}{B^2} + \frac{4\delta_2}{\sigma a^2}\left(\frac{C_T}{B^2}\right)^2 \tag{36}$$

The first two terms in equation (36) represent induced and climb power, while the remaining three terms represent profile-drag power.

By means of equation (36), the vertical-climb performance of a rotor may be calculated for any particular solidity, tip-loss factor ***B***, and operating at any thrust coefficient under the assumption of uniform inflow. Hovering performance is obtained by setting $V_v = 0$ in the equation.

Although hovering and vertical climb have been discussed, all remarks and formulas apply to small rates (i.e., between 0 and approximately 400 feet per minute) of vertical descent as well, if $-V_v$ is substituted for V_v. However, when the velocity of descent becomes of the same order of magnitude as the induced velocity v, the momentum theory no longer applies and another approach is indicated. The vertical descent condition as a whole will be considered in Chapter 6.

Rapid Performance Estimation

The performance of an ideally twisted, constant-chord rotor as developed in the preceding section may be used as a convenient baseline from which to estimate roughly and rapidly the characteristics of any rotor in the hovering condition. The procedure consists of calculating the performance of the basic rotor by means of equation (36) and then modifying the result to account for the actual twist, taper, and surface condition of the rotor considered. This procedure is summarized below together with a table of corrections for various values of twist and plan-form taper which were taken from the analysis developed in reference II–5 of Appendix IIA. (These corrections are discussed in detail in the following chapter.)

BASIC EQUATIONS.

$$C_T = \frac{T}{\pi R^2 \rho (\Omega R)^2}$$

$$C_Q = \frac{Q}{\pi R^2 \rho (\Omega R)^2 R} = \frac{\text{Power}}{\pi R^2 \rho (\Omega R)^3}$$

$$\sigma = \text{solidity} = \frac{bc}{\pi R}$$

where

T = rotor thrust, pounds
ρ = air density, slugs/cubic foot
Q = rotor torque, foot-pounds
ΩR = rotor tip speed, feet/second
R = rotor radius, feet
b = number of blades in rotor
c = blade chord

PROCEDURE

(1) For given weight, air density, radius, and tip-speed, calculate C_T and σ.

(2) Find tip-loss factor B from Fig. 4–4 or as $1 -$ (tip chord/$2R$); calculate $C_T/\sigma B^2$.

(3) Find C_{Q_o} from Fig. 4–9. Use this value without adjustment for smooth, well-contoured blades, such as plywood or metal blades of

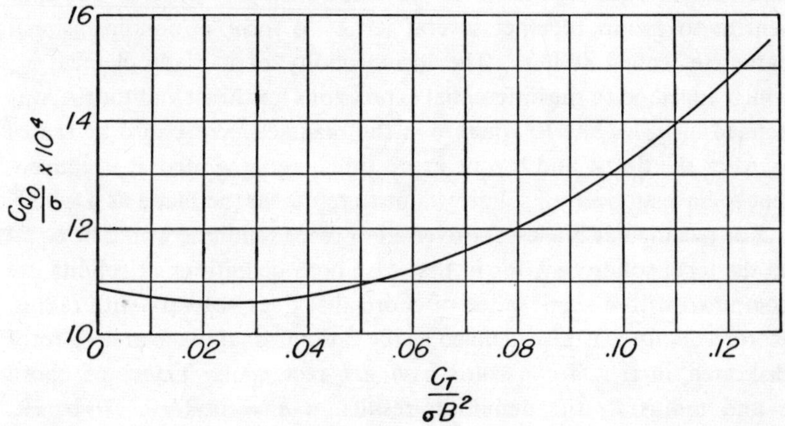

Fig. 4–9 Profile-drag torque of a rotor with ideally twisted, rectangular blades.

good construction. For blades having rough surfaces or inaccurate profile, increase C_{Q_o} by about 30 per cent.

(4) Calculate C_{Qi} as $C_T^{3/2}/\sqrt{2}B$.

(5) $C_Q = C_{Q_o} + C_{Q_i}$; calculate power from C_Q.

CORRECTIONS FOR PLAN FORM AND TWIST. Per cent power to be added to blades having taper, and linear or no twist is as follows: (Note dependence on C_T/σ values.)

Taper ratio	*Twist, degrees*	$C_T/\sigma = 0.067$	$C_T/\sigma = 0.100$
1:1	0	$5\frac{1}{2}$	$7\frac{1}{2}$
1:1	−8	3	$3\frac{1}{2}$
1:1	−12	$1\frac{1}{2}$	$1\frac{1}{2}$
3:1	0	$3\frac{1}{2}$	3
3:1	−8	0	$-\frac{1}{2}$
3:1	−12	0	$-\frac{1}{2}$

(Preceding chart gives per cent power to be added at fixed thrust. To find per cent thrust reduction at fixed power, multiply chart values by 0.8 for $C_T/\sigma = 0.067$ and by 0.4 for $C_T/\sigma = 0.100$.)

Equivalent Chords and Weighted Solidity

In connection with developing nondimensional expressions for rotor thrust and torque by means of the blade element theory, it was convenient to group together several terms to form a nondimensional parameter called solidity. The local solidity of a blade element σ_x, which appeared in the differential expressions for thrust and torque, was defined as $\sigma_x = bc/\pi R$, where c is the blade-element chord at radius x. After the thrust and torque expressions are integrated, it is convenient to have an over-all solidity value to represent the blade as a whole.

For rectangular blades, the over-all or rotor solidity σ would be equal to the local solidity σ_x, for in that case both definitions of solidity are computed with a single value of chord that does not vary with radius. Rotor solidity was also defined as the ratio of total blade area to rotor disk area, in that, for a rotor having b rectangular blades of chord c and radius R, the definition results in $\sigma = \mathrm{bcR}/\pi \mathrm{R}^2 = \mathrm{bc}/\pi \mathrm{R}$. If, however, the blade plan form is other than rectangular, the single, average chord which would yield the thrust and torque of the actual plan form must be determined. This chord is called the *equivalent chord*, c_e, and may be defined as the chord of a rectangular blade which would yield the same thrust and torque as the actual blade plan form.

One way of determining c would be to equate the expressions for the thrust and torque of a tapered blade to that for a rectangular blade and so find the relation between the chords of the two blades. Unfortunately, however, the value of σ as determined from the thrust and torque expressions differ. On a thrust basis, the outer portions of a blade are more important than the inner portion because the lift on an element depends on the square of its radial distance from the axis of rotation. The outboard elements, however, are still more effective with respect to torque, inasmuch as the torque on an element depends on the cube of its radial distance.

Consequently, the equivalent chord may be defined as follows:
On a thrust basis

$$c_e = \frac{\int_{x=0}^{x=1} cx^2\,dx}{\int_{x=0}^{x=1} x^2\,dx} \tag{37}$$

On a torque basis

$$c_e = \frac{\int_{x=0}^{x=1} cx^3\,dx}{\int_{x=0}^{x=1} x^3\,dx} \tag{38}$$

In order to be fully correct, each rotor of tapered plan form would have two solidities, equation (37) being used when thrust is calculated

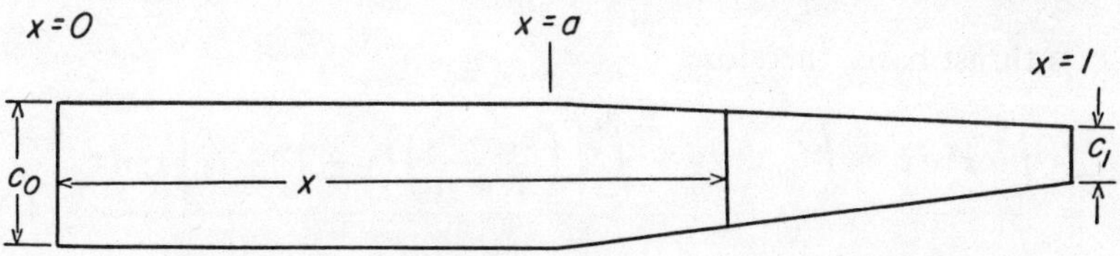

Fig. 4–10 Plan form of blade with partial, linear taper.

and equation (38) used for torque calculations. As this is cumbersome, however, and because the taper of conventional rotors is small enough so that the difference involved in using either definition is also small, one definition is usually used. Inasmuch as most helicopters are compared on a weight basis (which is relatively constant for a machine) rather than on power, the NACA and most agencies use a solidity involving an equivalent chord determined from thrust considerations. In comparing rotors, it is important to be consistent.

For a linearly tapered blade, equation (37) results in the chord at the 0.75 radius as being typical, compared to the chord at the 0.80 station resulting from the use of equation (38). Thus, on a thrust basis the equivalent chord of a linearly tapered blade is simply the chord at the 0.75 radius. If a rotor blade is, however, of odd shape or has large cutouts, the equivalent chord may be found by graphical integration. This is done by plotting the product of the chord and the square of the radial distance from the center of rotation (expressed in nondimensional form) against the distance, integrating the area beneath the curve, and multiplying the result by 3.0. (The factor of 3.0 arises from the integration of the denominator in equation (37), and is therefore perfectly general.)

Many current rotor designs have the partial taper plan form shown in Fig. 4–10 and it is, therefore, convenient to have an analytical expression for the equivalent chord of such a blade. Referring to the figure, it is seen that between $x = 0$ and $x = a$,

$$c = c_0 \tag{39}$$

and between $x = a$ and $x = 1$,

$$c = \left(\frac{c_1 - c_0}{1 - a}\right)(x - 1) + c_1 \tag{40}$$

On a thrust basis, therefore,

$$c_e = \frac{\int_0^1 cx^2\,dx}{\int_0^1 x^2\,dx} = \frac{\int_0^a c_0 x^2\,dx + \int_a^1 \left[\left(\frac{c_1 - c_0}{1 - a}\right)(x - 1) + c_1\right] x^2\,dx}{\int_0^1 x^2\,dx} \tag{41}$$

Integrating and substituting the limits in equation (41), the equivalent chord of a partially tapered blade is

$$c_e = (c_0 - c_1)a^3 - \left(\frac{c_0 - c_1}{1 - a}\right)\left(-\frac{1}{4} + a^3 - \frac{3}{4}a^4\right) + c_1 \tag{42}$$

If $c_0 = c_1$ (rectangular plan form), equation (42) reduces to

$$c_e = c_1$$

For a straight linear taper ($a = 0$),

$$c_e = c_0 - \tfrac{3}{4}(c_0 - c_1) \tag{43}$$

which further checks equation (42), inasmuch as it is already known from previous considerations that the 0.75 radius chord is typical for linear taper.

5

FACTORS AFFECTING HOVERING AND VERTICAL-FLIGHT PERFORMANCE

Several rather unrelated but important items that influence the hovering and vertical-flight performance of rotors are discussed in this chapter. Included are such factors as the detail design of the blades themselves, as well as the influence of such operating conditions as climb and hovering near the ground and at high altitudes.

Effects of Blade Twist and Taper

Methods whereby improved vertical-flight performance may be achieved without sacrifices in payload are of interest, particularly to designers of large, slow-moving, load-carrying helicopters. The helicopter can be made to carry a greater pay load or to climb at a greater speed either by improved structural design (which is dependent upon metallurgical advances and knowledge of the loads imposed on the helicopter rotor and fuselage), or by a reduction in the aerodynamic losses sustained by the rotor. As previously discussed, both the induced and profile-drag losses can be minimized in the hovering condition by operating with the largest diameter and slowest turning rotor compatible with structural criteria and operating considerations.

Additional improvement can be sought by the choice of the proper blade geometry, i.e., selection of the blade plan form and pitch distribution that would yield the maximum thrust for a given power input to the rotor. Improvements of this nature appear to be effective, for with the current ratio of pay load to gross weight, the percentage increase in pay load is approximately four times the percentage increase in thrust. In addition, the use of a more efficient pitch distribution might not entail additional production costs, for once the blade jigs are set up, it is just as easy (with most designs) to produce a blade having 8 degrees of linear twist, for example, as it is to produce a perfectly untwisted blade.

That part of the total induced loss arising from a nonuniformity of inflow may be minimized by twisting the blade so that its root end has higher pitch angles than the tip (washout), or by tapering the blade so that the root chord is greater than the tip chord. Over the major part of the operating range of thrust coefficients, the same expedients of twist and taper are beneficial with respect to the reduction of profile-drag losses. The reason for this fact is that profile-drag losses are dependent on the cube of the velocity, whereas thrust varies as the square of the velocity, thus making it advantageous for thrust to be produced by the low velocity inboard sections, while maintaining low angles of attack on the outboard section.

TWIST EFFECTS. The manner in which twist affects the inflow distribution and the blade loading (as represented by the section angles of attack) of an untapered blade operating at a representative $C_T = 0.0056$ and $\sigma = 0.060$ is shown in Fig. 5–1.[1] The induced velocities and section angles of attack shown in the figure are plotted against the square of the radius in order to put proper emphasis on each blade element, inasmuch as the thrust of the element is dependent on the square of the velocity at the element. The figure indicates that the application of increasing amounts of linear twist approaches the effects produced by ideal twist, in that both the induced velocity and the blade loading are increased near the inboard end of the blade. Although the effects of twist are such that the maximum section angle of attack occurring on the blade is increased, the effects of blade stall are minimized because the radial distance of the maximum angle of attack is decreased.

[1] The calculations discussed in this section were made by the strip-analysis method explained in the preceding chapter. This section is a condensation of reference I–5, in Appendix IIA, from which Figs. 5–1 through 5–6 were taken.

The reduction in induced and profile-drag losses that may be realized with a more uniform inflow distribution as obtained by linear twist may be seen from Fig. 5–2, in which both induced and profile-torque coefficients are plotted over a range of thrust coefficients for untapered

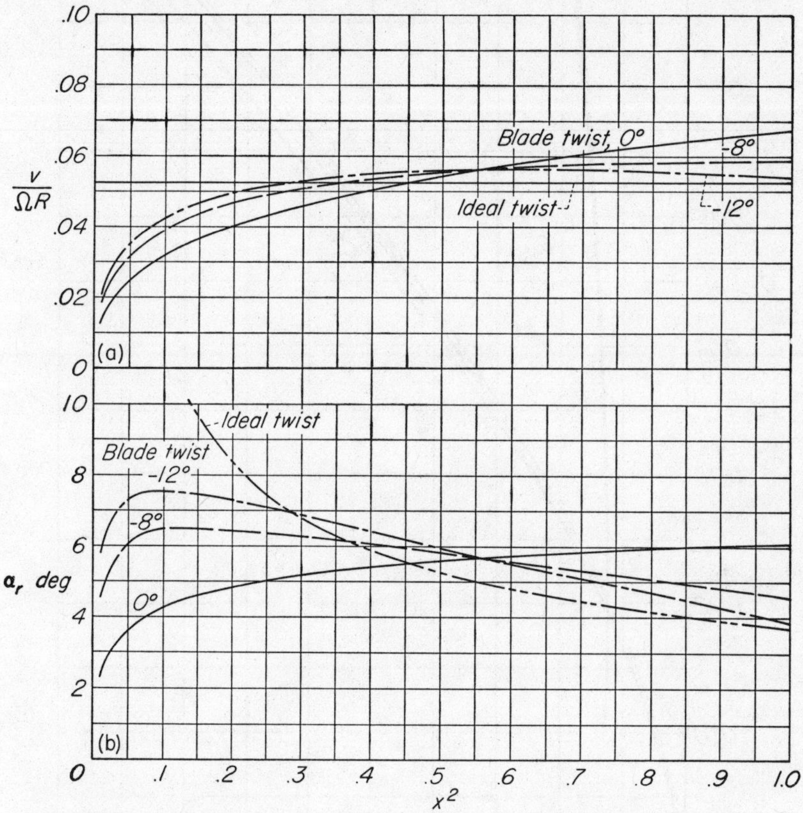

Fig. 5–1 Induced velocity and angle-of-attack distributions in hovering.

blades having varying amounts of twist (and 0.060 solidity). The figure indicates that twist is beneficial in reducing rotor losses over the range of thrust coefficients above the conventional minimum (i.e., above $C_T \cong 0.0030$), although the net change in profile-drag losses appears insignificant. As might be expected, twist is detrimental at thrust coefficients near zero, inasmuch as losses are incurred in producing negative thrust over the outer part of the blades. Consequently, the

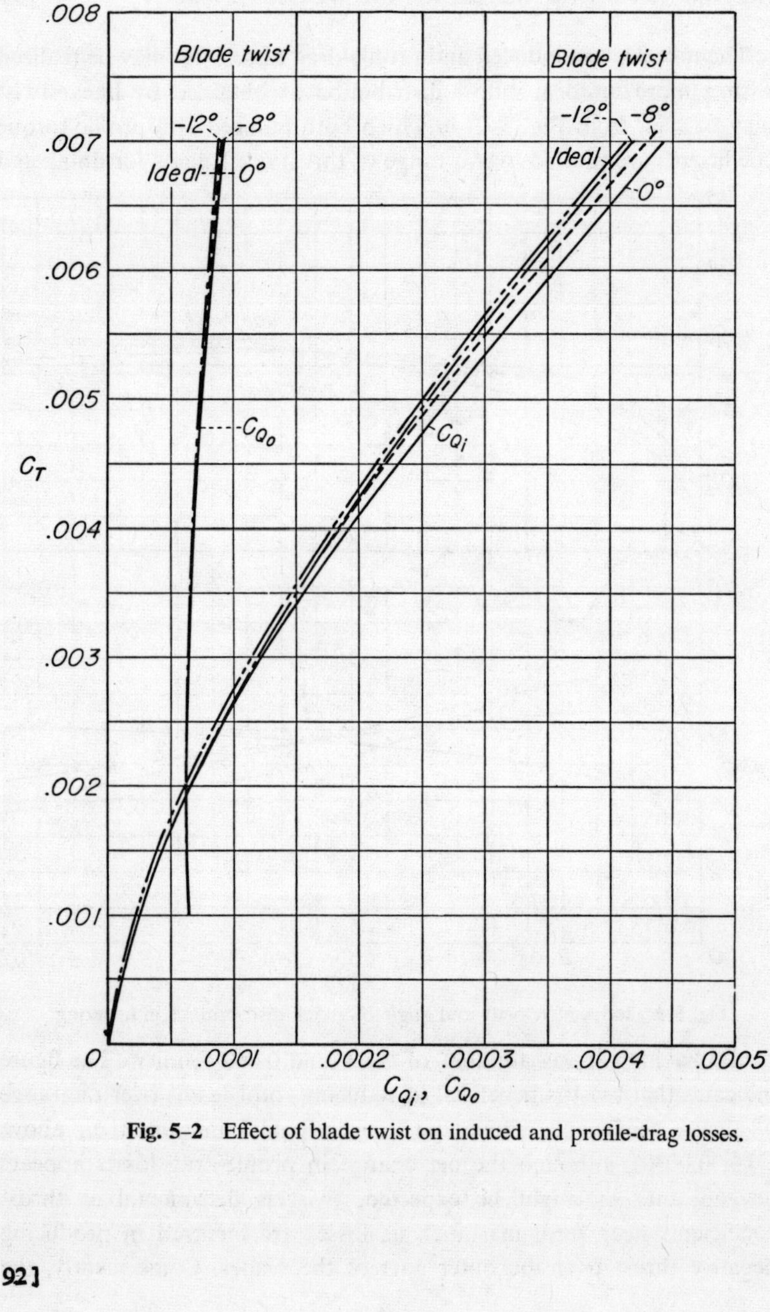

Fig. 5–2 Effect of blade twist on induced and profile-drag losses.

optimum amount of linear twist would decrease with lower operating mean-lift coefficients.

Figure 5–2 suggests that a linear twist of − 12 degrees realizes most of the reduction of induced loss made possible by ideal twist, while yielding almost identical profile-drag losses over the normal operating range. It appears probable that a few degrees more twist would produce slightly better results, but that amounts of twist very much above 12 degrees would result in decreasing gains. Highly twisted blades might also be detrimental in other flight conditions.

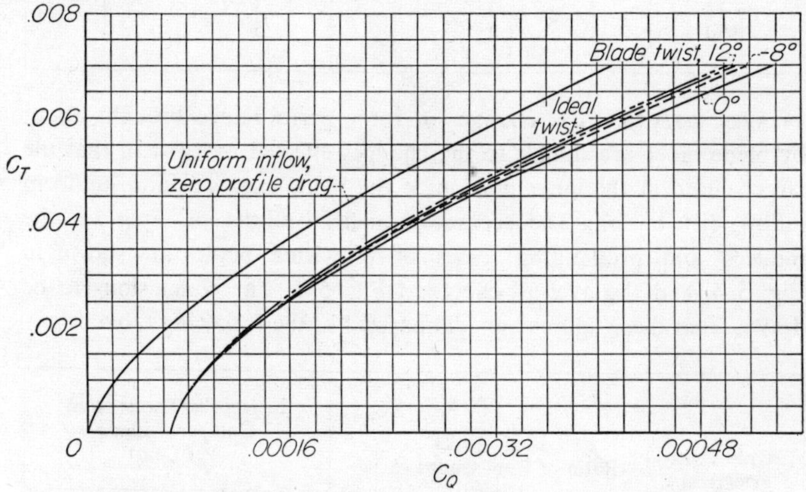

Fig. 5–3 Effect of blade twist on rotor-hovering performance.

The effects of twist illustrated thus far explain the comparative performance of the twisted, untapered 0.060 solidity rotors shown in Fig. 5–3.[1] The curves are compared in the figures with the performance of an ideal rotor having minimum induced loss (uniform inflow) and zero profile drag in order to determine the extent to which the ideal rotor can be approached by practical designs. The significant results from the figure are summarized in the following table which gives the percentage increase in thrust resulting from blade twist at fixed power.

[1] The blade airfoil section characteristics for the rotors discussed herein make use of the drag polar given by equation (29) of Chapter 4, and $c_l = 5.73\alpha_r$ (where α_r, is in radians).

The C_Q values were chosen to correspond to C_T's of 0.0040 and 0.0060 for the untwisted blade ($C_T/\sigma = 0.066$ and 0.10).

Blade Twist (Degrees)	Blade Taper (Ratio of Root Chord to Tip Chord	Increase in Thrust from Untapered Blade (Per Cent)	
		$C_Q = 0.00026$	$C_Q = 0.00044$
0	1	—	—
− 8	1	2	3
− 12	1	3	4
Ideal	1	5	5

TAPER EFFECTS. The changes in rotor performance brought about by blade taper are similar to the changes effected by twist in that the larger chord at the inner portion of the blade causes a more uniform inflow distribution. The separate beneficial effects of taper on the induced and profile-drag losses of untwisted blades are shown in Fig. 5–4, and the over-all effects in Fig. 5–5. The conclusions to be drawn from these figures are summarized in the following table:

Blade Twist (Degrees)	Blade Taper (Ratio of Root Chord to Tip Chord)	Increase in Thrust from Untapered Blade (Per Cent)	
		$C_Q = 0.00026$	$C_Q = 0.00044$
0	1	—	—
0	3	2	3

A several per cent increase in hovering performance may not justify the additional production costs of tapering the blades used on small helicopters, especially when the benefits of twist can be had at no additional production cost. Even small efficiency gains are significant with large helicopters, however, and the tapering of large diameter blades appears highly desirable for structural reasons.

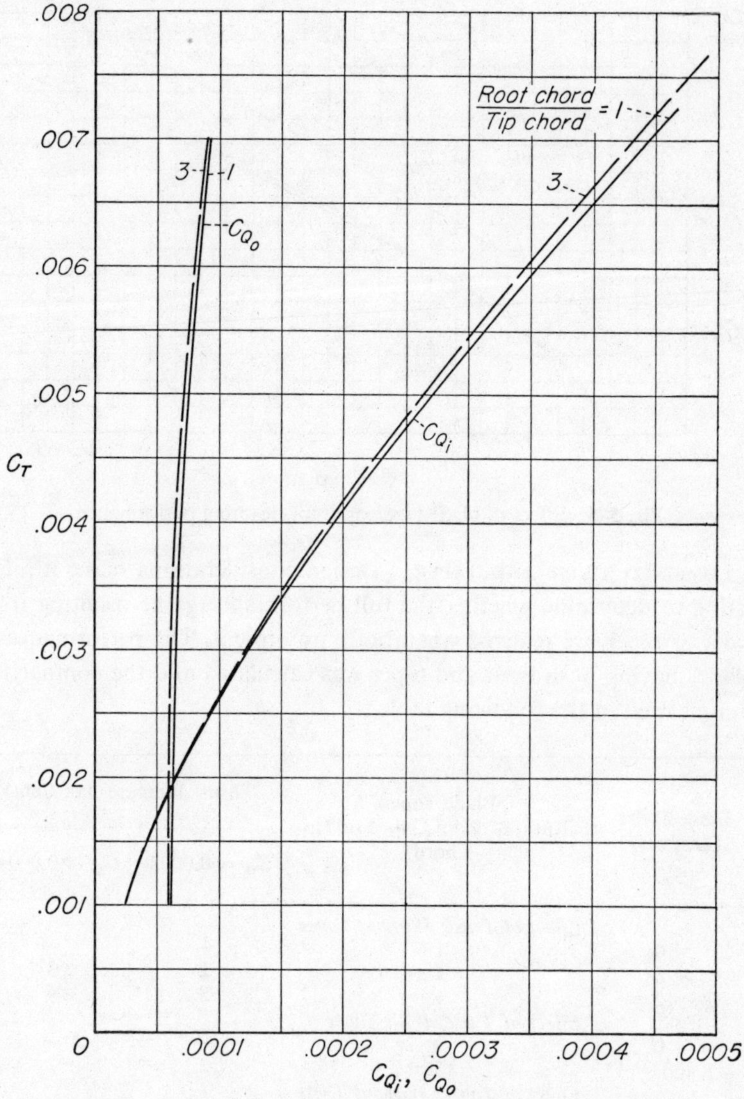

Fig. 5–4 Effect of blade taper on induced and profile-drag losses.

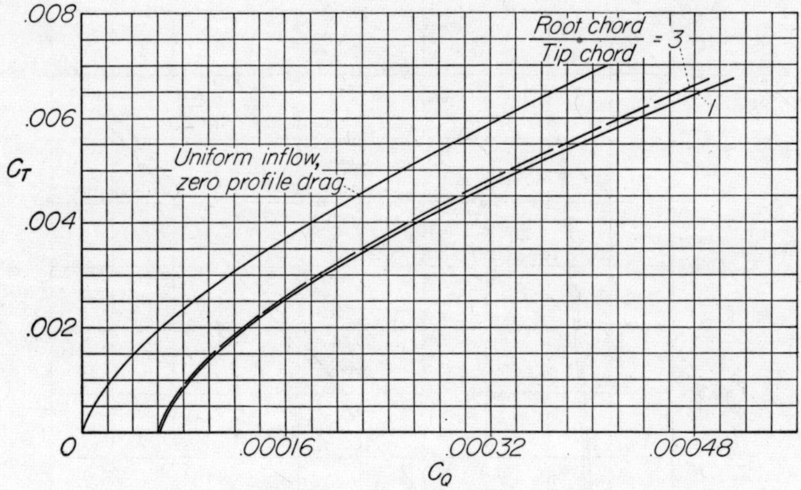

Fig. 5-5 Effect of blade taper on rotor-hovering performance.

COMBINED TWIST AND TAPER. Design considerations make it interesting to determine whether the full performance gains resulting from twist or taper are realized when both are present. The performance of blades having both twist and taper was calculated and the comparison summarized in the following table:

Blade Twist (Degrees)	Blade Taper (Ratio of Root Chord to Tip Chord)	Thrust Increase (Per Cent)	
		$C_Q = 0.00026$	$C_Q = 0.00044$
	Effect of Twist; Without Taper		
0	1	—	—
− 8	1	2	3
− 12	1	3	4
	Effect of Twist; With Taper		
0	3	—	—
− 8 and − 12	3	3	2
	Effect of Taper; Without Twist		
0	1	—	—
0	3	2	3
	Effect of Taper; With Twist		
− 12	1	—	—
− 8 and − 12	3	1	1

The increase in thrust obtained with various combinations of blade twist and taper as compared with the thrust of an untwisted rectangular blade is shown in the following table:

Blade Twist (Degrees)	Blade Taper (Ratio of Root Chord to Tip Chord)	Increase in Thrust from Untwisted Untapered Blade (Per Cent)	
		$C_Q = 0.00026$	$C_Q = 0.00044$
0	1	—	—
− 8	3	5	5
− 12	3	5	5
Ideal	Optimum	7	7

Note that a linearly twisted and tapered blade appears to produce only 2 per cent less thrust at the solidity considered than the optimum rotor, which very nearly represents the maximum increase in performance to be expected from twist and taper. (The optimum rotor is discussed in detail in the following section.)

SOLIDITY EFFECTS. In order to determine the extent to which the benefits of twist and taper can be obtained with lower solidity rotors than the 0.060 solidity treated, a similar analysis, covering equal amounts of twist and taper as well as the same thrust coefficient range, was made for a rotor having a solidity equal to 0.042. A comparison of these figures with the equivalent figures for the 0.060 solidity rotor shows that, in general, the same results were obtained with both rotors.

PARTIAL TAPER. In conventionally tapered rotor blades, the taper usually extends from the tip to approximately one-half of the radius, the remainder or inboard portion being rectangular in plan form. In order to determine whether the use of partial, rather than full, taper resulted in a loss of hovering efficiency, the performance of a blade tapered over its outer half and having a ratio of root-to-tip chord of 3 to 1 (the root chord being calculated by assuming the leading and trailing edges of tapered portion are extended to the root) was calculated and compared with the performance of a fully tapered blade having the same taper ratio and solidity. Little difference was found to exist between the two rotors. This conclusion is further substantiated by an inspection of Fig. 5–6, which shows that the inflow distribution (and consequently the power losses) of both the partially and fully tapered rotors operating at the same thrust coefficient is similar.

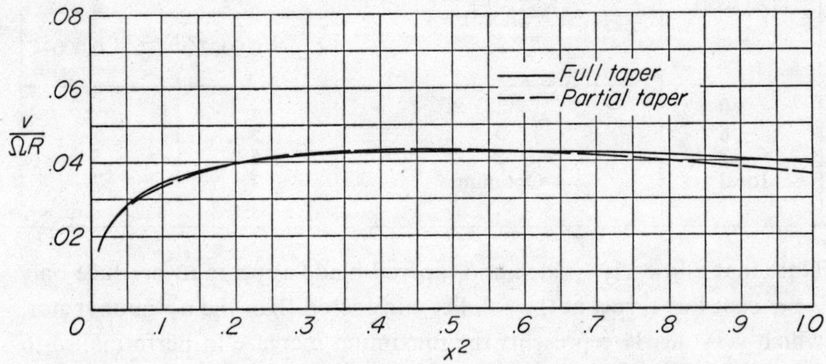

Fig. 5–6 Effect of partial, instead of full, taper on rotor inflow distribution.

It should be noted that tip losses have been neglected in all of the preceding calculations. This omission is justifiable in a comparative study, because it was found upon investigation that the effects of changes in plan form and twist were not influenced significantly by tip losses.

TWIST IN FORWARD FLIGHT. In connection with the use of twist in blade design, it will be subsequently shown in Chapters 9 and 10 that negative twist has been found to improve the efficiency of the rotor in forward flight as well as in hovering and may be expected to delay blade stalling at high forward speeds, and adverse compressibility effects at high tip speeds and high mean-lift coefficients. Blade stalling is delayed when twist is employed because twist unloads the tips by reducing the tip angles of attack. Compressibility losses are minimized for the same reason, in that the critical Mach number of the blade section is increased at reduced lift coefficients. Blade twist has been found experimentally to have little effect on the autorotative performance of rotors in forward flight and in vertical descent.

Optimum Hovering Rotor

The effects of various combinations of blade twist and plan form taper on hovering performance have just been discussed. The question then arises as to what the maximum benefits are that can be obtained by twist and taper, or in other words, what is the "optimum" rotor design?

With respect to induced losses, the most efficient rotor blade has been shown to be one that is twisted ideally. It is reasonable to assume that profile-drag losses will be a minimum when each blade element is operating at its most efficient angle of attack (i.e., the angle at which c_l/c_d is a maximum if tip speed is fixed, or the angle at which $c_l^{3/2}/c_d$ is a maximum if the disk loading is fixed). The problem, therefore, is to determine the twist and taper of a blade that will produce uniform downwash and each element of which is working at its optimum lift coefficient.

The optimum rotor for hovering is then defined as one that has uniform inflow over the disk and which has all its sections operating at constant angle of attack. Such a rotor is truly optimum only when rotational (i.e., angular momentum) and profile-drag losses are ignored in determining the criterion for minimum induced losses, namely, uniform inflow. It is expected, however, that the optimum rotor as defined approaches very closely the true optimum for the usual helicopter disk loadings. The performance of such a rotor may be rather easily and explicitly calculated.

THRUST. In hovering the blade-element pitch is

$$\theta = \alpha_r + \phi = \alpha_r + \frac{v}{\Omega r} \tag{1}$$

Thus the pitch distribution for the optimum rotor is composed of two parts; a constant part α_r, and a variable part ϕ, which varies inversely as the radius for a constant inflow v.

The differential thrust on a blade element is

$$dT = \tfrac{1}{2}\rho(\Omega r)^2 a\alpha_r c\, dr \tag{2}$$

In order to maintain α_r independent of r, the chord must be adjusted so that uniform downwash can be obtained. This latter condition exists when dT varies linearly with r; that is, when the thrust varies linearly from zero at the blade root to a maximum at the tip. (Such a loading is associated with uniform downwash inasmuch as momentum theory shows that the mass of air influenced by a blade element depends on the radial position of the element.)

Equation (2) shows that dT can be made a linear function of r by making c the following function of r

$$c = c_t \frac{R}{r} \tag{3}$$

Substituting equation (3) into equation (2) gives

$$dT = \tfrac{1}{2}\rho\Omega^2 r a\alpha_r c_t R\, dr \tag{4}$$

Integrating equation (4) and multiplying the resulting expression by the number of blades to give the total thrust produced by the rotor

$$T = \frac{b}{2}\rho\Omega^2 \frac{R^3}{2} a\alpha_r c_t \tag{5}$$

Expressed in nondimensional form, equation (5) reduces to

$$C_T = \frac{\sigma_t}{4} a\alpha_r$$

$$= \frac{\sigma_t}{4} c_l \tag{6}$$

where

$$\sigma_t = \frac{bc_t}{\pi R} \tag{7}$$

INDUCED TORQUE. The induced torque may be expressed as

$$Q_i = \int_0^R b\tfrac{1}{2}\rho\Omega^2 r^3 c_l \phi c\, dr \tag{8}$$

With the appropriate substitution for c and ϕ, equation (8) becomes after integration

$$Q_i = \frac{b}{2}\rho(\Omega R)\frac{R^2}{2} c_l c_t v \tag{9}$$

In coefficient form, equation (9) reduces to

$$C_{Q_i} = \frac{\sigma_t}{4} c_l \frac{v}{\Omega R} \tag{10}$$

From momentum considerations,

$$v = \sqrt{\frac{T}{2\rho\pi R^2}} = \Omega R \sqrt{\frac{C_T}{2}}$$

or

$$\frac{v}{\Omega R} = \phi = \sqrt{\frac{C_T}{2}} \tag{11}$$

Substituting equations (6) and (11) into equation (10), the expression for C_{Q_i} finally becomes

$$C_{Q_i} = \frac{C_T^{3/2}}{\sqrt{2}} \tag{12}$$

PROFILE-DRAG TORQUE. In a similar way the profile-drag torque may be expressed as

$$Q_0 = \int_0^R \tfrac{1}{2} b\rho(\Omega r)^2 c_{d_0} cr\, dr$$
$$= \tfrac{1}{6} b\rho(\Omega R)^2 R^2 c_t c_{d_0} \tag{13}$$

or, in nondimensional form, as

$$C_{Q_0} = \tfrac{1}{6}\sigma_t c_{d_0} \tag{14}$$

PERFORMANCE EQUATION. Adding equations (12) and (14), the performance of the optimum rotor (i.e., a rotor with twist and taper to give constant inflow and constant angle of attack along the span) is represented by

$$C_Q = \frac{C_T^{3/2}}{\sqrt{2}} + \frac{1}{6}\sigma_t c_{d_0} \tag{15}$$

In order to compare the optimum rotor with other rotors on the basis of the same solidity, it is necessary to determine the conventional weighted solidity of such a rotor. The equivalent chord of the optimum rotor as calculated from its definition and equation (3) is

$$c_e = \tfrac{3}{2} c_t$$

so that

$$\sigma = \tfrac{3}{2}\sigma_t$$

Thus, in terms of weighted solidity, equation (15) becomes

$$C_Q = \frac{C_T^{3/2}}{\sqrt{2}} + \frac{1}{9}\sigma c_{d_0} \tag{16}$$

DESIGN OF OPTIMUM ROTOR. The procedure for designing an optimum rotor for a particular operating C_T after the blade diameter has been decided upon would be as follows:

(1) For the airfoil section chosen, determine the angle of attack so that the profile drag is a minimum.

(2) Calculate σ_t for the design C_T as $\sigma_t = 4 \dfrac{C_T}{c_l}$.

(3) The proper twist and taper would then be

$$c = c_t \frac{R}{r} \left(\text{where } c_t = \sigma_t \frac{\pi R}{b} \right)$$

and

$$\theta = \alpha_r + \frac{R}{r} \sqrt{\frac{C_T}{2}}$$

The performance of the optimum rotor, as compared with the ideal rotor and with an untwisted, untapered rotor is given in Fig. 5–7.

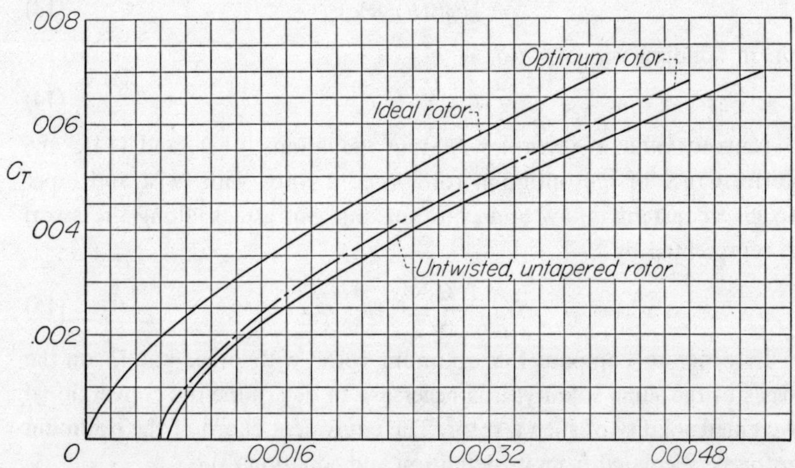

Fig. 5–7 Comparative performance of limiting rotor designs.

Effect of Climb on Induced-Power Loss

From previous considerations, it should be clear that the induced velocity produced by a rotor in the climb condition is less than that in hovering, inasmuch as the rotor handles a greater mass of air (because of the climb velocity) and consequently needs to accelerate the mass less to produce the same thrust. For similar reasons, a helicopter descending at a velocity less than the induced velocity at the rotor (i.e., at small rates of descent) suffers a greater induced loss than in hovering, because the descent velocity subtracts from the total flow through the rotor disk. At very large rates of descent, the induced velocity is again less than in hovering. Although the momentum theory does not cover the intermediate range of descent velocities (which is an operating condition treated in detail in Chapter 6), the change in the induced velocity in going from hovering to climb may be estimated from simple momentum considerations.

EQUATION FOR INDUCED VELOCITY IN CLIMB. Inasmuch as gross changes are sought, it can be assumed that the downwash is uniform across the disk. Let v_h = induced velocity in hovering and v_v = induced velocity in climb. From equation (31), Chapter 3,

$$v_h = \Omega R \sqrt{\frac{C_T}{2}} \tag{17}$$

Inasmuch as the total thrust is the same in both hovering and climb, the differential expressions for thrust obtained from momentum theory in the two conditions can be equated.

$$dT = 2\pi r\, dr\, \rho 2 v_h^2 = 2\pi r\, dr\, \rho (v_v + V_v) 2 v_v$$

From which

$$v_h^2 = V_v v_v + v_v^2$$

Solving for v_v

$$v_v = \frac{-V_v + \sqrt{V_v^2 + 4v_h^2}}{2} \tag{18}$$

Substituting equation (17) into the above, the following expression is obtained for the induced velocity in climb

$$v_v = \frac{-V_v + \sqrt{V_v^2 + 2C_T(\Omega R)^2}}{2} \tag{19}$$

EXPERIMENTAL CHECK OF INDUCED VELOCITY IN CLIMB. The validity of equation (19) will be checked by calculating the ratio of the power required by a helicopter climbing vertically at 450 feet per minute to the power required for hovering and comparing the answer with experimental data reported in reference I–10 (Appendix IIA).

Pertinent data are as follows:

$W = 2393$ pounds Engine speed = 2100 revolutions per minute
$R = 19$ feet Engine to main rotor gear ratio = 9.34

$$\rho/\rho_0 = 1$$

Then

$$\Omega R = \frac{2100}{9.34} \times \frac{2\pi}{60} \times 19 = 447 \text{ feet per second}$$

$$C_T = \frac{W}{\pi R^2 \rho(\Omega R)^2} = \frac{2393}{\pi(19)^2 \times 0.00238(447)^2} = 0.00442$$

From equation (17)

$$v_h = \Omega R \sqrt{\frac{C_T}{2}} = 447 \sqrt{\frac{0.00442}{2}} = 21.1 \text{ feet per second}$$

From equation (19)

$$v_v = \frac{-7.5 + \sqrt{(7.5)^2 + 2(0.00442)(447)^2}}{2} = 17.7 \text{ feet per second}$$

Neglecting fuselage parasite-drag losses, the power required by a helicopter in climb, P_v, is the sum of the induced and profile losses, $P_i + P_o$, as well as the power required to lift the aircraft at the velocity V_v, TV_v. Thus

$$P_v = P_i + P_o + TV_v$$
$$= Tv_v + \text{profile power} + TV_v$$

In order to calculate P_v, the profile-drag power must first be determined. Assuming ideally twisted blades, the profile-drag power in vertical climb is equal to that in hovering, because the spanwise angle of attack and spanwise velocity distributions are the same in both conditions. The magnitude of the profile-drag loss may therefore be estimated by subtracting a calculated induced power in hovering from the measured hovering power, which for the case under consideration is 125.0 horsepower. Thus, in hovering,

$$P_i = Tv_h = 2393 \times \frac{21.1}{550} = 91.5 \text{ horsepower}$$

and

$$P_o = 125.0 - 91.5 = 33.5 \text{ horsepower}$$

Writing the power ratio

$$\frac{P_v}{P_h} = \frac{Tv_v + TV_v + P_o}{Tv_h + P_o}$$

then

$$\frac{P_v}{P_h} = \frac{\frac{2393}{550}\left(17.7 + \frac{450}{60}\right) + 33.5}{125.0} = 1.14$$

This ratio checks the experimental value obtained from flight tests (Fig. 5–8) and shows that the simple momentum theory, with no

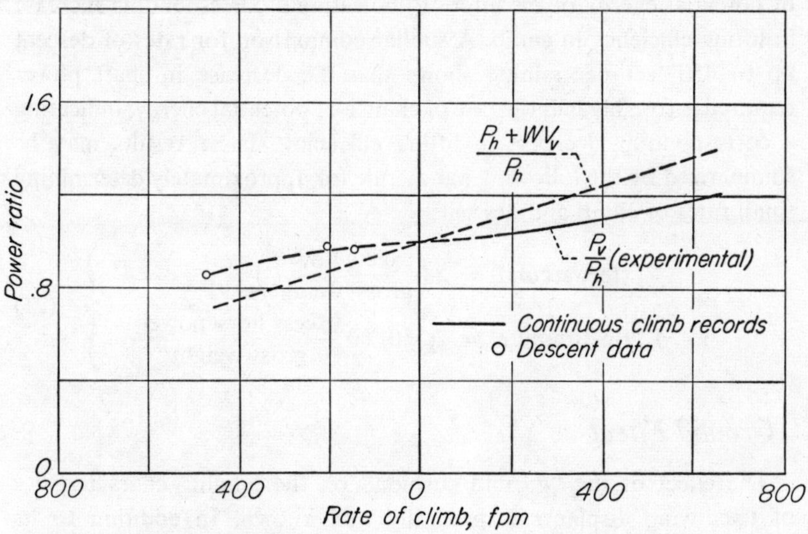

Fig. 5–8 Effect of climb velocity on rotor efficiency (from Appendix IIA, reference I–10).

refinements, correctly indicates the improvement of the induced efficiency in climb.

EFFECT OF CLIMB ON POWER REQUIRED. It is interesting to compare the actual power required in climb with the power required if there were no induced velocity changes. The comparison can be made by means of Fig. 5–8, using the curve labeled $(P_h + WV_v)/P_h$. The quantity WV_v is the power required to lift the aircraft at the velocity V_v, calculated from the known gross weight of the helicopter and the measured rate of climb. Thus the curve labeled $(P_h + WV_v)/P_h$ represents the climb power that would be expected if there were no changes in the induced or profile-drag power losses resulting from the climb velocity, while the curve P_v/P_h indicates the actual power required.

Figure 5–8 shows that the value of $(P_h + WV_v)/P_h$ at the rate of climb under consideration is 1.26, while the actual power required is represented by the value of 1.14, indicating a saving of $\frac{1.26 - 1.14}{1.26 - 1.00}$ = 46 per cent. In general, the figure shows that the increase in power actually required for climb is approximately half the rate of change of potential energy of the aircraft, indicating a corresponding increase in lifting efficiency in climb. A similar comparison for rates of descent up to 450 feet per minute shows that the decrease in shaft power required is roughly half the rate of change of potential energy, indicating a corresponding decrease in lifting efficiency. These results may be summarized by the following handy rule for approximately determining small rates of climb and descent:

$$\left.\begin{aligned} V_v \text{ (feet/second)} &= 2\left(\frac{\text{excess power}}{\text{gross weight}}\right) \\ V_v \text{ (feet/minute)} &= 2(550)\,60\,\frac{\text{excess horsepower}}{\text{gross weight}}, \end{aligned}\right\} \qquad (21)$$

Ground Effect

The effect of the "ground cushion" on the landing characteristics of fixed-wing airplanes is generally well known. In addition to its beneficial effect, the increased lift afforded by the presence of the ground often makes landing quite a problem with lightly loaded, high performance gliders, inasmuch as the pilot of the sailplane finds it difficult to lose sufficient airspeed so that he can land on a picked spot on the field. In general, ground effect on the helicopter rotor is beneficial for it enables overloaded helicopters to take off and hover by taking advantage of the lower power required for a given thrust by flying within the presence of the ground. Ground effect is also useful in checking the rate of descent of the helicopter in the flare-out near the ground, especially during power-off landings.

THEORETICAL TREATMENT. The ground influences the performance of a helicopter rotor through its restraint of the rotor downwash. As the helicopter approaches the ground, the induced velocity required to produce a given thrust is reduced, with a resultant decrease in induced power. Alternately, if a fixed amount of power is transmitted to the shaft, ground effect will produce an equivalent increase in thrust.

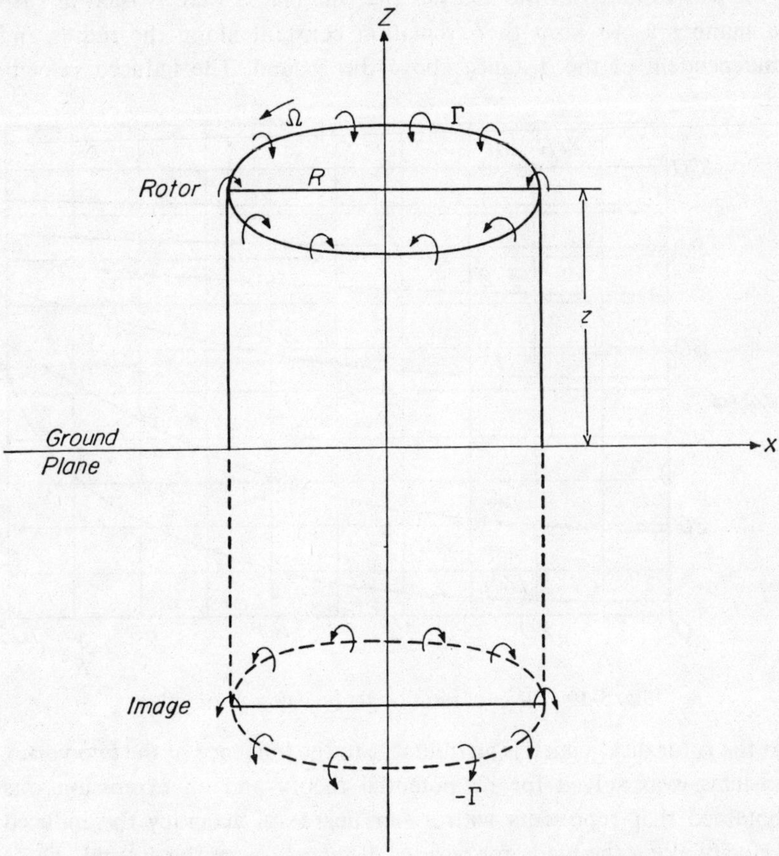

Fig. 5–9 Image vortex system for rotor within ground effect.

Ground effect was investigated mathematically in reference I–13 of Appendix IIA (from which Figs. 5–9 through 5–12 were adapted) by replacing the rotor by a cylindrical vortex and the ground by an image vortex cylinder (Fig. 5–9). It can be seen from the figure that the boundary condition of zero vertical velocity at the ground is satisfied, for the direction of the circulation along the blades of the two vortex systems is such as to make the induced velocities act in opposite directions. The velocities, being equal in magnitude, thus cancel each other at the ground.

It was assumed in the analysis that the blades were twisted in such a manner as to keep the circulation constant along the radius and independent of the distance above the ground. The induced velocity

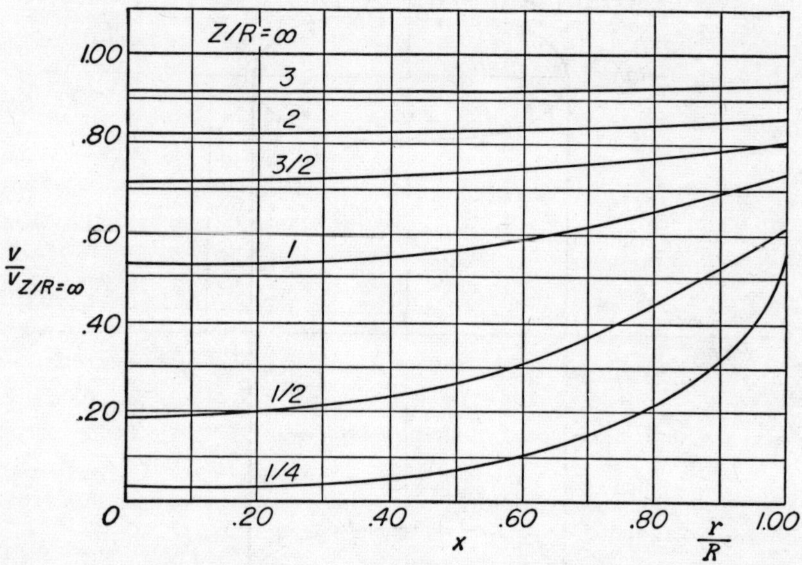

Fig. 5–10 Effect of rotor height on inflow distribution.

at the rotor disk, which is attributable to the influence of the two vortex systems, was solved for by potential theory and an expression was obtained that represents with a fair degree of accuracy the induced velocity along the blade for various distances above the ground. These curves are given in Fig. 5–10.

The results of the analysis of the ground effect on power may be summarized in a single plot representing the ratio of the power required in ground effect to that required in free air against height above the ground for the condition of constant thrust coefficient. The power ratio can be obtained in the following manner.

The total torque absorbed by the rotor is composed of a part required to overcome profile drag and an induced torque. From equation (36), Chapter 4, it can be seen that the profile-drag torque of a rotor hovering in free air (i.e., at an infinite distance above the ground plane) and having ideally twisted blades, is made up of terms that do not vary with thrust coefficient, $\sigma\delta_0/8$, and those that do, $\left[\frac{2}{3}\frac{\delta_1}{a}\frac{C_T}{B^2} + \frac{4\delta_2}{\sigma a^2}\left(\frac{C_T}{B^2}\right)^2\right]$. For hovering in free air, the induced torque as given by the same equation is $(C_T^{3/2})/(B\sqrt{2})$. If that part of the total torque coefficient which varies with thrust is denoted by ΔC_Q, then its value in free air is

$$\Delta C_{Q_\infty} = \frac{C_T^{3/2}}{B\sqrt{2}} + \frac{2}{3}\frac{\delta_1}{a}\frac{C_T}{B^2} + \frac{4\delta_2}{\sigma a^2}\left(\frac{C_T}{B^2}\right)^2 \tag{22}$$

For purposes of plotting, it is convenient to express ΔC_Q as a percentage of its free-air value, ΔC_{Q_∞}. The ratio may be symbolized by Λ, where

$$\Lambda = \frac{\Delta C_Q}{\Delta C_{Q_\infty}} \tag{23}$$

Figure 5–11 gives the variation of Λ with z/R for a range of values of C_T/σ^2. (The parameter C_T/σ^2 is used instead of C_T in order to make the plot independent of solidity.)

The use of Fig. 5–11 in calculating the power required by a rotor when hovering within ground effect can be illustrated by the following example.

Example: Determine the power required by a rotor having ideally twisted blades when hovering at a height of 15 feet above the ground. The following data are given:

R = 20 feet $\qquad C_T = 0.0060$

$\sigma = 0.060 \qquad \Omega$ = 25 radians/second

$a = 5.73 \qquad \rho$ = 0.002378 slugs/cubic foot

$c_{d_0} = 0.0087 - 0.0216\alpha_r + 0.400\alpha_r^2$

(i.e., $\delta_0 = 0.0087$, $\delta_1 = -0.0216$, and $\delta_2 = 0.400$)

Inserting the above data into equation (22) gives $\Delta C_{Q_\infty} = 0.000355$. Then entering Fig. 5–11 at $C_T/\sigma^2 = 1.67$ and $z/R = 0.75$, Λ is determined as 0.72. Using equation (23)

$$\Delta C_Q = 0.72 \times 0.000355 = 0.000256$$

The total torque coefficient may then be calculated as

$$C_Q = \Delta C_Q + \frac{\sigma \delta_0}{8} \tag{24}$$
$$= 0.000256 + 0.000065$$
$$= 0.000321$$

and

$$P = \frac{C_P \pi R^2 \rho (\Omega R)^3}{550} = 215 \text{ horsepower}$$

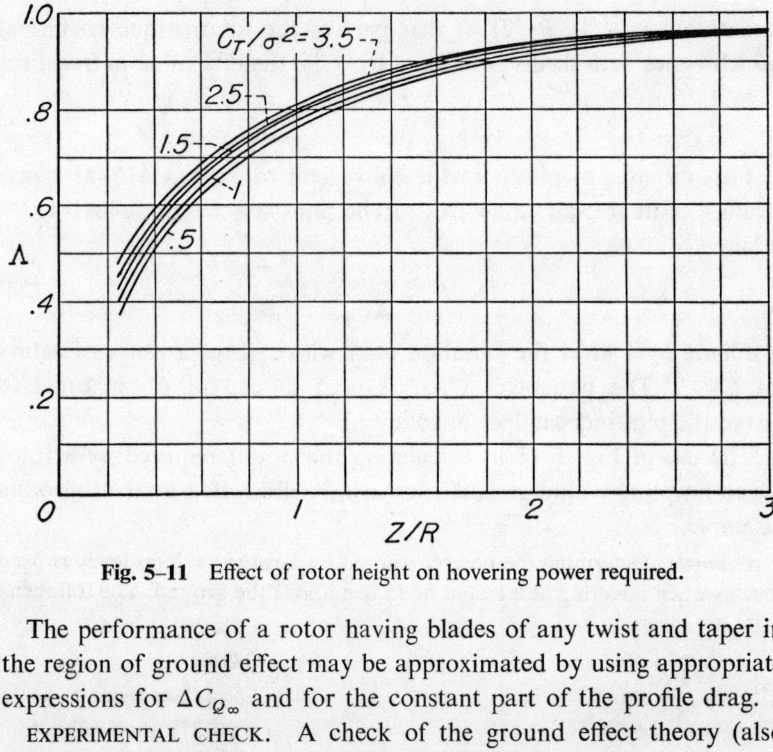

Fig. 5–11 Effect of rotor height on hovering power required.

The performance of a rotor having blades of any twist and taper in the region of ground effect may be approximated by using appropriate expressions for ΔC_{Q_∞} and for the constant part of the profile drag.

EXPERIMENTAL CHECK. A check of the ground effect theory (also presented in reference I–13 of Appendix IIA) was obtained from 5-foot diameter model tests employing untwisted blades. Samples of a comparison of the results of the tests with those of the mathematical analysis are given in Fig. 5–12. The rotor performance in free-stream operation ($z/R = \infty$) is also given in the figure in order to illustrate the magnitude of the ground effect. Figure 5–12 indicates that the

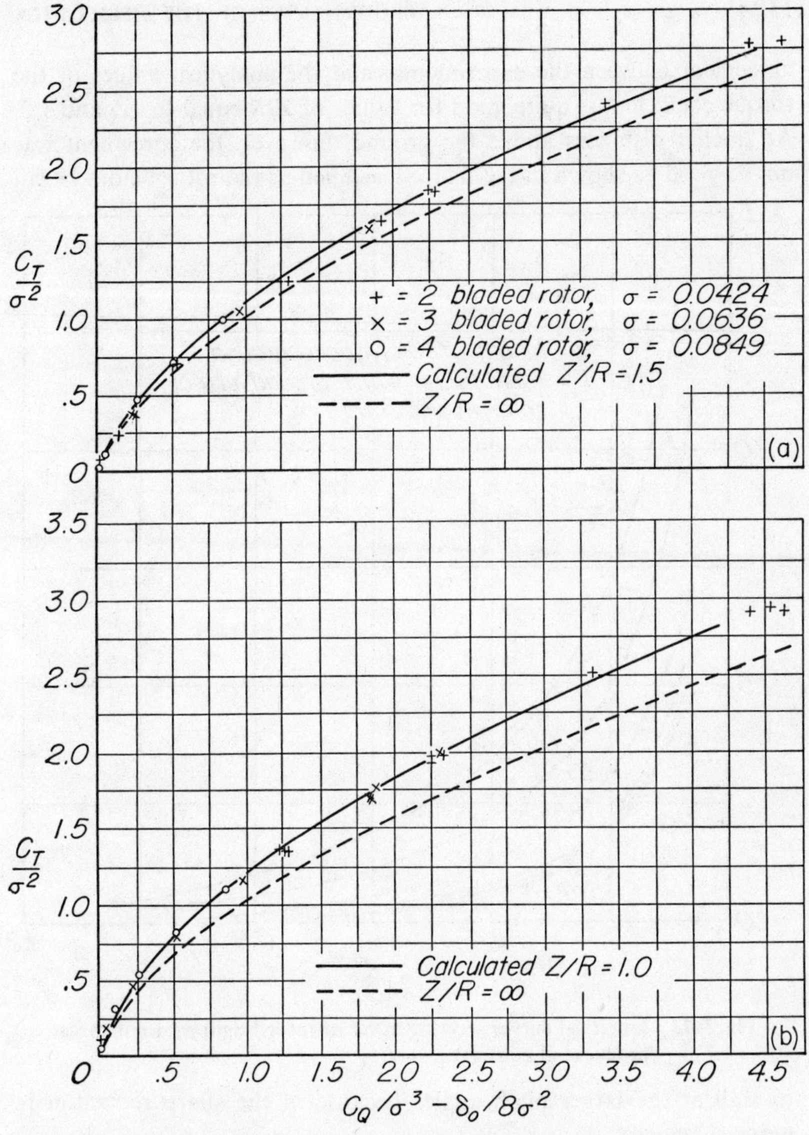

Fig. 5–12 Theory-data comparison of rotor performance within ground effect.
(a) $Z/R = 1.5$
(b) $Z/R = 1.0$

agreement between the experimental and the analytical values of the torque coefficient is quite good for values of Z/R equal to 1.5 and 1.0. At smaller distances above the ground, however, the agreement was not as good as shown in the figure inasmuch as the rotor blades began

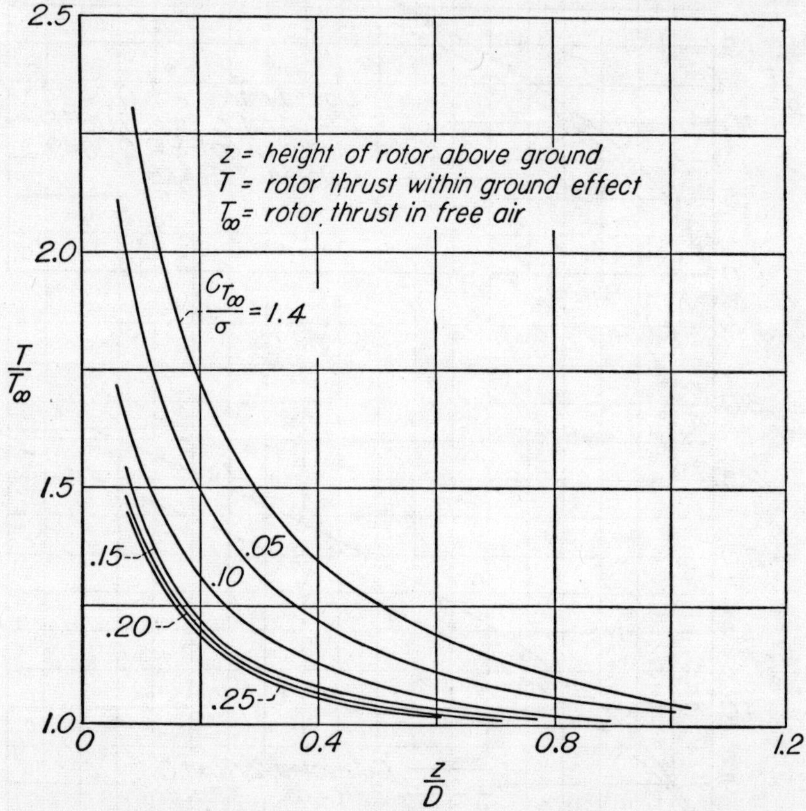

Fig. 5–13 Empirical curves showing effect of rotor height on rotor thrust produced at constant power.

to stall at the larger pitch angles because of the sharp reduction in induced velocity.

EMPIRICAL TREATMENT. An empirical means for calculating the effect of the ground on rotor hovering performance is presented in reference 14, (Appendix IIB). The paper (from which Figs. 5–13 and 5–14 were taken) contains an analysis of the results of about twenty-five different model tests (including the data of reference I–13 of Appendix IIA), the models having different blade shapes and numbers of blades, different blade twists and pitch settings, and operating at various rotor speeds. A summary of the results of the tests is given in Fig. 5–13, in which the ratio of rotor thrust in ground effect to that in free air is plotted against rotor height above the ground, expressed in terms of

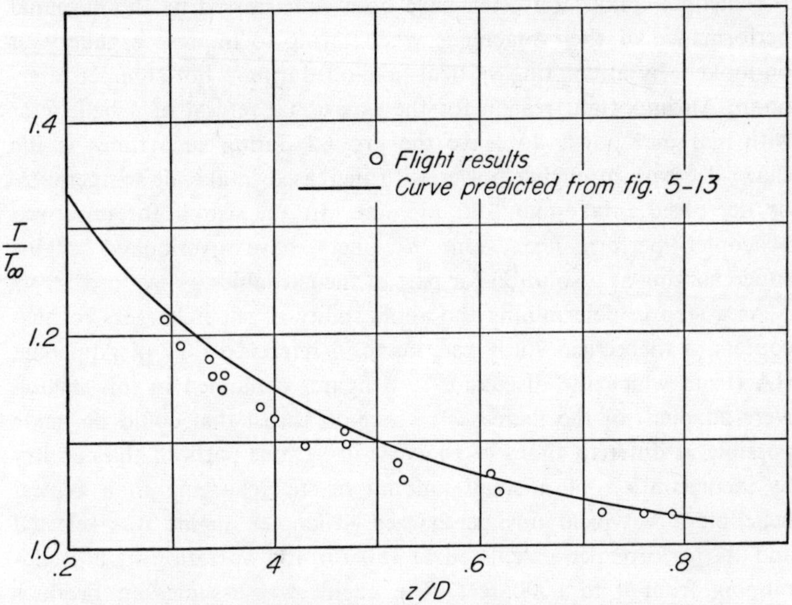

Fig. 5–14 Experimental check of ground-effect curves of Fig. 5–13.

the rotor diameter, for different values of free-air thrust coefficient-solidity ratios.

Figure 5–13 may be used to determine the increase in thrust obtained at constant power within ground effect for a rotor operating at various distances above the ground and at various values of free-air thrust and thrust coefficient-solidity ratio. For example, the figure shows that if a rotor operates at a height above the ground equal to about one-fourth of its rotor diameter, at a value of free-air C_T/σ corresponding to 0.10, it can hover with approximately 25 per cent greater load than it could lift in free air.

The curves of Fig. 5–13 were checked by means of flight tests in a typical single-rotor helicopter. The results of the tests, shown in Fig. 5–14, indicate good agreement with the predicted values of T/T_∞ as obtained from Fig. 5–13.

Effect of Engine Supercharging

Helicopter pilots too often have been embarrassed by the marginal performance of their machines when trying to impress a gallery of on-lookers by attempting vertical take-off during a hot summer afternoon. An important reason for the occasional refusal of a helicopter with marginal power to leave the ground during such times is the sharp decrease in engine power with increased intake-air temperature or decreased intake-manifold pressure. In the search for improved helicopter performance, some designers have overlooked engine supercharging as a solution for part of their troubles.

As a step in determining the applicability of superchargers to helicopters, a theoretical study was made in reference I–11 of Appendix IIA (from which the discussion and figures contained in this section were adapted) of the increases in take-off thrust that could be made possible at different times of the year in various parts of the country by incorporation of a small amount of supercharging in a typical helicopter. A typical unsupercharged helicopter engine was selected and its performance calculated at full throttle operation at altitudes ranging from 0 to 6000 feet. The engine was assumed to produce 200 indicated horsepower at standard sea-level conditions under continuous operation at 2100 revolutions per minute.

The engine was then assumed to be fitted with a single-stage, single-speed, gear-driven supercharger designed for a pressure ratio sufficient to give 230 indicated horsepower at 2100 revolutions per minute at 2600-feet altitude at full throttle and standard atmospheric conditions. The 15 per cent increase in the maximum permissible indicated horsepower was chosen because temporary increases in indicated mean effective pressure permitted for military take-off operation ranged from 10 to 20 per cent for a number of supercharged engines. The performance of the supercharged engine, calculated over the same temperature and pressure range as was done for the engine without

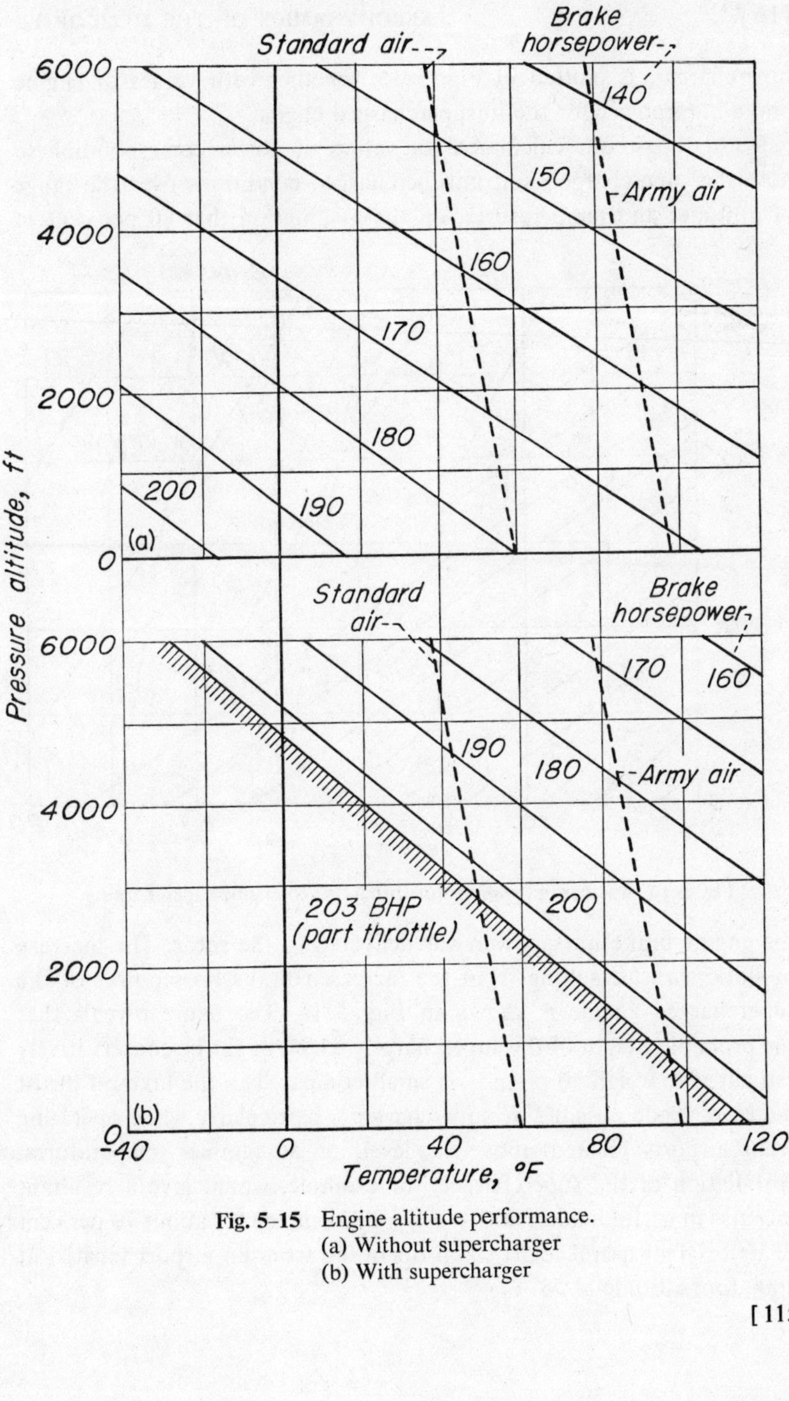

Fig. 5–15 Engine altitude performance.
(a) Without supercharger
(b) With supercharger

supercharger, is plotted in Fig. 5–15, together with values of engine brake horsepower for the unsupercharged engine.

Rotor thrust was calculated for values of power corresponding to both the supercharged and unsupercharged conditions over the range of altitudes and temperatures, on the assumption that 80 per cent of

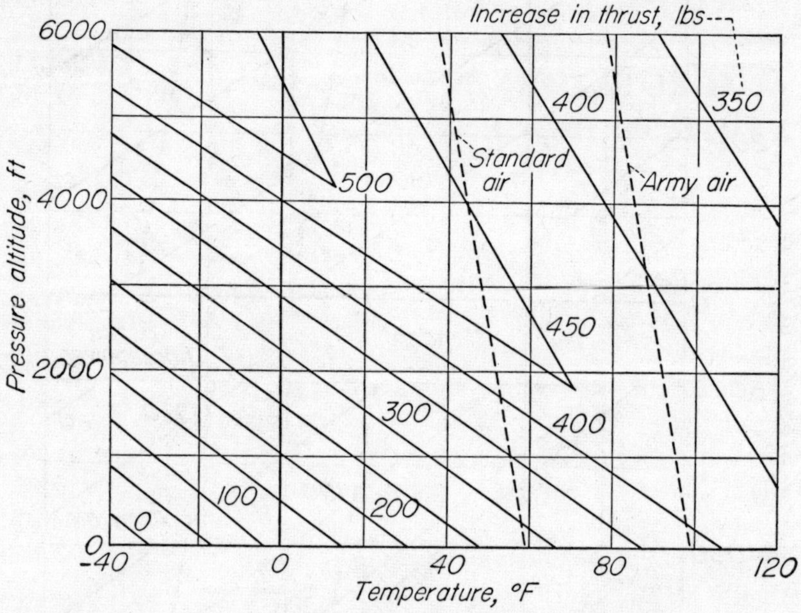

Fig. 5–16 Increase in static rotor thrust due to engine supercharging.

the engine brake horsepower was delivered to the rotor. The increase in static thrust resulting from the increased brake horsepower of the supercharged engine is shown in Fig. 5–16. The figure reveals that the probable weight of the supercharger, which might be conservatively estimated to weigh 50 pounds, is small compared to the take-off thrust increases made possible by supercharging, particularly when operating from airports located above sea-level, or at summer temperatures. Installation of the supercharger, for example, would give a resultant increase in useful load of 400 − 50 = 350 pounds (or about 70 per cent of its original useful load) when operating from an airport located at 1000-foot altitude at 75° F.

6

AUTOROTATION IN VERTICAL DESCENT

Autorotation may be defined as the condition of flight where the lifting rotor is driven in rotation by air forces, with no power being supplied through the rotor shaft. Inasmuch as by definition no power is expended through the rotor shaft, it is clear that the power required to produce thrust (induced power) and the power required to drag the blades through the air (profile-drag power) must be supplied from some external source. The autogyro supplies power through a propeller at the front of the fuselage, pulling the rotor through the air. The autogyro or helicopter in the power-off condition glides in autorotation and the rotor is pulled through the air by the force of gravity.

That a rotor will autorotate in descent is not at all surprising when one considers the rotor as a windmill. Every boy has whittled out a propeller and watched it spin in the wind. Closer observation would have shown that the spinning propeller offered resistance to the wind and that the higher the wind velocity the greater that resistance, or drag, became. For the power-off helicopter, this phenomenon may be interpreted to mean that as the machine descends, the rotor spins in the wind and that equilibrium is reached when the resistance of the rotor is equal to the weight of the helicopter. The question for a given rotor is not "Will it autorotate?" for any rotor will autorotate if the pitch is low enough. Rather the question is "What will be its minimum rate of descent in steady autorotation?" The surprising fact about autorotating rotors is not that they spin in the wind but rather that they are capable of producing such high resistance; that is, capable of supporting the helicopter in descent at such low rates of descent. Indeed, a rotor in vertical autorotation is as effective in producing resistance as a parachute of the same diameter.

The Energy Balance in Autorotation

Before considering the mechanism by which a rotor blade element achieves autorotative equilibrium, it is well to establish one further concept of the autorotating rotor. It has already been pointed out that the power to overcome blade profile drag and to produce lift must be supplied in power-off flight by the force of gravity pulling the rotor at a rate of descent.

Analytically, then (neglecting parasite drag)

$$WV_v = Wv + \text{profile-drag power}$$

where W = weight of the helicopter

V_v = rate of descent

v = mean effective induced velocity

Once the weight of a given helicopter is fixed, the induced loss is determined. For normal rotors in autorotation, profile power represents 25 per cent to 50 per cent of the total rotor losses. The rate of descent thus depends heavily on the profile power. It should be quite clear, then, that the blades must be as smooth as possible and of good airfoil section in order to achieve low rates of descent. Thus, in order for the rotor as a whole to produce the required thrust at the lowest rate of descent, the blades should have as little profile drag as possible.

Forces on the Blade Element in Autorotation

Consider the forces acting on an element of a windmilling rotor, as in Fig. 6–1. The velocity at the blade element is made up of two components, as in hovering; the rotational speed Ωr and the inflow velocity u. The section lift force is perpendicular to the relative wind and therefore a component of the lift vector exists that tends to accelerate the element in the direction of rotation, while at the same time the profile drag tends to decelerate the rotor. As drawn in Fig. 6–1 a net force exists in the plane of rotation which tends to accelerate the blade element.

Figure 6–1 is drawn with the blade at a negative pitch angle, as one would normally think of a windmilling rotor. It may be seen, however, that no fundamental reason exists to prevent autorotation at positive

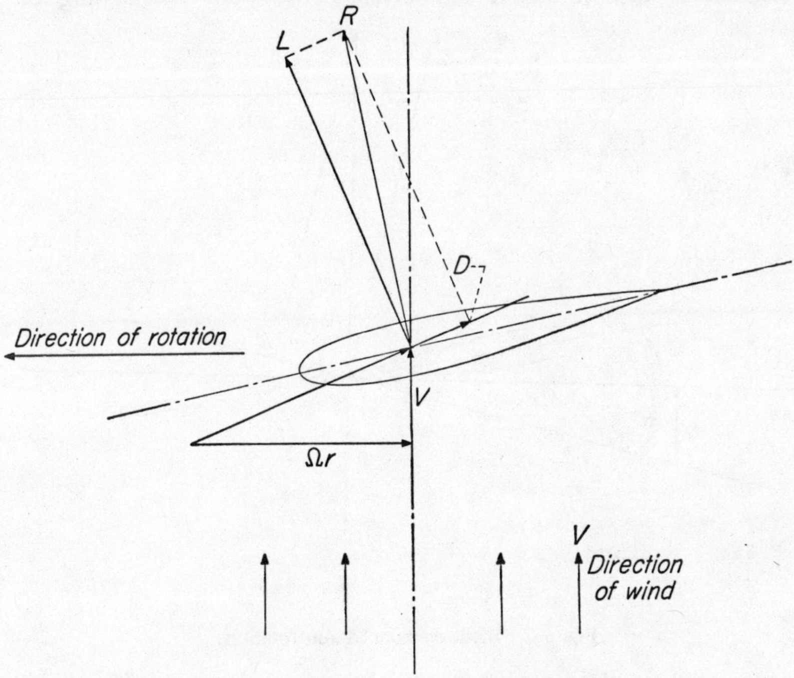

Fig. 6–1 Forces acting on a windmilling blade element.

pitch angles, as long as the ratio of lift to drag is high enough to permit equilibrium. It will be shown that optimum autorotative performance is obtained at moderate positive angles.

It is well to point out the distinction between the terms *autorotation* and *windmilling*. As generally accepted, these terms differ only in degree. Windmilling implies blade settings which produce a maximum of torque, regardless of the thrust produced. Autorotation implies blade settings that produce maximum axial resistance to the wind at zero torque. Moderate positive angles are best for autorotation; negative angles are best for producing torque. Any windmilling rotor does produce resistance, however, and in this respect windmilling may be considered as defining an operating condition during which the rotor is primarily a torque-producing device rather than a thrust-producing device.

A normal autorotating blade element is shown in Fig. 6–2. The element is in autorotative equilibrium because no accelerating or

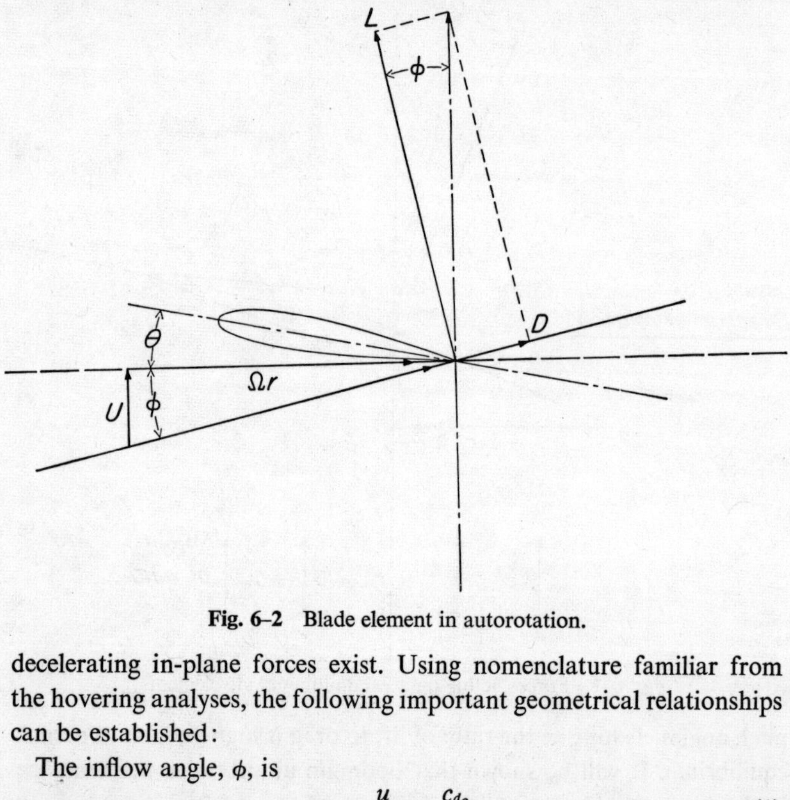

Fig. 6–2 Blade element in autorotation.

decelerating in-plane forces exist. Using nomenclature familiar from the hovering analyses, the following important geometrical relationships can be established:

The inflow angle, ϕ, is

$$\phi = \frac{u}{\Omega r} = \frac{c_{d_o}}{c_l} \tag{1}$$

where c_{d_o} and c_l are the blade-element drag and lift coefficients. Thus, in autorotative equilibrium,

$$c_{d_o} = c_l \phi \tag{2}$$

The blade angle of attack, α_r is

$$\alpha_r = \theta + \phi \tag{3}$$

Autorotation Diagram

A convenient diagram for studying the mechanism of autorotation and determining the autorotative characteristics of a particular airfoil section is given in Fig. 6–3. (This diagram and its explanation first appeared in British *R & M No. 1108*, 1926.) The figure consists of the airfoil section characteristics, c_{d_0}/c_l, plotted against the blade section angle of attack, α_r, both quantities being drawn to the same scale,

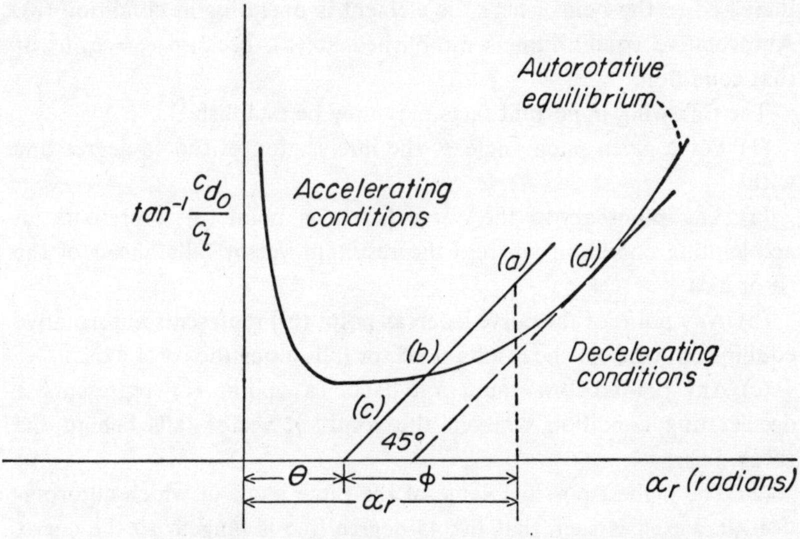

Fig. 6–3 Autorotation diagram for investigating equilibrium conditions at a blade element.

such as radians or degrees. [The vertical scale may be considered as an angle whose tangent is c_{d_0}/c_l (i.e., $\tan^{-1} c_{d_0}/c_l$).]

The diagram is applied to a particular section by first marking off the pitch angle of the section from the origin of the plot along the α_r axis, and then by constructing a 45-degree line from the pitch angle as shown in Fig. 6–3. A perpendicular dropped from any point on this line to the horizontal axis will then define the blade-element inflow angle ϕ on the horizontal axis, inasmuch as the inflow angle may be obtained by subtracting the pitch angle from the blade section angle of attack [equation (3)]. Also, because of the equality of the two legs of the right triangle, the vertical leg is also equal to ϕ.

It may now be seen that the condition of operation as represented by point (a) in Fig. 6–3 is one in which $\phi > c_{d_0}/c_l$. For this condition, the resultant force on the airfoil is displaced from the axis of rotation in such a manner as to cause the blade element to accelerate, with a consequent increase in Ωr. (See Fig. 6–2.) As Ωr increases, ϕ decreases. The element continues to accelerate until its rotational speed has increased to the point where the element is operating at condition (b). Autorotative equilibrium is established at (b), because $\phi = c_{d_0}/c_l$ at that condition.

The following important facts may now be established:

(1) For a given pitch angle, θ, the intersection of the 45-degree line with:

(a) Any point above the curve [such as point (a)] represents an accelerating condition wherein the resultant vector falls ahead of the rotor axis.

(b) Any point of the curve [such as point (b)] represents autorotative equilibrium wherein the resultant vector falls along the rotor axis.

(c) Any point below the curve [such as point (c)] represents a decelerating condition wherein the resultant vector falls behind the rotor axis.

(2) The highest possible value of the pitch angle at which autorotation may exist is such that the 45-degree line is tangent to the curve, as at point (d).

It is important to note that autorotation is a stable phenomenon as long as the pitch angle is less than the maximum as defined by point (d). Any disturbance which slows down the rotor increases ϕ and accelerates the rotor to autorotative equilibrium. Similarly, if a disturbance causes the rotor to speed up, ϕ is decreased, thereby tilting the resultant vector rearward and decelerating the rotor to equilibrium.

The autorotation diagram also demonstrates some important concepts concerning variations of rotational speed with blade pitch. If changes in the inflow velocity u are neglected, then ϕ varies inversely with Ωr. On the diagram, then, the highest rotational speed corresponds to the lowest ϕ. The pitch for maximum rotor speed is therefore the pitch defined by the intersection of a 45-degree line through the minimum of the c_{d_0}/c_l curve and the horizontal axis. Operating at higher pitches means operating at lower rotational speeds. As the pitch is increased, the rotational speed will decrease more and more rapidly until the highest possible pitch for autorotation is reached [point (d)]. This point represents a discontinuity. Physically, any slight increase in pitch would result in the rotor stopping, and then turning in the opposite direction, which of course would be catastrophic in practice with rotor blades which depend on centrifugal forces for flapping equilibrium.

Changes in rotational speed with blade pitch may be summarized by saying that as the pitch is decreased from the maximum value of autorotation, the rotational speed increases until the minimum c_{d_0}/c_l is reached, then decreases again as the pitch is decreased further until finally, at $-$ 90 degrees, the rotational speed becomes zero.

The principal value of the autorotation diagram lies in helping to visualize autorotative relationships. It must be remembered however that the diagram applies only to a single blade element inasmuch as each element of a rotor blade in a given flight condition operates at different velocities and angles of attack. Some elements may encounter accelerating forces while others may encounter decelerating forces. It is therefore necessary to integrate forces on the elements along the blade in order to determine the behavior of the blade as a whole.

Most Efficient Angle of Attack for Autorotation

The aim in the design of an autorotating rotor is to obtain a minimum rate of descent at a given helicopter gross weight and horizontal velocity. While the autorotation diagram clarifies the mechanism of autorotation it deals only with the resultant rotor inflow, and therefore not with the rate of descent itself. The relationships employed in the diagram are therefore inadequate to determine the angle of attack for minimum rate of descent.

The optimum operating angle of attack may be determined from energy considerations. It has already been pointed out that the rate of descent of a given rotor is influenced by the profile-drag power.

The optimum blade angle of attack is therefore determined as the angle for minimum profile-drag power. The profile-drag power of a blade element may be expressed as

$$d(\text{profile-drag power}) = \tfrac{1}{2}c_{d_o}\rho(\Omega r)^2 c\, dr(\Omega r) \tag{4}$$

For a given element, then,

$$\text{profile-drag power} \sim c_{d_o}(\Omega r)^3$$

Also

$$dT = c_l \tfrac{1}{2}\rho(\Omega r)^2 c\, dr \tag{5}$$

For a given thrust, then,

$$(\Omega r) \sim \frac{1}{\sqrt{c_l}}$$

Therefore

$$\text{profile power} \sim c_{d_o}(\Omega r)^3 \sim \frac{c_{d_o}}{c_l^{3/2}} \tag{6}$$

It is apparent from the preceding relationships that for a minimum profile-drag power loss, and therefore for a minimum rotor sinking

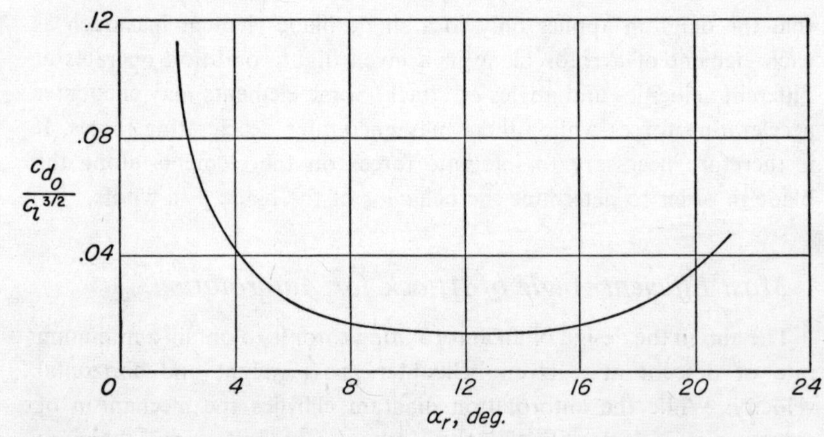

Fig. 6–4 Section characteristics of a practical-construction NACA 23012 airfoil.

speed, the $c_{d_o}/c_l^{3/2}$ ratio of each blade element should be as low as possible. This conclusion may be tempered somewhat by an examination of Fig. 6–4, in which the $c_{d_o}/c_l^{3/2}$ ratio of an NACA 23012 airfoil section of practical construction is plotted against angle of attack α_r. The figure shows that for good efficiency the operating angle of attack could be anywhere in a reasonably high range below the stall angle, inasmuch as the optimum part of the curve is very flat.

Statement of the Performance Problem

The aim of the performance analysis for the hovering condition was to find relationships between the power required to hover and the major variables—thrust, rotational speed, pitch, solidity, and profile

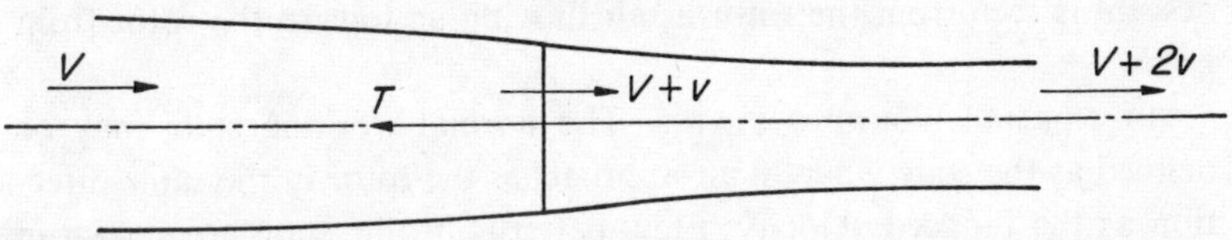

Fig. 6–5 Slipstream of an airscrew as derived from momentum considerations.

drag. The problem in vertical autorotative descent is to find relationships between these same variables and the rate of descent. As in hovering, three fundamental physical relationships are available:

(1) The blade element thrust equation
(2) The blade element torque equation
(3) The momentum equation

The first two of these equations express the thrust and torque of an element in terms of the inflow velocity, and the third relates the thrust of the rotor as a whole to the inflow. These three relations are sufficient to define completely the operating conditions of the rotor. The blade-element thrust and torque expressions are general and are free of qualifying assumptions. The simple momentum equation, however, requires a definite physical condition for the airflow. Specifically, the momentum equation assumes that a definite slipstream exists with respect to the rotor and that air moves in the same direction far ahead and far behind the rotor, as shown in Fig. 6–5. Inasmuch as the airflow direction is downward in hovering and vertical climb and upward in rapid vertical descent, it is clear that there must be intermediate states where no definite slipstream exists. In view of these facts it is well to examine the nature of the flow through the rotor in vertical flight and to determine the conditions under which the momentum equation involves irrational assumptions.

Flow States of the Rotor

Three basic flow states exist for the rotor in vertical flight; the normal working state, the vortex-ring state, and the windmill-brake state. (The various flow states were first described in detail in reference 9, Appendix IIB.) These flow states will now be discussed from the point of view of an observer moving with the rotor. Climb, from this point of view, is seen as air moving toward the rotor from above, while descent is seen from the rotor as air moving up toward the rotor from below.

THE NORMAL WORKING STATE. The normal working state may be defined as the state wherein air approaches the rotor in the same direction as the induced velocity. Flow patterns in this state are shown in Fig. 6–6a. The flow is downward through the disk and the flow at the disk is always equal to or greater than the induced velocity. The normal working state includes conditions from infinite rate of climb to hovering.

THE VORTEX-RING STATE. As soon as the rotor starts to descend from hovering, a definite slipstream—extending from far above to far below the rotor—ceases to exist. The flow pattern is as shown at the left of Fig. 6–6b. The resultant flow through the disk is still downward, because of the large induced velocity, but the flow far above the rotor is upward. The limits of the vortex-ring state are hovering and the condition where the rate of descent is equal to twice the average induced velocity at the rotor. This limiting condition is shown at the right of Fig. 6–6b. The vortex-ring state is characterized by the absence of a definite slipstream and large recirculating flows.

THE WINDMILL-BRAKE STATE. At large rates of descent, the flow again becomes smooth, and as shown in Fig. 6–6c, a definite slipstream exists. The flow is up through the rotor in this condition, and its velocity decreases as it approaches and passes through the rotor as a result of the induced velocity which opposes the direction of motion of the main flow. The slipstream thus expands above the rotor. For low rates of descent near the limiting conditions of the windmill-brake state, the expansion of the slipstream is very large and considerable recirculation and turbulence exist, as shown at the left of Fig. 6–6c. Inasmuch

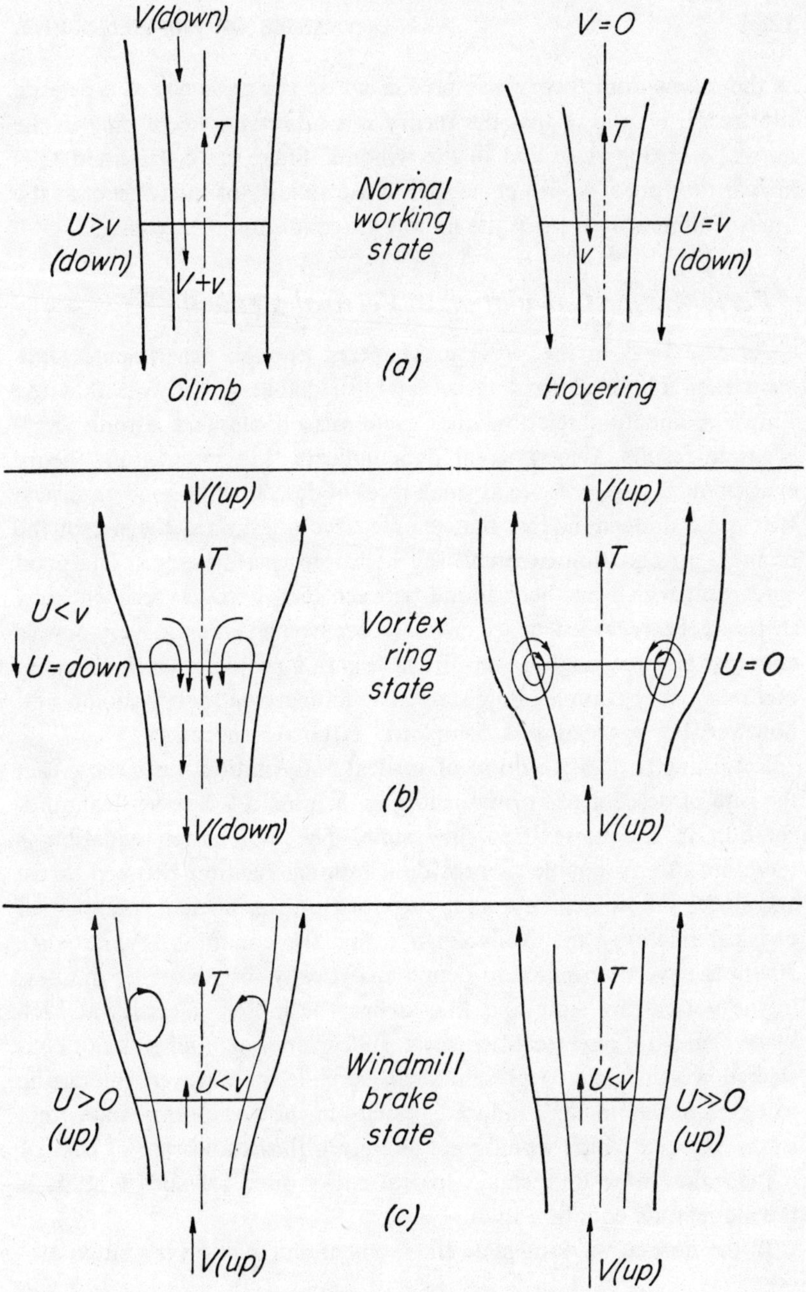

Fig. 6–6 Flow states of an airscrew.

as the momentum theory was predicated on the existence of a definite slipstream, it follows that the theory is a rational concept only in the normal working state and in the windmill-brake state. It would thus appear that proof would be needed of the validity of the concept at the limit of these states, or in the important condition of hovering.

Performance Calculation in Vertical Descent

As explained in the previous chapters, enough experimental data have been obtained to date to establish definitely the fact that the simple momentum relation does yield reasonable and astonishingly accurate results. Experimental data indicate that momentum theory even predicts performance at small rates of descent with good accuracy. It is quite understandable that people have questioned the use of the momentum equation in predicting helicopter performance. The good agreement which has been found between theory and experiment may therefore be regarded as a fortunate occurrence wherein very simple concepts give reasonable over-all answers to a problem which in reality involves a very complex flow state. The momentum theory should not, however, be regarded as a completely rational concept.

Returning to the condition of vertical autorotation, tests show that the rate of descent of normal rotors is such as to locate vertical autorotation in the vortex-ring flow state. The momentum equation is therefore utterly unable to provide a rational relation between thrust and local inflow velocity and therefore an insufficient number of physical relations are available to define the condition. While some attempts have been made to define analytically the nature of the flow in the vortex-ring state and thus define the inflow velocities at each blade element, these treatments are quite elaborate, and it is not considered worth while to present them here. It is, however, interesting to estimate the "mean" induced velocity in the vortex-ring state; that is, the velocity which would yield the given thrust and rate of descent if the inflow were uniform. A logical background for such a study is the momentum equation itself.

In the normal working state the momentum equation is written as

$$T = \pi R^2 \rho (v + V_v) 2v \qquad (7)$$

where both the induced velocity v, and the rate of climb V_v are positive when directed downward with respect to the rotor. In the windmill-brake state equation (7) becomes

$$T = 2\pi R^2 \rho(-V_v - v)v \tag{8}$$

It is convenient to make the velocity terms nondimensional. The following definitions are therefore introduced:

$$\text{Nondimensional induced velocity,} \quad \bar{v} = \frac{v}{\sqrt{\dfrac{T}{2\rho\pi R^2}}} \tag{9}$$

$$\text{Nondimensional rate of climb,} \quad \bar{V}_v = \frac{V_v}{\sqrt{\dfrac{T}{2\rho\pi R^2}}} \tag{10}$$

For a fixed weight, diameter, and air density, $\bar{v}$ is proportional to v, and $\bar{V}_v$ is proportional to V_v. The nondimensional velocities permit a single plot to represent all possible values of disk loading and density. In terms of $\bar{v}$ and $\bar{V}_v$ equation (7) becomes

$$1 = \bar{v}(\bar{v} + \bar{V}_v)$$

or

$$\bar{V}_v = \frac{1}{\bar{v}} - \bar{v} \tag{11}$$

for the normal working state, and

$$1 = -\bar{v}(\bar{v} + \bar{V}_v)$$

or

$$\bar{V}_v = -\frac{1}{\bar{v}} - \bar{v} \tag{12}$$

for the windmill-brake state.

These relations are plotted in Fig. 6–7. As previously discussed, the momentum equation loses significance when a definite slipstream does not exist. For the normal working state this limit occurs when the induced velocity in vertical descent becomes greater than the hovering induced velocity, or when $\bar{v} > 1$. In the windmill-brake state no slipstream exists if $\bar{v} > -\frac{1}{2}\bar{V}_v$ or again when $\bar{v} > 1$. The regions of the momentum equation which have no physical significance are represented by broken lines in Fig. 6–7.

Using Fig. 6–7 as a background, it is now possible to formulate a method for determining the mean effective induced velocity in the vortex-ring state from flight measurements and of predicting the performance of any rotor in partial-power and power-off descent.

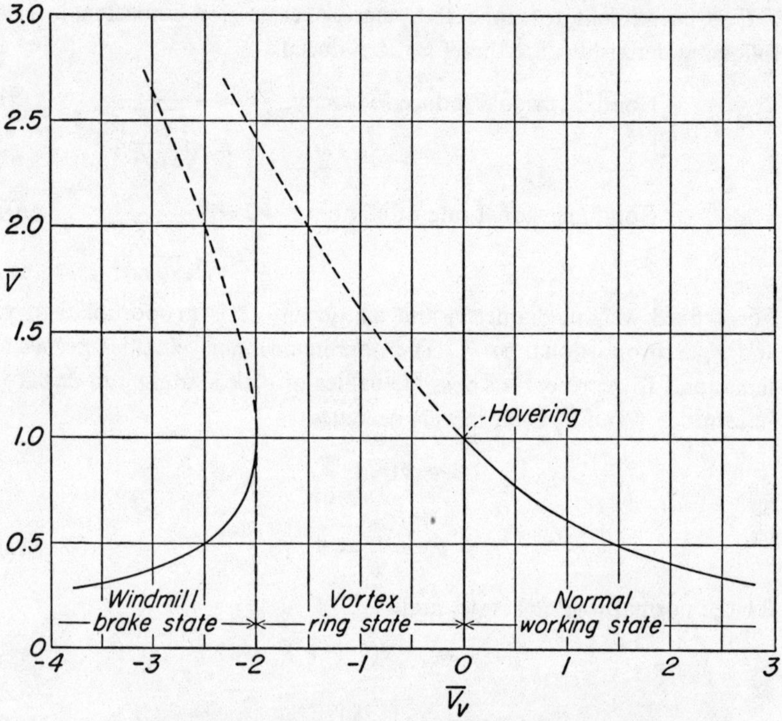

Fig. 6–7 Theoretical induced-velocity parameter. (Dashed lines indicate region wherein momentum concepts do not apply.)

It is necessary to deduce, rather than exactly determine, the mean effective induced velocity from flight measurements, inasmuch as it is almost impossible to measure the induced velocity directly.

The mean effective induced velocity is defined by the following equation which states that the power at the shaft is the sum of the climb, induced, and profile-drag powers.

$$\text{Shaft power} = TV_v + Tv + \frac{\delta\rho}{8}(\Omega R)^3\sigma\pi R^2 \qquad (13)$$

where δ is the mean blade section drag coefficient, or the drag coefficient of the airfoil at the mean blade angle of attack, which is given as the angle at which the lift coefficient is $c_{l_{\text{mean}}} = (6C_T)/\sigma$.

Equation (13) may be rewritten in terms of the following nondimensional quantities: C_T, C_Q, $\bar{V}_v$, and $\bar{v}$. It then becomes

$$C_Q = \frac{\sigma\delta}{8} + (\bar{v} + \bar{V}_v)\frac{C_T^{3/2}}{\sqrt{2}} \tag{14}$$

It is apparent from equations (13) and (14) that with measured quantities for shaft power, gross weight, rotor speed, rate of descent,

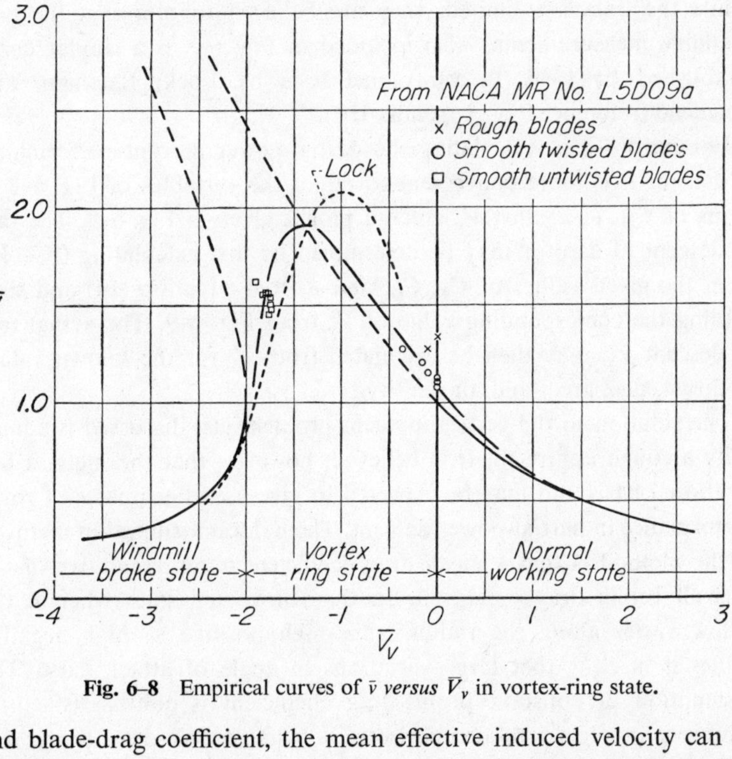

Fig. 6–8 Empirical curves of $\bar{v}$ *versus* $\bar{V}_v$ in vortex-ring state.

and blade-drag coefficient, the mean effective induced velocity can be determined. Furthermore, once a curve of effective induced velocity against rate of descent is established in the vortex-ring state it may be used to predict the rates of descent of rotors of different solidity and profile-drag characteristics. Little reliable data, however, are presently available with which to determine the relationship in the vortex-ring state.

Vertical-flight data are difficult to obtain, for small deviations from a vertical-flight path may cause sizable reductions in the rate of descent. Some data are available from NACA tests (reference I–3 of Appendix IIA) in which instrumentation was provided to give the pilot a means of holding a vertical-flight path and to record the exact angle of descent. Data from these flight tests have been reduced as previously described and are presented in Fig. 6–8. These data represent conditions of flight where the angle of descent was less than 5 degrees from the vertical. While the data are limited, they are believed to represent the best available measurements. Also included in Fig. 6–8 is a similar curve established by early British model tests by Lock, Bateman, and Townend (reference 9 of Appendix IIB).

For purposes of calculating the partial-power descent performance of a rotor it is more convenient to replot the variables of Fig. 6–8 in terms of $\bar{v} + \bar{V}_v$ against $\bar{V}_v$. Such a plot is given in Fig. 6–9. The rate of descent of a rotor may be determined by first calculating $(\bar{v} + \bar{V}_v)$ from the given values of C_T, C_Q, and δ from equation (14) and then reading the corresponding value of $\bar{V}_v$ from Fig. 6–9. The actual rate of descent V_v, may then be calculated from $\bar{V}_v$ for the known values of thrust, disk area, and air density.

The solution to the vertical-descent problem just discussed is admittedly a rough approach. It is believed, however, that the method has a rational basis and may be expected to give useful estimates of rotor performance in partial-power descent. The boldest assumption involved in the method is that a mean drag coefficient is representative of the over-all profile-drag condition. In the vortex-ring state wherein the inflow varies along the radius from high positive to high negative values it is clear that large variations in angle of attack exist. The assumption of constant profile-drag coefficient is quite satisfactory, however, in regard to comparisons of total profile-drag power for different rotors with conventional airfoil sections, for it is approximately true that blades of varying roughness have proportionate profile drags throughout the most significant angle-of-attack range. The assumption is questionable in that it may underestimate the profile

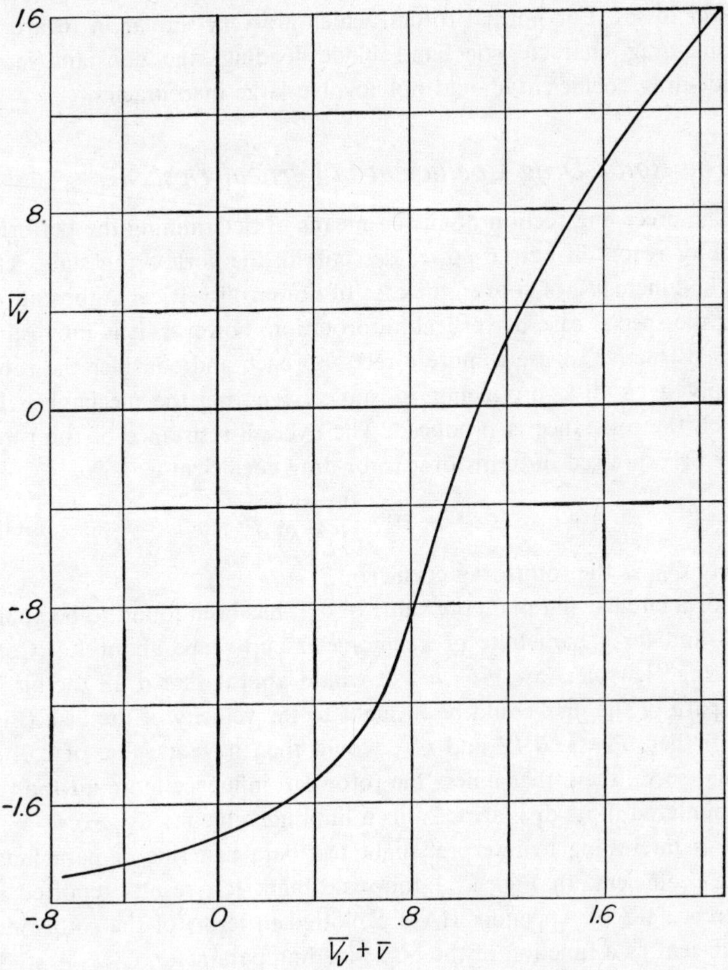

Fig. 6–9 Empirical curve for calculating vertical-descent velocities.

losses and thus may not properly proportion the profile and induced-power losses. For normal rotors, which are very similar in regard to profile-drag characteristics and blade loading, the constant mean blade-drag coefficient should not involve large inaccuracies.

The Rotor Drag Coefficient in Vertical Descent

The preceding section discussed means of determining the performance of rotors in partial-power descents in the vortex-ring state. The method includes, of course, the case of power-off vertical autorotation. For the special case of vertical autorotation, however, it is interesting and instructive to use a more direct approach and consider the rotor simply as a disk producing resistance, ignoring the mechanism by which the resistance is produced. The over-all resistance of the rotor may be expressed in terms of a rotor-drag coefficient as

$$C_{D_R} = \frac{\text{thrust}}{\frac{1}{2}\rho V_v^2 \pi R^2} \tag{15}$$

where C_{D_R} is the rotor-drag coefficient.

For a circular flat plate the value of C_{D_R} has been found to be about 1.28 and for a parachute or anemometer cup shape about 1.4. Considering Newton's law, $F = ma$, it would appear that if all the air in the path of the disk could be brought to the velocity of the disk (Fig. 6–10) then $T = \rho\pi R^2 V^2$ and C_{D_R} would then have a value of 2.0. It would seem, then, that unless the rotor can influence more air than is encountered in its disk area, 2.0 is a limiting value for C_{D_R}.

It is interesting to interpret flight test data in terms of parachutal drag coefficient. In Fig. 6–11 full-scale flight test results reported in reference I–3 of Appendix IIA are plotted in terms of the rotor-drag coefficient as a function of the blade-loading parameter, C_T/σ. A study of the figure indicates that values of about 1.2 may be expected for rotors at normal pitch angles and which have good profile-drag characteristics.

Also shown in Fig. 6–11 are theoretical curves of C_{D_R} plotted against C_T/σ for two different solidity values. It may be seen that the theory overestimates the rotor-drag coefficient by about 15 per cent, which means that the helicopter rate of descent (which varies inversely as the square root of C_{D_R}) is underestimated only by about half that amount. Thus vertical rates of autorotative descent may be predicted with satisfactory accuracy by theoretical methods.

EFFECT OF PITCH. The parameter C_T/σ is a measure of the blade-pitch angle, because the rotational speed of a rotor will vary in autorotative descent as the pitch angle is changed. The data presented in

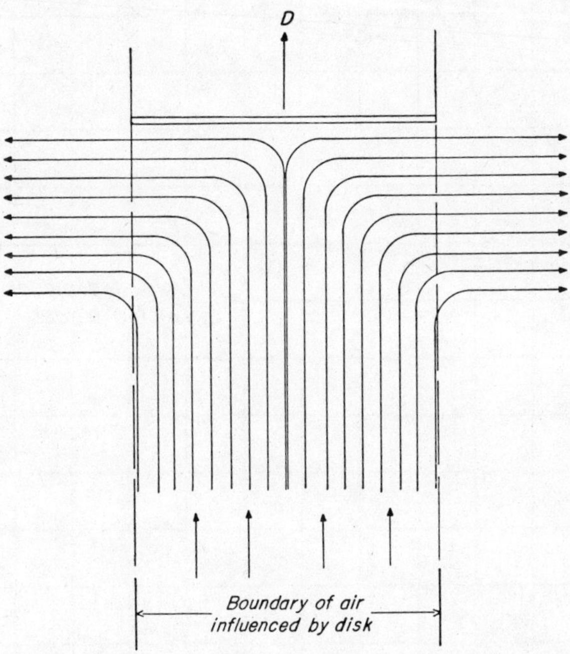

Fig. 6–10 Flow picture for calculating drag of disk by Newton's law.

Fig. 6–11 were obtained at several values of rotational speed in the range of normal helicopter operating rotor speeds. It may be concluded from a study of the data that:

(1) The rotational speed of a helicopter rotor in vertical descent can be kept in the range of normal operating values.

(2) Within the scatter of the data, the rate of descent does not vary with pitch—rotor speed combinations in the normal working range of rotor speeds. (This conclusion is substantiated by the theoretical

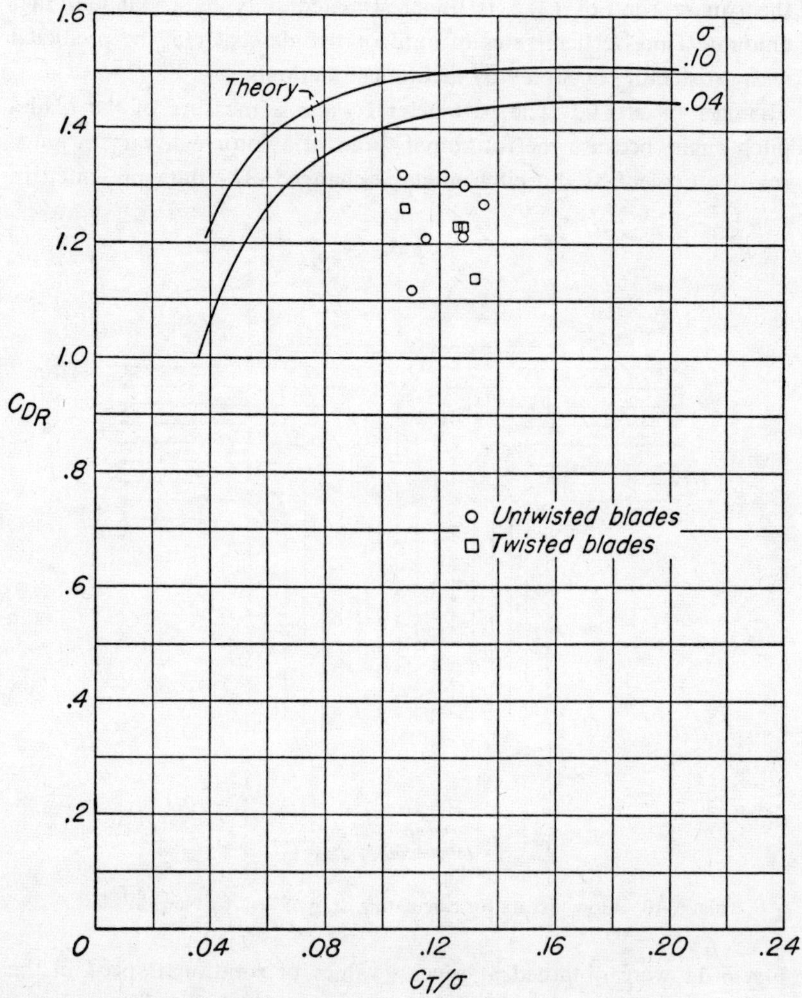

Fig. 6–11 Experimental vertical-autorotative-descent data and a comparison with theory.

curves shown in Fig. 6–11, in that the curves are relatively flat over the normal operating range.)

EFFECT OF TWIST. As noted in Fig. 6–11, the data include measurements for both twisted and untwisted plywood-covered blades of identical plan form. It should be noted that within the accuracy of the measurements, the twisted blades (low angles at the tip for good hovering and forward-flight performance) gave the same rate of descent as untwisted blades. It appears, therefore, that "helicopter" twist is not detrimental to vertical-autorotative performance as has often been suspected from considerations of high inboard angles of attack.

7

PHYSICAL CONCEPTS OF BLADE MOTION AND ROTOR CONTROL

Before developing quantitative expressions for the forces and moments acting on a lifting rotor moving edgewise through the air, it is well to gain a physical understanding of the rotor in forward flight. As shown in Fig. 7–1, the advancing blades of the rotor encounter higher velocities than the retreating blades as the rotor moves forward. Considering first a rigid propeller, it is seen that a sizable rolling moment would be present in forward flight as a result of the inequality in lift produced on the advancing and retreating blades. While it is possible to utilize two side-by-side rigid propellers rotating in opposite directions to cancel the rolling moments, high alternating blade loads are involved.

Two standard means are available to overcome the dissymmetry of lift in forward flight:

(1) The blades may be hinged at their roots so that no moments can be transmitted through the hub. Control is then achieved by tilting the hub axis until the resultant rotor vector points in the desired direction.

(2) The blades may be rigidly attached to the shaft but cyclically feathered, decreasing the pitch on the advancing (high velocity) side and increasing the pitch on the retreating (low velocity) side so as to equalize the lift around the disk.

It will be shown that the two systems are fundamentally equivalent, differing only in the axis of reference. Historically, the first successful rotating-wing aircraft was of the flapping type. Consequently, rotor theory was developed for the hinged system. Because no sacrifice in generality is involved and because the assumption of a hinged-blade system is in accordance with the bulk of the literature on rotors, the

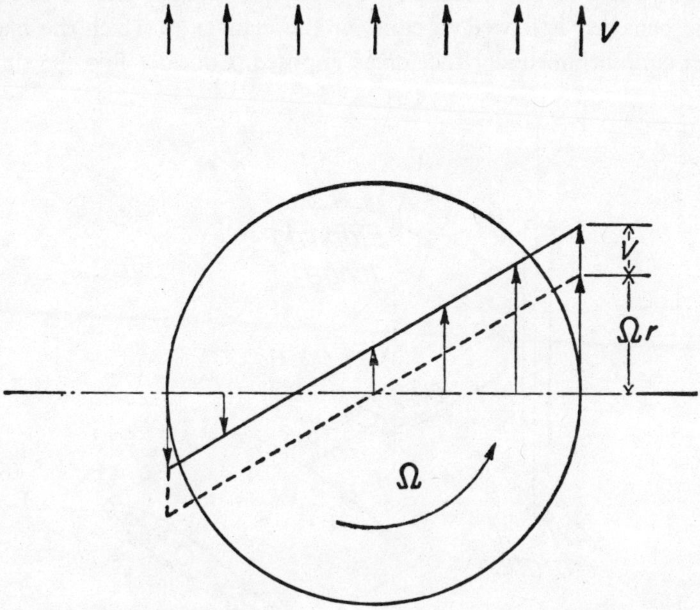

Fig. 7–1 Velocity distribution on advancing and retreating blades of a rotor in forward flight.

forward-flight theory developed in later chapters and most of the physical discussions of this chapter will deal with the pure flapping rotor.

Equilibrium of Hinged Blades

With blades free to flap, it is clear that the rolling-moment problem is solved, for no moments (other than blade-pitching moments) can be transmitted to the hub. The blades will move in such a manner as to seek equilibrium; that is, in such a way as to make the summation of the moments about the flapping hinge zero. It was found that the introduction of flapping hinges usually necessitated hinging the blade in the dragging or in-plane direction (for reasons which will be discussed later). The normal flapping blade is thus effectively mounted to the hub on a universal joint—free to flap, lead, or lag but always fixed in pitch. As an initial step toward the physical picture of rotor blade behavior it is well to consider the manner in which the blades reach equilibrium under the forces imposed. Consider first the simple

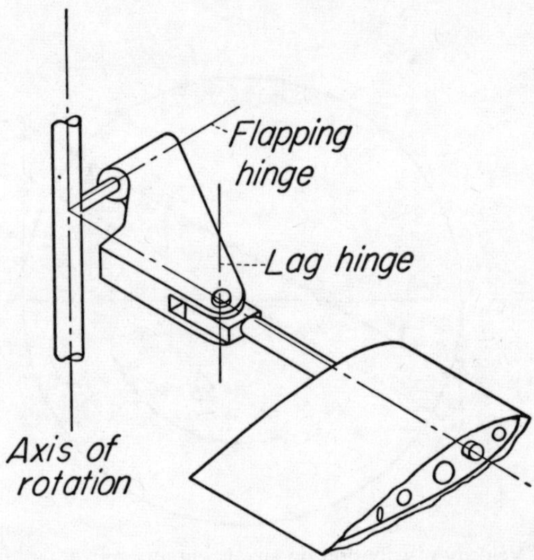

Fig. 7–2

case of a rotor in hovering flight with blades hinged in the manner shown in Fig. 7–2.

EQUILIBRIUM ABOUT THE FLAPPING HINGE. The forces acting on the blade in the flapping direction in steady hovering flight are air (lift) forces, centrifugal forces, and blade weight. These forces produce moments about the flapping hinge, and the summation of these moments must be zero. Blade weight is a small factor relative to the centrifugal and lift forces and for the present purpose it shall be considered negligible.

To establish the nature of flapping equilibrium in hovering, consider the simple case of a blade having uniform mass distribution. The centrifugal force distribution may be derived as follows. Referring to Fig. 7–3, the elemental centrifugal force is given by

$$d(\text{C.F.}) = (m\,dr)\Omega^2 r \cos \beta \tag{1}$$

where m = mass per unit length of blade
Ω = the rotational speed
r = the radius of the element
β = the blade flapping angle

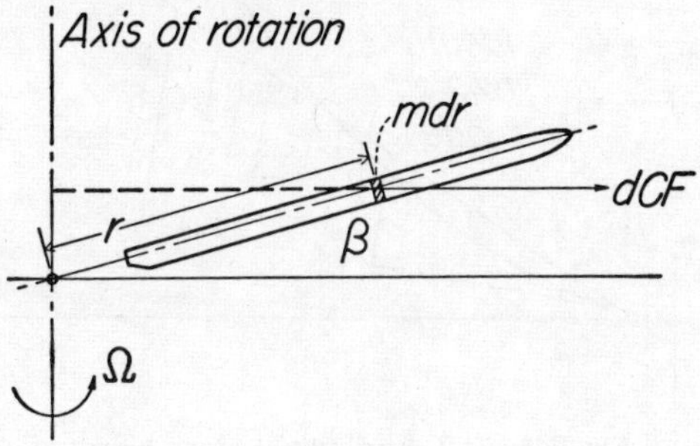

Fig. 7–3

The component of the centrifugal force perpendicular to the blade is given by

$$d(\text{C.F.}) \sin \beta = m\,dr\,\Omega^2 r\beta \tag{2}$$

(inasmuch as β is always a small angle). The centrifugal force distribution perpendicular to the blade thus varies linearly with the radius for a uniform mass blade, as shown in Fig. 7–4.

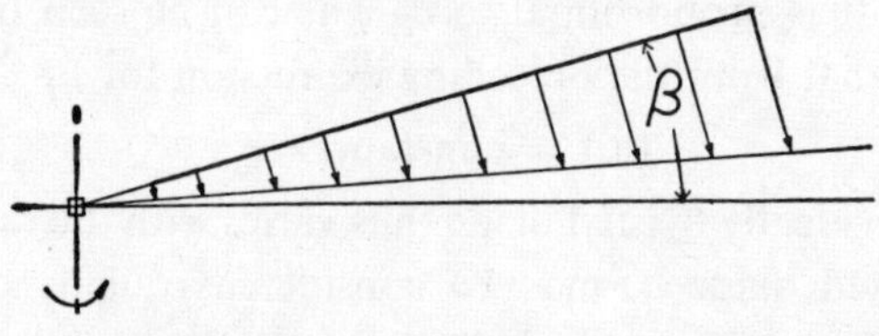

Fig. 7–4 Centrifugal force distribution perpendicular to blade.

The moment exerted by the centrifugal forces about the flapping hinge is then

$$\text{C.F. moment} = \tfrac{1}{3}R(M\Omega^2 R\beta) = \tfrac{2}{3}(\text{CF})R\beta \tag{3}$$

where $M = mR$, the blade mass

The lift force distribution along the blade in hovering depends on the particular blade plan form and twist. As extremes, consider the untwisted constant-chord blade and the ideally twisted constant-chord blade. For the untwisted blade the inflow varies approximately linearly with the radius whereas for an ideally twisted blade the inflow is

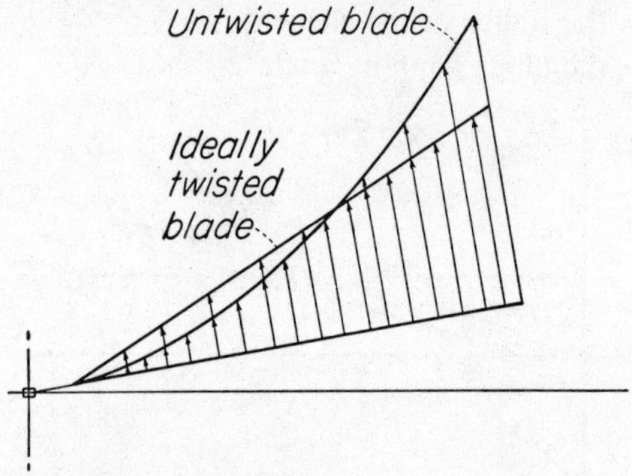

Fig. 7–5 Lift distributions along blade.

constant along the radius. The lift per foot of span for a blade of any twist is always given by

$$\frac{dL}{dr} = c_l \frac{\rho}{2} \Omega^2 r^2 c \tag{4}$$

For an ideally twisted constant-chord blade,

$$c_l = \alpha_r a = a\left(\theta_t \frac{R}{r} - \frac{v}{\Omega r}\right) \tag{4a}$$

Inasmuch as lift is proportional to $c_l r^2$, as can be seen from equation (4), it follows that from the preceding expression for c_l

$$\text{Lift} = \text{constant} \times r$$

The lift of an ideally twisted blade thus varies with the radius, whereas for an untwisted blade α_r may be considered roughly constant, and thus the lift of an untwisted blade varies as the radius squared. (See Fig. 7–5.)

The moment of the lift about the flapping hinge is then

$$\text{Lift moment} = \tfrac{2}{3}R \times \text{lift (for “ideal” twist)} \tag{5}$$

and

$$\text{Lift moment} = \tfrac{3}{4}R \times \text{lift (for no twist and no taper)} \tag{6}$$

The coning angle β is determined by equating the lift and centrifugal force moments so that

$$\beta = \frac{\text{blade lift}}{\text{C.F.}} \text{ (for an ideally twisted blade)} \quad (7)$$

$$= \frac{\frac{9}{8}\text{ blade lift}}{\text{C.F.}} \text{ (for an untwisted, constant-chord blade)} \quad (8)$$

The above expressions for β are only approximate. Tip losses, for example, have been neglected. The important point to remember is

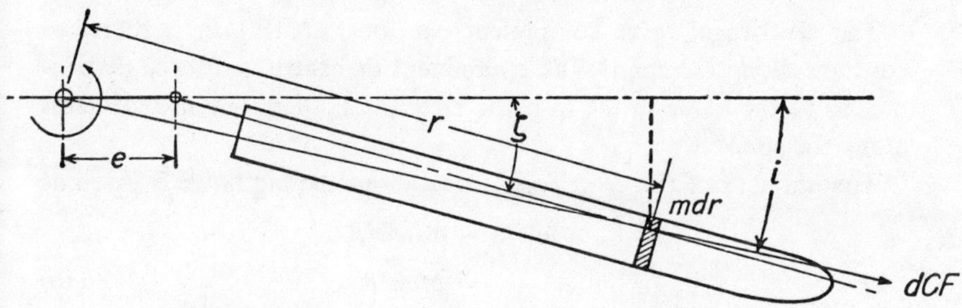

Fig. 7–6 Geometry of blade about lag hinge.

that the coning angle varies with the rotor thrust and inversely with centrifugal force; that is, with the square of the rotor speed. The coning angle in hovering is, therefore, essentially proportional to the thrust coefficient C_T.

EQUILIBRIUM ABOUT THE DRAG HINGE. The hovering equilibrium about the drag hinge is also determined by equating the vertical hinge moments to zero. The geometry of the lagging blade is shown in Fig. 7–6. The lag angle ζ is defined as the angle between the blade and a line passing through the lag hinge and the axis of rotation. The angle ζ is considered positive in the lagging direction. The component of centrifugal force perpendicular to the blade, tending to rotate the blade toward zero lag, is given as

$$d(\text{C.F.}) = m\Omega^2 r\, dr(\zeta - i)$$

where ζ = lag angle

i = angle between no lag position and line of action of centrifugal force

From the geometry of Fig. 7–6,

$$ir = \zeta(r - e)$$

or

$$i = \zeta\left(1 - \frac{e}{r}\right)$$

Thus

$$d(\text{C.F.}) = m\Omega^2 r\, dr\, \zeta \left[1 - \left(1 - \frac{e}{r}\right)\right]$$
$$= m\Omega^2 e\zeta\, dr \tag{9}$$

The centrifugal force component per foot, $d(\text{C.F.})/dr$, is therefore constant along the span. The component of centrifugal force perpendicular to the blade in the in-plane direction is therefore also constant along the span.

The moment of the centrifugal force about the lag hinge is given as

$$\text{C.F. moment} = mRe\Omega^2 R_{\text{c.g.}}\zeta$$
$$= Me\Omega^2 R_{\text{c.g.}}\zeta \tag{10}$$

where $R_{\text{c.g.}}$ = distance from axis of rotation to blade c.g.

The aerodynamic forces acting on the blade in the plane of rotation include both induced drag and blade profile drag. For a rough physical picture, the spanwise aerodynamic force distribution may be considered proportional to the spanwise lift distribution. Denoting the resultant of these forces as F, and the point of application as $R_{d.f.}$, then the aerodynamic moment about the lag hinge is $= FR_{d.f.}$. Equating the aerodynamic and centrifugal force moments,

$$FR_{\text{d.f.}} = M\Omega^2 R_{\text{c.g.}} e\zeta$$

or

$$F = M\Omega^2 R_{\text{c.g.}} e\zeta / R_{\text{d.f.}} \tag{11}$$

The resultant in-plane aerodynamic force F may be expressed in terms of known quantities by equating the shear forces at the lag hinge. Resolving forces perpendicular to the no-lag axis, it is seen from Fig. 7–7 that

$$\text{Torque}/e = F\cos\zeta + M\Omega^2 R_{\text{c.g.}} \sin\zeta$$
$$= F + M\Omega^2 R_{\text{c.g.}}\zeta \tag{12}$$

Substituting in equation (11)

$$\text{Torque}/e = M\Omega^2 R_{\text{c.g.}} \zeta \frac{e}{R_{\text{d.f.}}} + M\Omega^2 R_{\text{c.g.}} \zeta$$
$$= \zeta (M\Omega^2 R_{\text{c.g.}}) \left(\frac{e}{R_{\text{d.f.}}} + 1 \right)$$

or

$$\zeta = \frac{\text{torque}}{M\Omega^2 R_{\text{c.g.}} e \left(\frac{e}{R_{\text{d.f.}}} + 1 \right)} \tag{13}$$

Equation (13) indicates that for a given rotor, the mean drag angle is essentially a function of torque/Ω^2. Because the radial position of the resultant aerodynamic force is very large with respect to e ($R_{\text{d.f.}}$ is of the order of $0.7R$), the drag angle should be relatively insensitive to changes in $R_{\text{d.f.}}$. For example, when $R_{\text{d.f.}} = 0.7R$, a variation in $R_{\text{d.f.}}$ of $\pm$ 30 per cent in equation (13) produces a change in ζ of only $\pm$ 2 per cent.

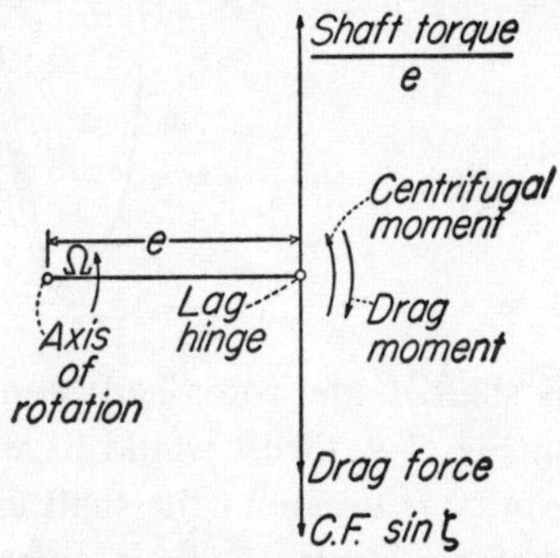

Fig. 7–7 Force and moment equilibrium at lag hinge.

At a given altitude the torque coefficient C_Q is also proportional to torque/Ω^2, and therefore the lag angle (at fixed altitude) is proportional to the torque coefficient, C_Q.

Control of the Hinged Rotor in Hovering

To gain a physical understanding of the behavior of a flapping rotor, it is convenient to consider the action of a simple model rotor. Imagine a small rotor with freely-hinged blades (as shown in Fig. 7–2) set at some positive pitch angle. If the shaft is held vertically (Fig. 7–8) and the model started, the blades will cone upward about the shaft axis. Because no wind is involved, the model represents a hovering rotor. The tip-path plane is perpendicular to the shaft, and the resultant

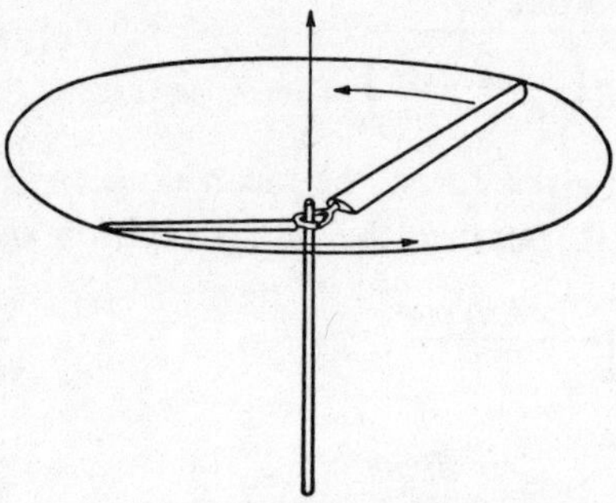

Fig. 7–8

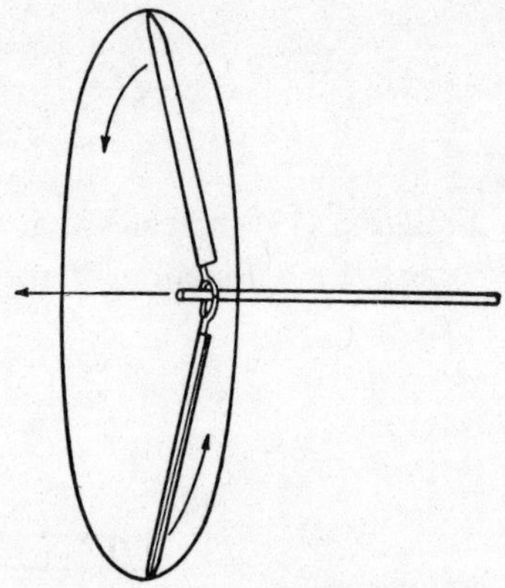

Fig. 7–9

vector lies along this shaft. If the rotor had been started up with its shaft horizontal as in Fig. 7–9, thrust would have been produced and the blades would again cone up about the shaft axis. It follows, then, that anywhere the shaft, or control axis, is pointed, there the thrust vector will point. Consider a rapid maneuver—such as a sudden change in the control-axis position in space—as in Fig. 7–10. Because no

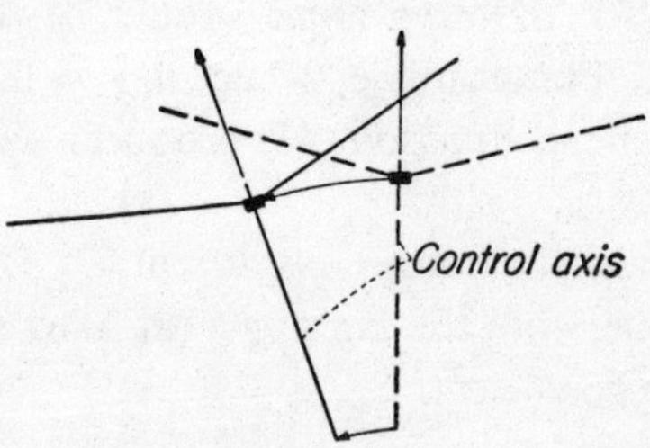

Fig. 7–10

moments can be transmitted from hub to blade, the blades cannot be forced into new positions (and conversely, no gyroscopic forces can be thrown back into the hub). Consider then, the mechanism by which the blades again align themselves to rotate about the control axis.

Suppose the model is held with its control axis vertical, as indicated by the broken lines of Fig. 7–11. Then suppose that the control axis is suddenly rotated forward in the plane of the paper to the position shown by the solid lines in Fig. 7–11. Note that the motion changed the pitch angle of the blade perpendicular to the plane of the paper, and the blade, which intended to rotate on in the direction shown by the dashed line now finds itself operating at a higher angle of attack. Because of the increased lift the blade moves, or "flaps," upward.

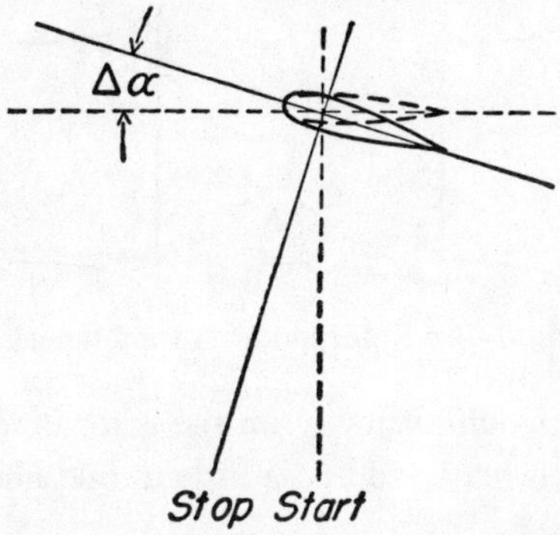

Fig. 7–11 Change in blade angle of attack due to shaft tilt.

Similarly, the blade on the opposite side finds itself with a decreased angle of attack, and it moves downward. This process continues until the plane of the blades is again perpendicular to the control axis, at which position no cyclic-pitch changes occur. Thus, although by tilting the axis it was physically impossible to force the hinged blades to align themselves with the control axis, the tilt produced a cyclic change in the blade angle of attack such that the air forces brought the blades into proper alignment.

It is evident that there will be some delay between a rapid control angle change and the realignment of the rotor disk. With conventional blade mass characteristics, this lag is extremely small and its effect cannot be detected by the pilot as far as response of the rotor disk to applied control is concerned.

Up to this point only the hovering rotor has been considered. It has been seen that in hovering the thrust vector always lies along the control axis. Consider now the model moving edgewise through the air. For convenience, imagine the model held in the jet of a wind tunnel, its shaft, or control axis, held vertically as in Fig. 7–12.

Considering the air-flow conditions under which the model is operating, it is not hard to believe that the blade motion (the flapping of the blade) will be influenced by the horizontal air flow. The following paragraphs will attempt to explain blade motion physically and will point out the factors that influence this motion. It will be shown that

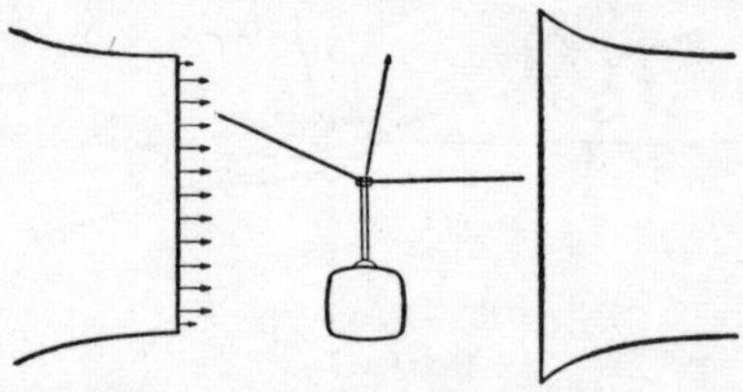

Fig. 7–12 Rotor model in wind tunnel.

the effect of the horizontal velocity on the rotor is to "blow" the cone back; that is, to tilt it rearward and a little to one side.

Blade Flapping Motion

FLAPPING AS REPRESENTED BY A FOURIER SERIES. In studying flapping motion in forward flight, it is convenient to express the variation of the flapping angle β with azimuth angle ψ in terms of simple sinusoidal motions about the longitudinal and lateral axes. The axis of reference in this study is the control axis, which by definition is the axis about which there is no cyclic-pitch change, or the axis about which the blade pitch is constant. (The control axis is thus similar to the shaft of the simple model flapping rotor.) The flapping motion is written as a sum of simple harmonic motions as follows:

$$\beta = a_0 - a_1 \cos \psi - b_1 \sin \psi - a_2 \cos 2\psi - b_2 \sin 2\psi - \ldots \quad (14)$$

In equation (14) β is the angle between the control axis and the blade, and ψ is the azimuth angle as measured from the downwind position in the direction of rotation (Fig. 7–13). Because the Fourier coefficients in equation (14) find such extensive use in rotor theory it is important to understand their physical significance. Accordingly, an interpretation of each coefficient is given below.

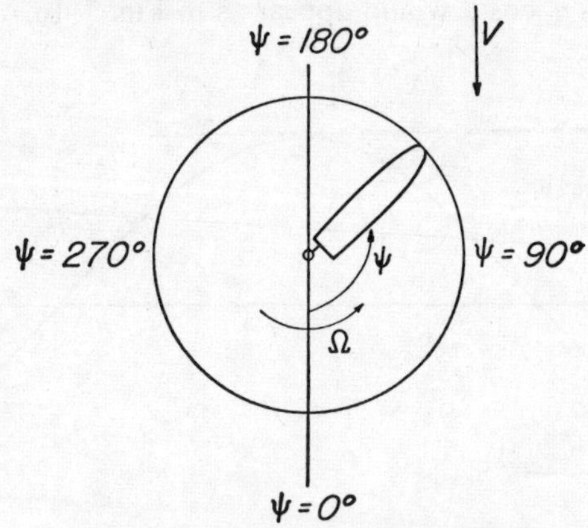

Fig. 7–13 Azimuth angle measurement.

GEOMETRICAL INTERPRETATION OF THE FOURIER COEFFICIENTS

a_0 The coefficient a_0 represents that part of the flapping angle which is independent of the blade azimuth angle, ψ. In hovering, $\beta = a_0$.

A motion represented by $\beta = a_0$ is shown in Fig. 7–14.

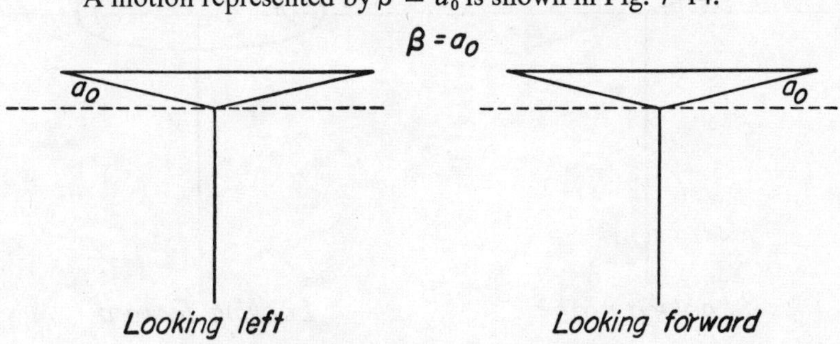

Fig. 7–14 Hovering case wherein $\beta = a_0$.

a_1 The coefficient a_1 represents the amplitude of a pure cosine motion. If the motion represented by $\beta = -a_1 \cos \psi$ is plotted as β against ψ, the motion is as shown in Fig. 7–15. The figure shows that β is a maximum at $\psi = 180$ degrees, or at the upwind position, and is a minimum at $\psi = 0$ degrees, or at the downwind position. When viewed from the side and rear, a rotor executing a periodic blade motion represented by $\beta = -a_1 \cos \psi$ would appear as in Fig. 7–16.

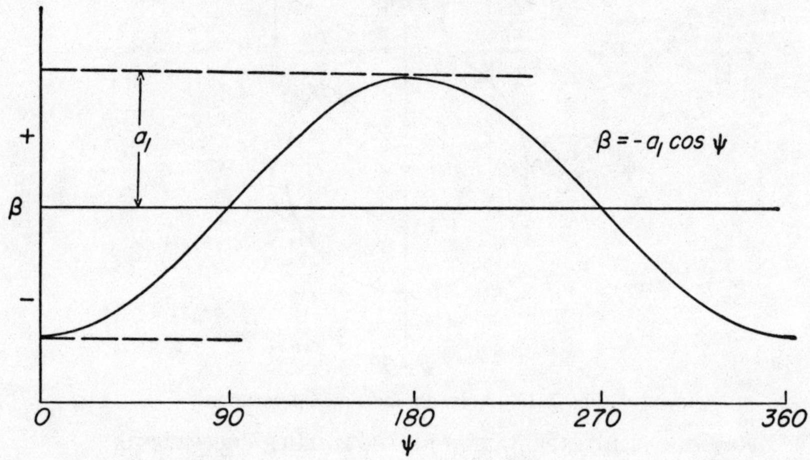

Fig. 7–15 Graphical representation of first harmonic cosine blade motion.

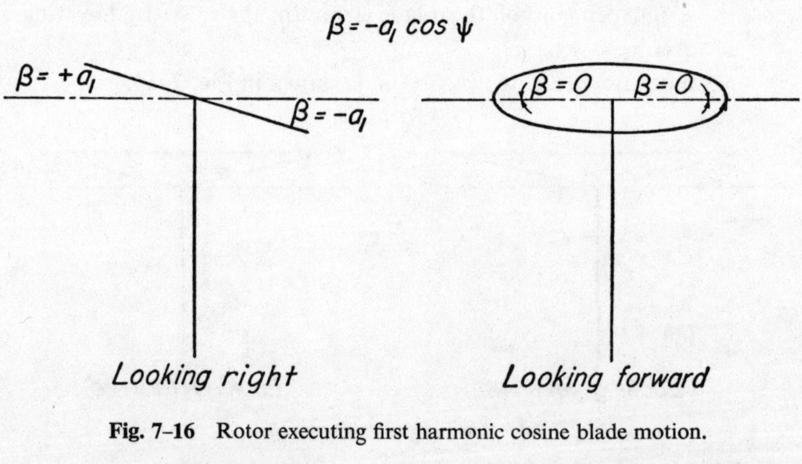

Fig. 7–16 Rotor executing first harmonic cosine blade motion.

b_1 The coefficient b_1 represents the amplitude of a pure sine motion. If the blade motion represented by $\beta = -b_1 \sin \psi$ is plotted against ψ, the result is as shown in Fig. 7–17. Thus, flapping is zero at the fore and aft positions, is a maximum at $\psi = 270$ degrees, and a minimum at $\psi = 90$ degrees. A rotor executing

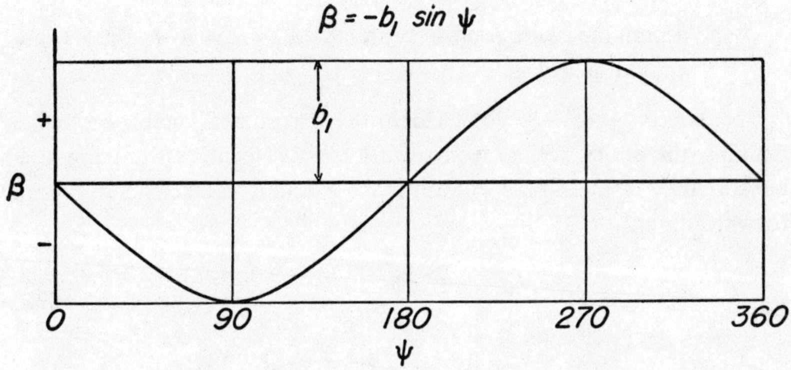

Fig. 7–17 Graphical representation of first harmonic sine blade motion.

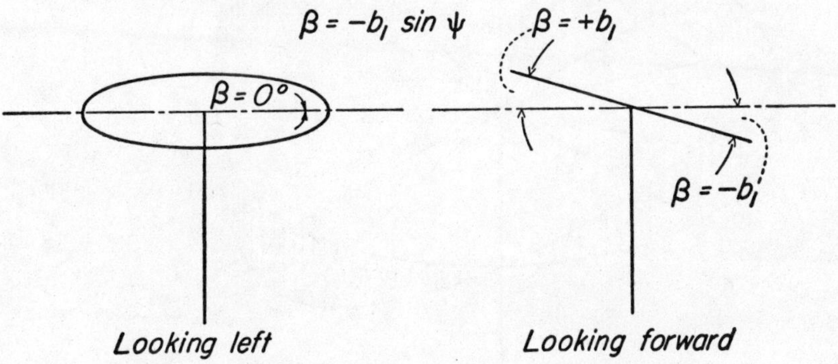

Fig. 7–18 Rotor executing first harmonic sine blade motion.

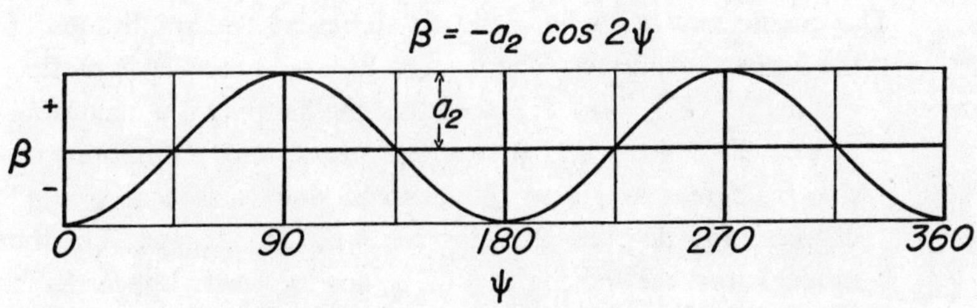

Fig. 7–19 Graphical representation of second harmonic cosine blade motion.

a periodic blade motion represented by $\beta = -b_1 \sin \psi$ would appear as in Fig. 7–18.

The signs (+ or −) with which the terms are established when defining the series are, of course, arbitrary. Negative signs are used because they result in plus values for the a_1 and b_1 coefficients in normal forward flight.

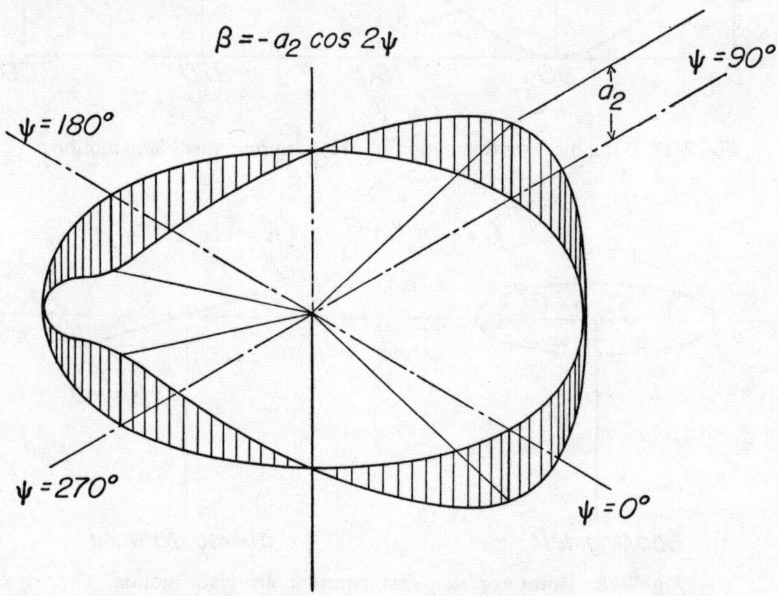

Fig. 7–20 Rotor executing second harmonic cosine blade motion.

a_2 The coefficients a_2, b_2, and so on represent the amplitudes of the higher harmonics. The motion $\beta = -a_2 \cos 2\psi$ is plotted against ψ in Fig. 7–19. It is seen that the flapping is a maximum at $\psi = 90$ degrees and $\psi = 270$ degrees, and a minimum at $\psi = 0$ degrees and $\psi = 180$ degrees, and is zero at $\psi = 45$ degrees, 135 degrees, 225 degrees, and 315 degrees. On the model rotor, the motion $\beta = -a_2 \cos 2\psi$ would appear as in Fig. 7–20.

PHYSICAL EXPLANATION OF THE EXISTENCE OF THE COMPONENT MOTIONS. Any periodic blade motion may be expressed as a superposition of the motions previously described. The accuracy with which the series describes the motions depends entirely on how many terms are considered—an infinite number of terms exactly describes any arbitrary motion. Actually, only a few terms are necessary to represent blade flapping motions to a high degree of accuracy. The a_0, a_1 and b_1 terms usually represent the actual flapping motion to within 1 degree and inclusion of the second and third harmonics (a_2, b_2, a_3, b_3) increases the accuracy to well within 0.1 degree. The magnitude of a typical

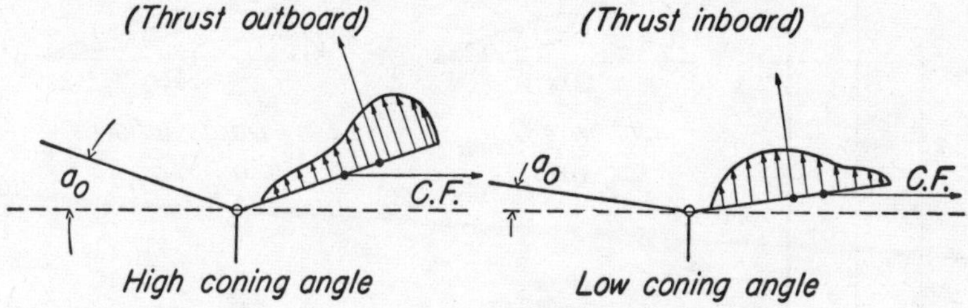

Fig. 7–21 Effect of thrust distribution on coning angle.

flapping motion in forward flight is given by the following example. [These figures were measured in flight on a small, single-rotor helicopter at 70 miles per hour (reference III–4, Appendix IIA).]

Coning angle	a_0 = 8.7 degrees
Backward tilt	a_1 = 6.1 degrees
Tilt to right	b_1 = 3.9 degrees
In-plane weaving	a_2 = 0.5 degree
	b_2 = − 0.1 degree

Having defined the terms in which the flapping motion may be expressed, the physical reason for the existence of each coefficient will now be pointed out.

The coning angle, a_0. The existence of a_0 has been discussed earlier under blade equilibrium conditions in the hovering case. There it was seen that a_0 depended upon the magnitudes of the two primary moments about the flapping hinge—the thrust moment and the centrifugal moment. It was also pointed out that if the thrust distribution along the blade changed, thus changing the radial position of the resultant force, coning was affected. (See Fig. 7–21.)

The inflow velocity through the rotor has a significant effect on the spanwise thrust distribution. Referring to Fig. 7–22, it can be seen that the loading on the blade shifts toward the tips with increased inflow through the disk. A given inflow velocity is much more effective in reducing the angle of attack near the center of the disk, where the tangential velocity is small.

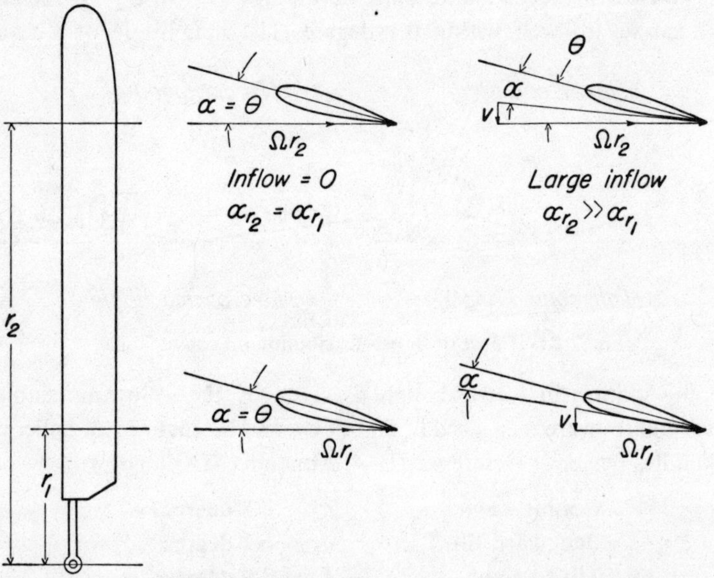

Fig. 7–22 Effect of a change of inflow through the disk on coning angle.

The inflow through the disk varies essentially as the power-required curve. In hovering, where induced flow is large, the inflow through the disk is considerable; the mean velocity is of the order of 25 feet per second for normal disk loadings, whereas velocities near the tip are as high as 35 feet per second. As the rotor moves forward and encounters a larger mass of air, the induced flow, and thus the total rotor inflow, decreases rapidly. At higher speeds, however, the rotor must tilt forward in order to provide forward thrust to drag along the fuselage. Thus a component of the forward speed acts through the tilted disk and the inflow increases. The following statements summarize the inflow and coning angle changes for a given rotor:

Flight Condition	*Inflow*	*Load*	*Coning*
Hovering	Large (induced)	Toward tips	Larger (9 degrees)
Minimum power	Small (small induced, small parasite)	More inboard	Smaller (8 degrees)
High speed	Large (parasite)	Toward tips	Larger (9 degrees)

The backward tilt, a_1. As the simplest approach to a_1, consider the rotor in forward flight, assuming that no inertia forces exist on the blades

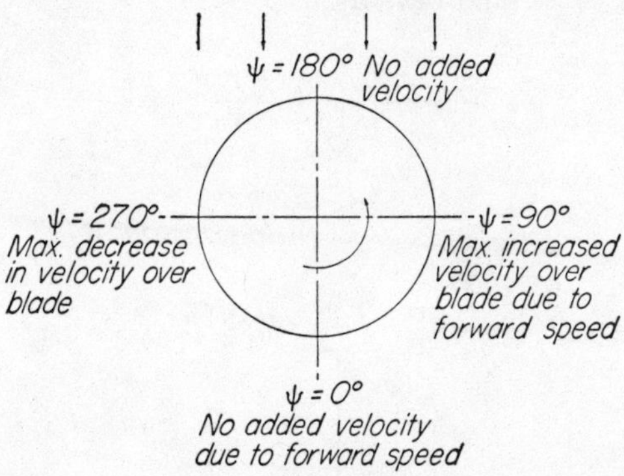

Fig. 7–23 Effect of forward speed on velocity distribution.

to retard their flapping motion when a force is applied. As shown in Fig. 7–23, a blade in the downwind position encounters no added velocity due to forward speed. As the blade moves forward, however,

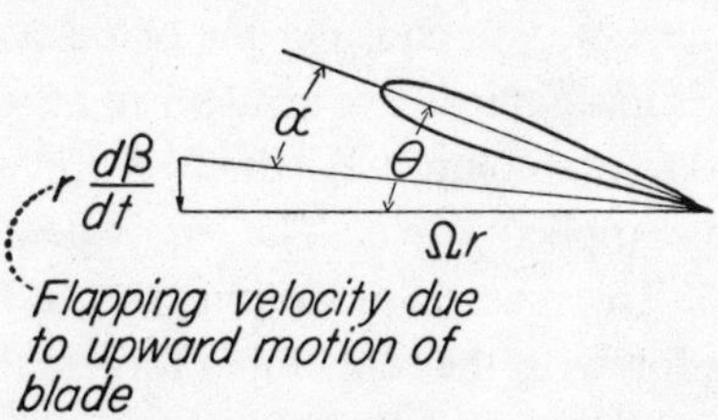

Fig. 7–24 Effect of blade flapping on element angle of attack.

the velocity and thus the lift are increased. As the blade experiences this first increased force it immediately moves upward. In flapping upward, the direction of the relative wind changes so as to decrease the angle of attack and thus the lift (Fig. 7–24). With no inertia forces the blade must at all times be in equilibrium; that is, no unbalanced forces may exist. The flapping velocity must therefore be of such a magnitude as to decrease the angle of attack enough so that the lift will remain constant. It follows then that the maximum upward flapping velocity occurs at $\psi = 90°$ where it is needed most. Following the blade around, then, the flapping velocity would be zero at $\psi = 0°$, maximum upward at $\psi = 90°$, zero at $\psi = 180°$, maximum downward at $\psi = 270°$ and again zero at $\psi = 360°$. The rotor is therefore high in front and low in back, a $+a_1$ kind of motion.

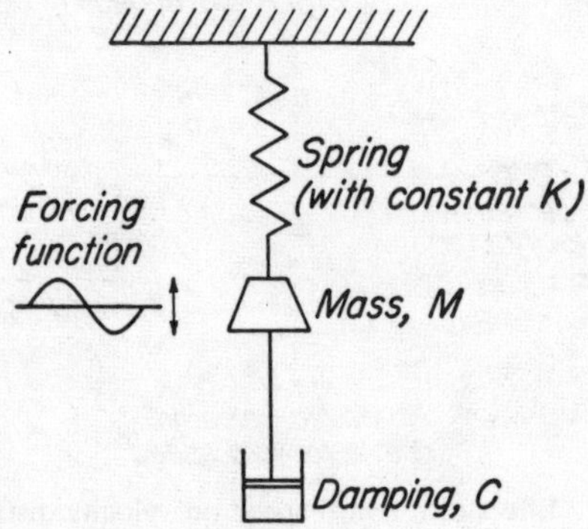

Fig. 7–25 Mechanical analogy to a flapping blade.

It was assumed in the preceding discussion that the blade had no inertia. To examine the effects of blade mass and of air damping on the a_1 motion, it is necessary to consider the blade as a dynamic system. The flapping blade is mechanically equivalent to a system with 1 degree of freedom with some damping and which is forced to vibrate by a sinusoidally varying applied force. The equivalent simple system is shown in Fig. 7–25. The mass is analogous to the flapping blade and the spring restoring forces to the centrifugal forces; the forced vibration is supplied by the air force, and the damping is air damping (proportional to the flapping velocity).

For such a system, the force-displacement phase is related to the frequency of the forced vibration as shown in Fig. 7–26. In the figure, ω_n is the natural frequency of the system, ω is the frequency of the forced vibration, and ϕ is the phase angle between the maximum applied force and the maximum displacement (that is, the angle by which the force leads the displacement). The ratio C/C_c is the ratio of actual damping to critical damping, critical damping being defined as the amount of damping just sufficient to let the displaced mass return to equilibrium without overshooting.

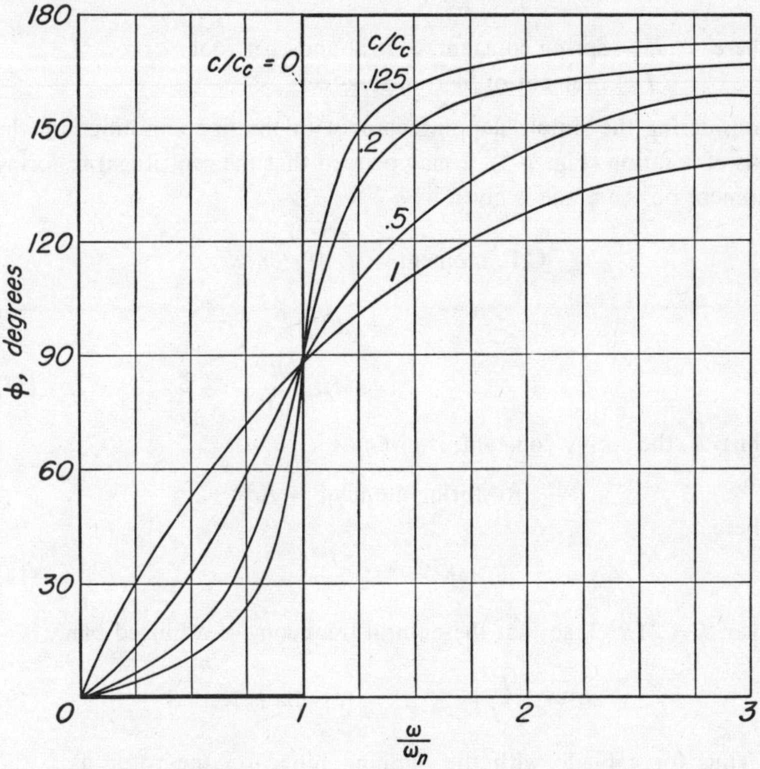

Fig. 7–26 Relation between force-displacement phase and frequency for system with various amounts of damping.

It is seen from the figure then, that the displacement-force phase angle at any value of ω/ω_n is a function of the amount of damping, except when the force is applied at the natural frequency of the system. When the exciting force is applied at the natural frequency, the phase angle is 90° and is independent of the amount of damping.

The natural frequency of a hinged blade is expressed by the general equation for rotational oscillations,

$$\omega_n = \sqrt{\frac{K}{I}} \text{ radians/second} \tag{15}$$

or

$$\omega_n = \frac{1}{2\pi}\sqrt{\frac{K}{I}} \text{ cycles/second} \tag{16}$$

where K = spring constant, foot-pounds per foot
I = moment of inertia, slug-feet^2

Considering the simple flapping rotor with the flapping hinge on the axis of rotation (Fig. 7–2), it may be seen that the centrifugal restoring moment on the blade is given by

$$\begin{aligned} \text{C.F. moment} &= \int_0^R \Omega^2 r^2 \beta m\, dr \\ &= m\Omega^2\beta \frac{R^3}{3} \\ &= M\Omega^2\beta \frac{R^2}{3} \end{aligned} \tag{17}$$

Thus, K, the spring constant, is given as

$$\text{Restoring moment} = K\beta$$

where

$$K = M\Omega^2 \frac{R^2}{3} \tag{18}$$

Also, $I = MR^2/3$, so that the natural frequency of a hinged blade is

$$\omega_n = \sqrt{\frac{K}{I}} = \sqrt{\Omega^2} = \Omega \text{ radians/second} \tag{19}$$

Thus for a blade with the flapping hinge on the rotor axis, the natural frequency equals the rotational frequency. (This simple result is modified slightly if the flapping hinge does not lie on the axis of rotation and if hinge angularities exist so that a blade pitch change accompanies a flapping motion.) When the flapping hinge is at a distance h from the axis of rotation, the natural frequency is given as

$$\omega_n = \Omega\sqrt{1 + \frac{3}{2}\frac{h}{R}} \tag{19a}$$

Because the exciting air force also has a frequency of 1 per revolution, then $\omega/\omega_n = 1$ and the force-displacement phase is 90 degrees and is independent of the amount of damping present. Also, the ratio K/I has been shown to be independent of the blade mass for a uniform mass distribution and can be shown independent for any mass distribution, so that for the simple rotor neither blade mass nor damping affect the 90° phase shift between force and displacement. Thus, the results of the initial approach to a_1 (assuming no inertia forces and

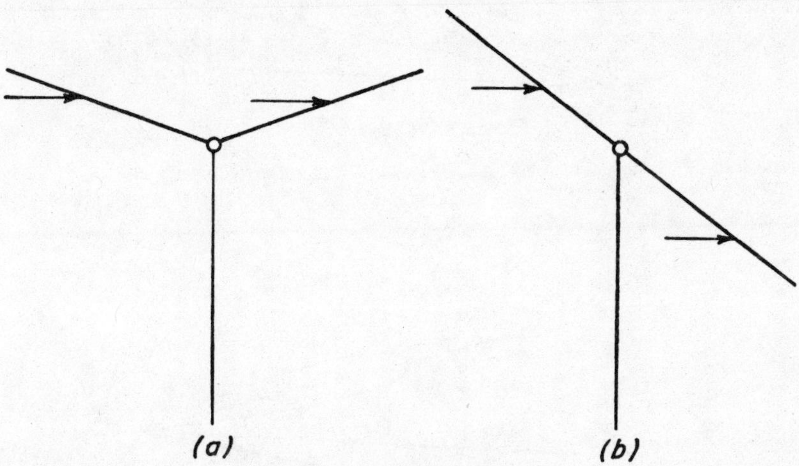

Fig. 7–27 (a) With coning
(b) Without coning

no damping) are unchanged in the real case, and on the basis of the forces and effects considered, maximum flapping will occur at $\psi = 180°$ and minimum flapping at $\psi = 0°$.

The sideward tilt, b_1. The sideward tilting of the cone, b_1, may be viewed as arising from coning, a_0. For the coned rotor in Fig. 7–27a, it may be seen that there is a difference in angle of attack of the blades at the front and rear of the rotor because of the forward speed. Note that with no coning (Fig. 7–27b), the effect of the forward velocity is identical in the fore and aft positions, but that for the coned rotor a periodic air force is produced because of the forward velocity, V.

As shown in Fig. 7–28, this force is a maximum at $\psi = 180°$, the front of the rotor, and is a minimum at $\psi = 0°$, the rear of the rotor.

Here then is another periodic force, and just as the lateral forces produced a_1 motion, with a force-displacement phase of 90°, the longitudinal force results in a flapping motion which is a maximum at $\psi = 270°$, the retreating side, and a minimum at $\psi = 90°$, the advancing side. Therefore, a $+ b_1$ motion results because of coning. The b_1 flapping is of the same order of magnitude as the a_1 flapping,

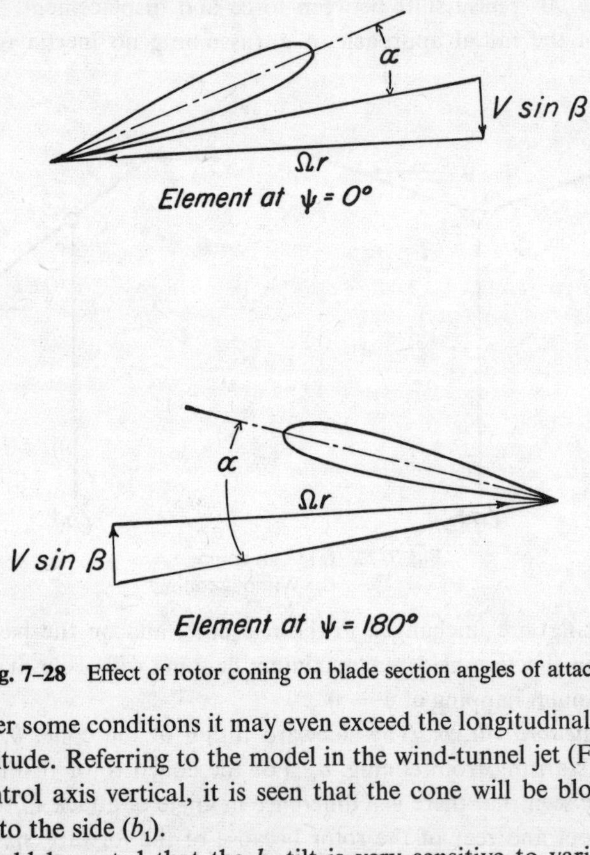

Fig. 7–28 Effect of rotor coning on blade section angles of attack.

and under some conditions it may even exceed the longitudinal flapping in magnitude. Referring to the model in the wind-tunnel jet (Fig. 7–12) with control axis vertical, it is seen that the cone will be blown back (a_1) and to the side (b_1).

It should be noted that the b_1 tilt is very sensitive to variations in inflow of air from the front to the rear of the disk. It is usually assumed in forward-flight performance analyses that the induced flow is uniform across the disk, this being a reasonable assumption for most cases. At low forward speed, however, where the induced velocities are still quite large, the air is bent downward as it crosses the disk, so that the inflow is greater at the rear of the disk. The angle-of-attack differences at the front and rear of the disk that are a result of coning are thus accentuated by inflow variations and b_1 is increased. Indeed, the nonuniform inflow may be thought of as an effective coning angle insofar as its effect on b_1 motion is concerned. The asymmetry of the inflow increases from zero in hovering to a maximum at some low forward speed, then decreases steadily as speed increases. At high speeds the inflow is again almost uniform inasmuch as the bulk of the inflow is due to the flight velocity acting through the tilted disk.

The higher harmonics. The higher harmonics, a_2, b_2, a_3, b_3, etc., may be viewed as a weaving of the blade in and out of the surface of the cone formed by the first harmonic motions of the blades. The physical cause of their existence cannot be shown as simply as with a_o, a_1, and b_1, but it should suffice to point out the presence of forces which can produce higher harmonic motions. The most obvious source is in the asymmetrical flow pattern of the rotor. On the retreating side of the disk there is a region of reversed velocity which is not present on the advancing side. Simple, symmetrical first harmonic force variations cannot be expected from such a situation and a source of higher harmonic motion is evident.

Another source for higher harmonic flapping lies in the very forces which have been shown to produce a_1 and b_1. It was explained that the blade was forced to flap to relieve the forces produced by periodic changes in velocity. To keep constant lift on the blade, however, the product of angle of attack and velocity squared must be constant. In order to keep constant lift, the blade would then have to have a flapping velocity proportional not to $\sin \psi$ but to $\sin^2 \psi$. Thus the blade moving in a first harmonic pattern still has forces acting on it and these forces result in higher harmonic motions of the blade. Higher harmonic blade motions are of little importance in problems of rotor control and rotor performance but are extremely important in problems of rotor vibration and blade stress.

THE EFFECTS OF BLADE MASS ON FLAPPING MOTION. The coning angle, a_0, is directly affected by blade mass because the centrifugal forces are increased with respect to the lift forces when blade mass is increased. The fore and aft flapping, a_1, has been shown to be independent of blade mass, for the exciting forces act on a system in resonance, which by definition is a system in which the mass forces are in equilibrium with the spring forces, and in which external forces are in equilibrium with the damping forces.

The sideward flapping, b_1, also involves a system in resonance and in this respect is independent of blade mass. The exciting forces for the b_1 motion, however, have been shown proportional to the coning angle, a_0, which is proportional to blade mass. The amplitude of any oscillation of a system with damping and in resonance is, of course, proportional to the magnitude of the excitation. Therefore, as blade mass increases to infinity, the coning angle decreases to zero, and b_1 decreases to zero inasmuch as the exciting forces decrease to zero.

The higher harmonic motions do not involve a system in resonance, but rather a forced vibration well above resonance, wherein the exciting force is opposed almost entirely by the mass forces. The amplitudes of motion which the higher harmonic air forces produce are therefore inversely proportional to blade mass and go to zero as blade mass becomes infinite.

A rotor with very heavy blades thus performs a nearly pure a_1 flapping motion; the coning angle, sideward tilt, and higher harmonic flapping become greater as blades become lighter.

Rotor Control in Forward Flight

The preceding paragraphs have dealt with the behavior of a rotor when its control axis, or axis of no cyclic-pitch change, is held fixed and the velocity over the rotor increased (as for the model in the wind tunnel). It has been shown that the plane of the tips tilts backwards and to one side (by the amount of a_1 and b_1) with respect to the control axis. It has also been stated that the resultant thrust of a rotor is approximately perpendicular to the plane of the tips.

Control of the helicopter in any flight condition involves the proper orientation of the rotor thrust vector, and therefore of the tip-path plane, in space. In the preceding discussion of hovering control it was shown that the tip-path plane is perpendicular to the control axis and remains so as the control axis assumes different positions in space. Control in forward flight is obtained in a similar manner, except that a given increment in control-axis tilt does not correspond to exactly the same increment of tip-path plane tilt. For example, Fig. 7–29 shows the rate of change of a_1 flapping with the angle of tilt of the control axis α for a typical rotor. It is seen that as the control axis is tilted forward, the rearward tilt of the tip-path plane with respect to the control-axis plane decreases. The tip-path plane tilts forward faster

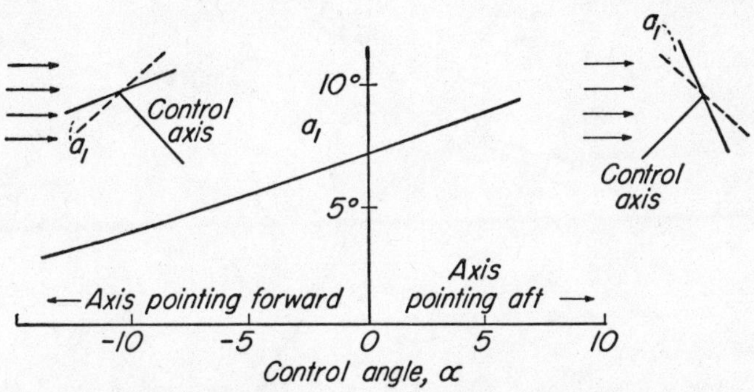

Fig. 7–29 Variation of longitudinal tip-path plane tilt with control axis tilt.

than the control axis is tilted forward, and tilts rearward faster than the control axis is tilted rearward. This effect constitutes an instability of the rotor with angle of attack and is very important in stability considerations. (See Chapter 11.) In regard to static control, however, the important point to establish is that the only difference between control in hovering and in forward flight is one of degree, control being somewhat more sensitive as forward speed increases.

Having shown that the location of the control axis in space is the fundamental problem of helicopter control, there remains only the mechanical problem of how best to achieve the desired control-axis tilt. On autogyros of the so-called *direct control* type, the control axis was located in space by physically tilting the rotor shaft and hub to the desired position with respect to the fuselage. With the helicopter, where power is transmitted through the shaft, it becomes mechanically awkward to obtain control by shaft tilt. To solve the control axis problem in the helicopter, two means have come into common use:

(1) The rotor hub may be tilted with respect to the shaft, as in Fig. 7–30. This method divorces the shaft from the control axis because the hub axis then becomes the control axis—or the axis of no feathering.

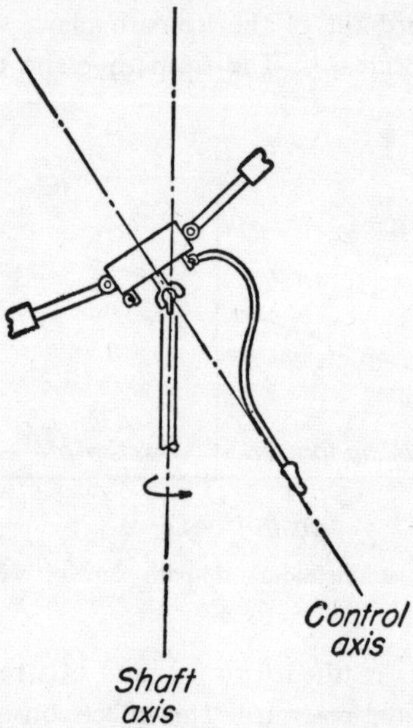

Fig. 7–30 Control by tilting hub with respect to shaft.

(2) The hub may remain fixed to the shaft and a means provided for cyclically varying the blade pitch with respect to the hub. Such a system is shown in Fig. 7–31. The system consists of blades attached to the hub which are free to flap and free to change their pitch angles. The blade is held in pitch through a linkage connecting it to the swash plate. If simple feathering control is defined as a swash-plate system with a 1 : 1 linkage ratio, then the pitch of the blade is always constant with respect to the plane of the swash plate. The feathering system provides a convenient means of moving the control axis.

It is well to point out at this time the basic equalities of the flapping and feathering blade. Consider a rotor in forward flight as in Fig. 7–32. The rotor shown is a simple flapping rotor. The control axis is vertical and the tip-path plane tilts rearward by an amount a_1. (The rotor under

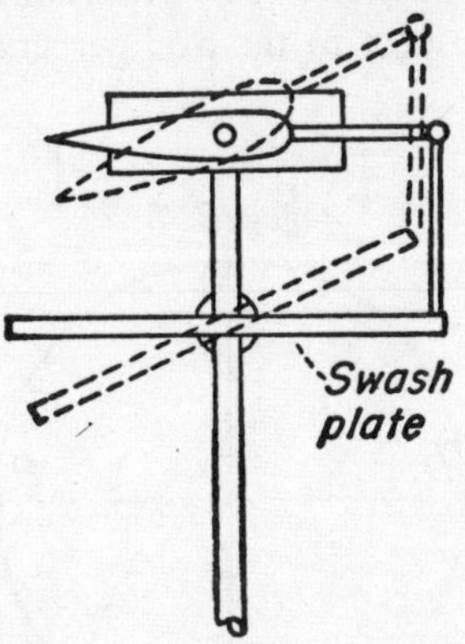

Fig. 7–31 Control by cyclic pitch.

consideration is one with infinitely heavy blades so that only a_1 motion exists.) An observer riding on the control axis and rotating with the blades observes that the blades flap up and down each revolution but that they are fixed in pitch. At the same time an observer who sits

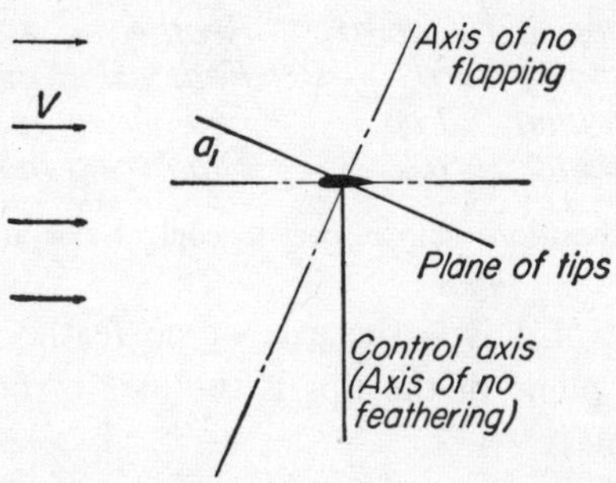

Fig. 7–32 Flapping rotor in forward flight.

in the plane of the tips, rotating with the blades, observes that the blades do not flap at all but do change their pitch—high, then low—once each revolution. The pitch is low on the advancing side of the rotor and high on the retreating side.

Examining Fig. 7–33, it is seen that the amount of blade feathering with respect to the plane of the tips is equal in degrees to the amount of blade flapping with respect to the control axis. Fore and aft (a_1) flapping with respect to the control axis is therefore equal to lateral (B_1) feathering with respect to the axis perpendicular to the plane of

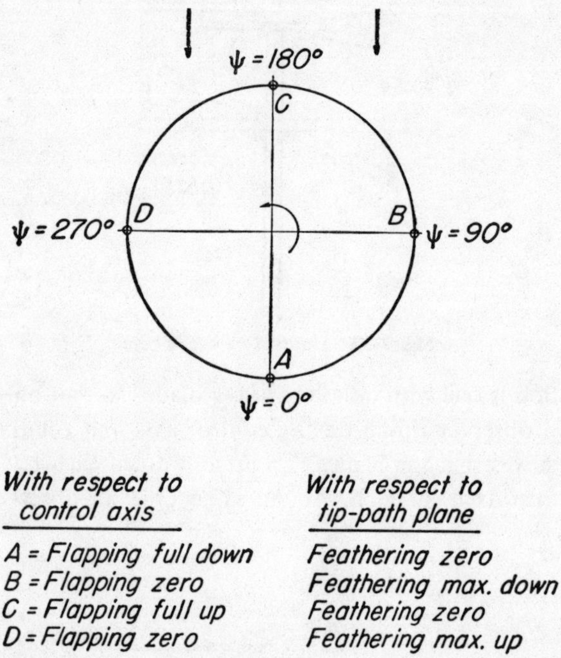

Fig. 7–33 Blade positions with respect to control axis and tip-path plane.

the tips. The control axis is the axis of no feathering; the axis perpendicular to the plane of the tips is the axis of no flapping (except for higher harmonics).

The normal helicopter appears to be quite complicated in its blade motion because both flapping and feathering motions exist with respect to the physical, powered shaft. If, however, the feathering is thought of as simply a means of tilting the control axis through a given angle, then the system is reduced to one in which only pure flapping exists with respect to this control axis.

It is desirable to derive geometrical relations that can be used to transfer the axis of reference to the axis of no feathering (control axis) from the axis of no flapping (tip-path plane axis), or from an inter-

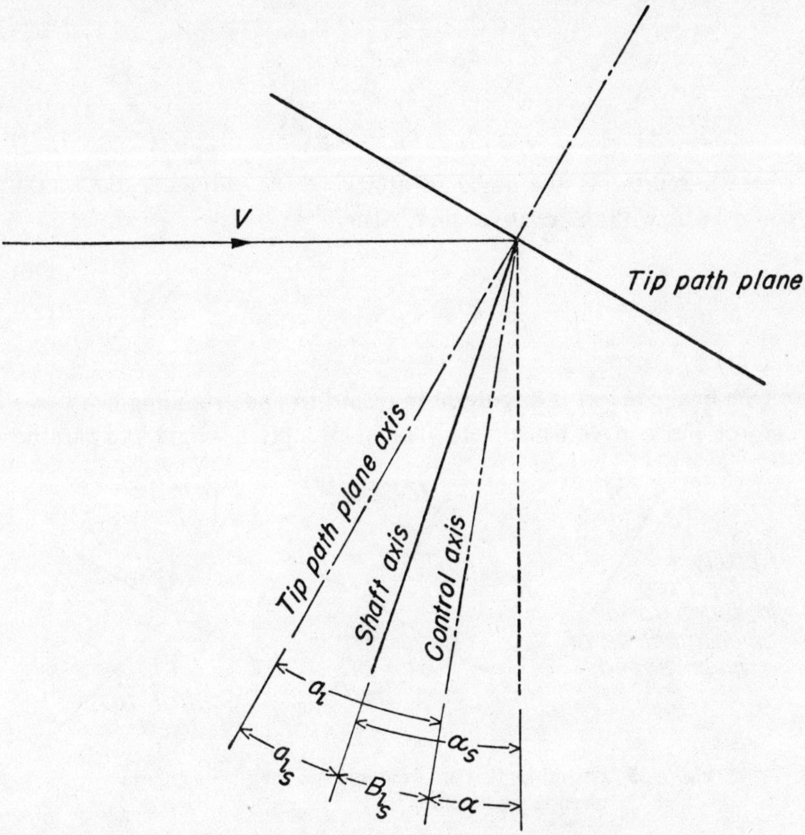

Fig. 7–34 Flapping—feathering relationships.

mediate shaft axis. If the blade motion with respect to the control axis is given by

$$\beta = a_0 - a_1 \cos \psi - b_1 \sin \psi - a_2 \cos 2\psi - b_2 \sin 2\psi \ldots \quad (20)$$

and the feathering motion with respect to the plane of the tips is given by

$$\theta = A_0 - A_1 \cos \psi - B_1 \sin \psi - A_2 \cos 2\psi - B_2 \sin 2\psi \ldots \quad (20a)$$

and if flapping and feathering motions with respect to the shaft are denoted by the subscript s, then from Fig. 7–34,

$$\alpha = \alpha_s - B_{1_s} \tag{21}$$

$$A_0 = A_{0_s} \tag{22}$$

$$a_0 = a_{0_s} \tag{23}$$

$$a_1 = a_{1_s} + B_{1_s} \tag{24}$$

$$b_1 = b_{1_s} - A_{1_s} \tag{25}$$

where α represents the angle of attack of the perpendicular to the control axis with the relative wind. Also

$$a_2 = a_{2_s} \tag{26}$$

$$b_2 = b_{2_s} \tag{27}$$

etc. . . .

One final concept is important in regard to understanding helicopter control. For a given helicopter—that is, for a given weight and parasite

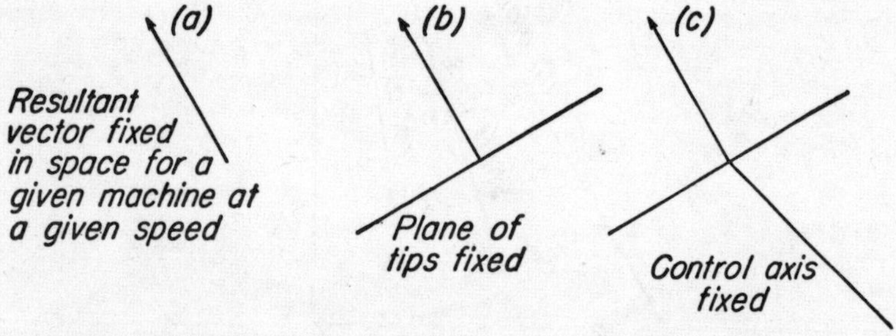

Fig. 7–35 Quantities that are fixed in space for a helicopter in a given flight condition.

drag, moving at a given speed—the resultant rotor vector is fixed in space (Fig. 7–35). Because the resultant vector is essentially perpendicular to the plane of the tips, the plane of the tips is also fixed in space. It will be proven later, when equations for forward flight are developed, that for a given condition of operation—a given thrust, velocity, parasite drag, and a given rotor turning at a given angular velocity—the flapping motion of the rotor is completely determined. The control axis is thus fixed with respect to the plane of the tips. In a given condition, then, the three items which are fundamentally oriented in space are the resultant vector, the plane of the tips, and the control axis. The identities presented in equations (21) through (27) show that a given control axis inclination may be achieved with various combinations of shaft angle and feathering control. Assuming that the fuselage angle has no effect on their pitching moment, lift, or

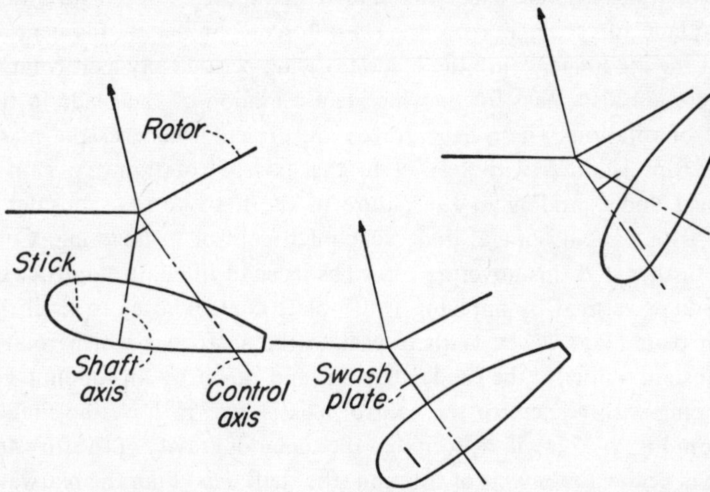

Fig. 7–36 Helicopter in a given flight condition with different combinations of shaft angle and feathering control.

parasite drag, the three helicopters shown in Fig. 7–36 are in identical flight conditions as far as the rotor is concerned.

Actually, fuselage attitude and control position at a given speed may change from day to day, because of differences in fuselage center-of-gravity position. The important point to remember is that fuselage attitude and control position combinations have no effect on the attitude of the rotor in space, except insofar as they may secondarily affect the magnitude and direction of the resultant force (through changes in fuselage moments, lift, or parasite drag). No gains in rotor performance can therefore be expected from different combinations of fuselage attitude and control position (i.e., different combinations of flapping and feathering).

Blade Motion in the Plane of the Disk

SOURCE OF IN-PLANE BLADE MOTION. Periodic blade motion in the plane of the disk arises from two sources: periodically varying air forces in the plane of the disk and periodically varying mass forces. In forward flight it is not surprising that the blade should encounter varying air forces as it rotates. In view of the variations in velocity and angle of attack that are encountered it would indeed be odd if variations in in-plane force did not occur. These forces—or better, moments—about the lag hinge, cause the blade to swing periodically as it rotates.

There are also mass forces which cause motion of the blade in the plane of rotation. These mass forces are present whenever the plane of the tips is tilted with respect to the axis of rotation. To gain a physical understanding of the nature of the mass forces, consider a rotor with flapping hinges and cyclic-pitch control in hovering. Consider further that this hovering rotor has its shaft tilted, its control axis being kept vertical by applying swash-plate control so as to level the swash plate (Fig. 7–37). With respect to the shaft axis which rotates at constant velocity, the blades flap up and down by an amount a_{1_s}. (Flapping with respect to the control axis is zero in hovering flight.)

From Fig. 7–37 it will be seen that the center of gravity of the forward blade is nearer to the axis of rotation (the shaft axis) than the rearward blade. In order to maintain its angular momentum the forward blade must accordingly move faster than normal and the rearward blade must move slower than normal. Thus, an observer viewing the rotor from the powered shaft would see that the blades move back and forth as they rotate.

On the other hand, an observer viewing the rotor from above, standing on, and rotating with, the control axis is utterly unaware that anything unusual is occurring. From his point of view there is no flapping, no cyclic-pitch change, no cause of any sort for the blades to move in the plane of rotation. It is not odd, then, that he finds that there is no in-plane motion with respect to the plane of the tips. Apparently, then, the existence of this in-plane motion depends on the axis of reference used. With respect to the shaft axis, the blades move toward and away from each other as they rotate, while with respect to the tip-path plane axis no in-plane motion occurs.

It is a worth while exercise to reconcile these two points of view by determining the forces and resulting motions with respect to one axis system and showing that the results are compatible with motions found for a second axis system.

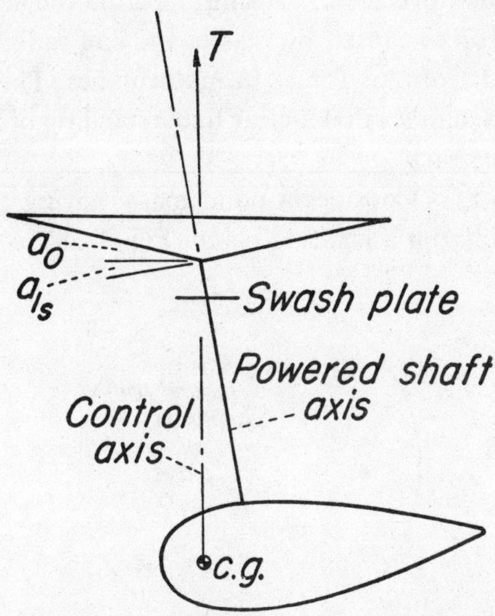

Fig. 7–37 Hovering helicopter illustrating the source of in-plane blade motion.

Consider, first, motion with respect to the shaft axis. For this case then, the flapping motion with respect to the shaft axis is given by

$$\beta_s = a_0 - a_{1_s} \cos \psi \tag{28}$$

In this case, the blades experience two periodic torques about the shaft axis.

(1) A periodically varying torque exists because of the component of lift force which acts in the plane perpendicular to the shaft. This is given as

$$\text{Lift torque component in plane perpendicular to shaft} \quad \mp T_b r_{a.f.} a_{1_s} \sin \psi \tag{29}$$

where T_b = thrust per blade
$r_{a.f.}$ = radius of resultant lift force on blade
a_{1_s} = fore and aft flapping with respect to shaft
(A positive torque is taken as one increasing the lag angle.)

(2) A periodic "mass force" torque exists because of the fact that the blades move toward and away from the axis of rotation, periodically changing their moment of inertia. In the absence of external torques, a system must have constant angular momentum, $I\Omega$; then as I increases, Ω must decrease, causing a periodic change in angular velocity. These forces caused by masses moving radially in a rotating plane are termed Coriolus forces in mathematics. They are so fundamental in rotor dynamics that a clear understanding of their occurrence is considered necessary.

CORIOLUS FORCES. Consider a point mass moving radially outward at constant velocity on a rotating plane (Fig. 7–38). A man getting off

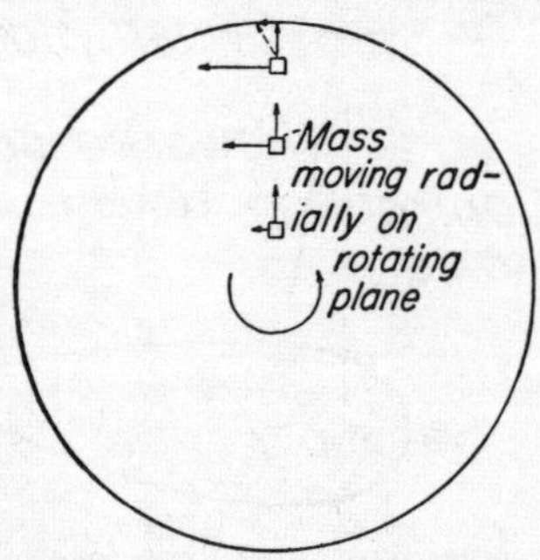

Fig. 7–38 Origin of Coriolus forces.

a moving merry-go-round is a practical analogy. It is clear that as the point moves radially outward its tangential velocity is increasing. Because any mass resists changes in velocity, according to Newton's law, the mass resists the tangential acceleration to the left by exerting a force to the right on the rotating plane. The tangential acceleration of the mass due to its increasing tangential velocity is given by

$$d(\Omega r)/dt = \Omega dr/dt = \Omega V_{\text{radial}}$$

A second less obvious source of tangential acceleration caused by the radially moving mass lies in the swinging of its radial velocity vector in space. Because the plane is rotating, the radial velocity vector is continually changing its direction in space. Acceleration is the rate of change of velocity per unit time, and therefore the swinging of the velocity vector must represent an acceleration. The mass is again being accelerated to the left and resists by pushing on the plane to the right. The tangential acceleration arising from the swinging of the velocity vector is given as

$$(\text{Vector length}) \times d\theta/dt$$

where $d\theta$ is the angle of swing of the vector in time dt. But $d\theta/dt = \Omega$, so that the acceleration is given by ΩV_{radial}. Thus, the resultant acceleration of the mass point moving radially in the rotating plane is $2\Omega V_{\text{radial}}$ and because the force exerted on a mass is given by $F = ma$, the Coriolus force is given as

$$F_{\text{Coriolus}} = 2mV_{\text{radial}}\Omega \tag{30}$$

To summarize the above:

(1) Whenever a mass moves radially in a rotating plane it experiences a tangential force which is proportional to the rotational velocity and the radial velocity of the mass. This force is given as $F_{\text{Coriolus}} = 2mV_{\text{radial}}\Omega$.

(2) The direction of the force exerted by the mass on its surroundings is opposite to the direction of rotation when the mass is moving outward and in the direction of rotation when the mass is moving radially inward.

The Coriolus torque acting on the hovering rotor of Fig. 7–37 can now be evaluated. With respect to the rotor shaft, a blade element moves outward with a velocity

$$\frac{d(r\cos\beta_s)}{dt} = -r\sin\beta_s\frac{d\beta_s}{dt}$$
$$= -r\beta_s\dot{\beta}_s \tag{31}$$

Because $\beta_s = a_0 - a_{1_s}\cos\psi$ and $\dot{\beta}_s = a_{1_s}\Omega\sin\psi$, the radial velocity of the blade is given by

$$V_{\text{radial}} = r\beta_s\dot{\beta}_s = -r\Omega\left(a_0a_{1_s}\sin\psi - \frac{a_{1_s}^{\,2}}{2}\sin 2\psi\right) \tag{32}$$

If $a_{1_s}^{\,2}$ is neglected as small compared to $a_0a_{1_s}$, then the Coriolus torque is given as

$$\text{Torque}_{\text{Coriolus}} = -\int_0^R 2r^2\Omega^2 a_0 a_{1_s}\sin\psi m\,dr$$
$$= -\tfrac{2}{3}MR^2a_0a_{1_s}\Omega^2\sin\psi \tag{33}$$

Remembering that for a uniform mass blade,

$$a_0 = 3r_{a.f.}T_b/M(\Omega R)^2 \tag{34}$$

where $r_{a.f.}$ = radius of resultant air force

The Coriolus torque then becomes

$$\text{Torque}_{\text{Coriolus}} = -2T_b r_{a.f.} a_{1_s} \sin \psi \tag{35}$$

The Coriolus torque, therefore, depends on the thrust loading and the a_{1_s} flapping but is independent of blade mass.

EQUATION OF MOTION FOR BLADE IN LAG. Expressions have now been developed for the exciting forces acting on a blade in the in-plane direction in hovering with respect to the shaft axis. In order to find what motion results from these applied forces, the mass and spring characteristics of the system must be known, whence the forces may be equated to the motion by $F = ma$.

The spring restoring torque on the blade is given as

$$\text{C.F. torque} = \frac{MR}{2} e\Omega^2 \zeta \tag{36}$$

The equation of motion of the blades in lag is formed by equating all torques to the angular acceleration according to the equation, $F = ma$, or because this case involves angular motions, to $T = I\ddot{\zeta}$. Thus,

$$T_{\text{air}} - T_{\text{Coriolus}} - T_{\text{spring}} = I\ddot{\zeta} \tag{37}$$

This equation states that the torques acting on the system in the positive direction are equal to the blade moment of inertia multiplied by the acceleration in the positive direction. Substituting in the expressions for the torques,

$$T_b r_{a.f.} a_{1_s} \sin \psi - 2T_b r_{a.f.} a_{1_s} \sin \psi - Me\Omega^2 \frac{R}{2} \zeta = I\ddot{\zeta} \tag{38}$$

The signs may be checked physically as follows. The air force torque, as seen in Fig. 7–37, tends to increase the lag angle for $+ a_1$ motion on the advancing side of the disk where $\sin \psi$ is positive. As the blade moves from the rear to the front of the disk, the blade center of gravity moves toward the axis of rotation for $+ a_1$ motion. According to the previous discussion of Coriolus forces, a mass moving toward the axis of rotation results in a torque applied by the mass on the rest of the blade in the direction of rotation. Because the equation expresses the torques acting opposite to the direction of rotation, or in the positive direction, the Coriolus term bears a negative sign. The spring torque always resists the motion, and so exerts a force in a negative direction.

Combining and rearranging terms,

$$I\ddot{\zeta} + Me\Omega^2 \frac{R}{2} \zeta = - T_b r_{a.f.} a_{1s} \sin \psi \qquad (39)$$

It will be noted that damping has been neglected in establishing the equation of motion. Damping actually arises from two sources: (1) air damping resulting from the increased drag of the blade as it moves forward in its motion, and (2) damping at the blade root, imposed by physical dampers. Unless large friction dampers are employed at the blade roots, damping has small effect on the blade motion and can be neglected.

The solution to an equation of motion such as equation (39) is determined by assuming a solution of the form $\zeta = \zeta_0 \sin \omega t$, where ζ_0 is the amplitude of the forced motion. Thus,

$$- I\omega^2\zeta_0 + M \frac{R}{2} e\Omega^2\zeta_0 = - T_b r_{a.f.} a_{1s} \qquad (40)$$

and

$$\zeta_0 = \frac{- T_b \, r_{a.f.} \, a_{1s}}{\frac{MR}{2} e\Omega^2 - I\omega^2}; \quad \Omega = \omega \qquad (41)$$

For a uniform mass blade, $I = MR^2/3$. Substituting this expression for I, and equation (34) for a_0,

$$\zeta_0 = \frac{\frac{2}{3} a_0 a_{1s}}{\frac{2}{3} - \frac{e}{R}} \qquad (42)$$

This result has a definite physical significance. For the limiting case of zero lag-hinge distance, the blade motion is given by

$$\zeta = \zeta_0 \sin \psi = a_0 a_{1s} \sin \psi \qquad (43)$$

It may be seen from Fig. 7–39 that from purely geometrical considerations the in-plane motion about the shaft is given by $\zeta = a_0 a_{1s} \sin \psi$ when the blades move at constant velocity with respect to the plane of the tips. Thus, for a rotor with blades hinged at the center of rotation, the Coriolus forces cause the blades to move always at constant velocity

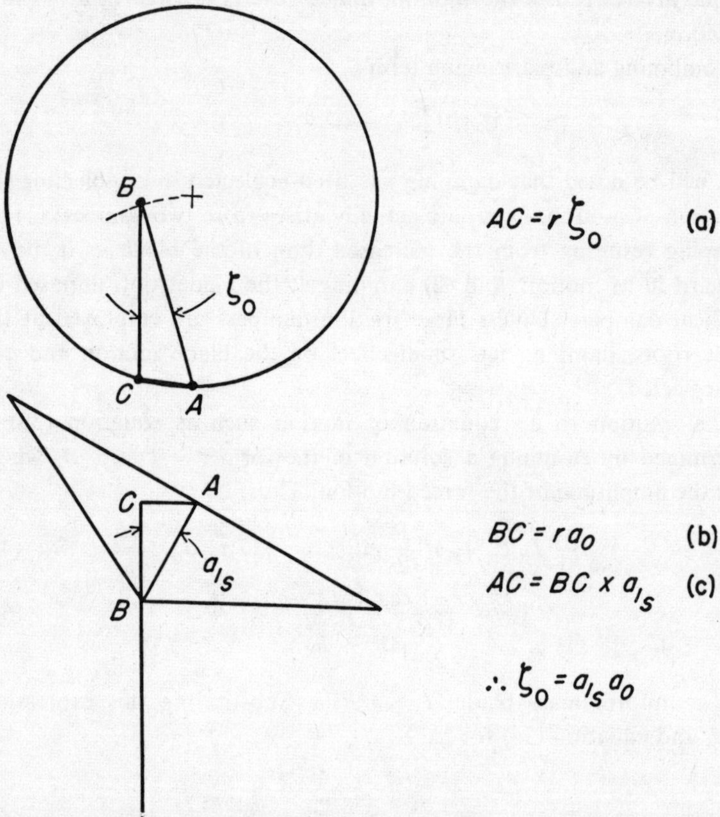

Fig. 7–39 Geometrical concept of in-plane motion.

with respect to the tip-path plane. As the lag hinge is moved outward, the blade motion in the plane of the tips is given as

$$\zeta_{TPP} = a_0 a_{1s} \frac{\frac{3}{2}\frac{e}{R}}{\left(1 - \frac{3}{2}\frac{e}{R}\right)} \tag{44}$$

It is important to note that while blade flapping motion is essentially a resonant phenomenon, lagging motion has a natural frequency well below the rotational speed for normal lag-hinge offsets. The natural frequency of the lagging motion, wherein spring forces are always in equilibrium with mass forces, is found by setting the forcing torque in equation (40) to zero and solving for the frequency. Thus, for uniform mass distribution

$$-I\omega_n^2 = \frac{MR}{2} e\Omega^2$$

and

$$\omega_n = \Omega \sqrt{\frac{3}{2}\frac{e}{R}} \tag{45}$$

A nonuniform mass distribution has small effect on this result. A blade which is twice as heavy per unit length at the root as at the tip

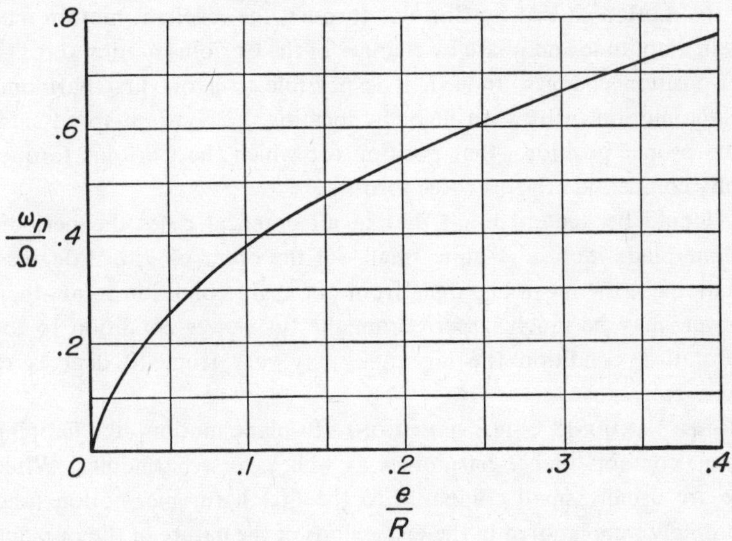

Fig. 7–40 Variation of blade natural lagging frequency with lag-hinge offset.

differs in natural frequency from a blade with uniform mass by less than 5 per cent. (See Appendix IIB, reference 27.) For normal lag-hinge distances, the natural frequency of the blade in lag is about one-third to one-fourth of the rotational speed. The variation of blade natural frequency with lag-hinge distance for a uniform mass blade is shown in Fig. 7–40.

LAG MOTION IN FORWARD FLIGHT. In the preceding discussion it has been pointed out that for hovering flight the in-plane motion depends entirely upon the flapping motion of the blades with respect to the shaft. If no flapping exists, no in-plane motion exists. In forward flight, however, an additional exciting torque is applied to the system as a result of the periodic variations in blade drag. This excitation and the resulting blade motion depend on the condition of flight but are independent of the particular combination of shaft flapping and feathering that is used to locate the control axis. The flapping relative to the shaft may be positive or negative, depending on the fuselage center-of-gravity position and the tail setting. Because the resultant in-plane motion is the sum of the motion caused by periodic air forces and the motion caused by Coriolus torques, the resultant motion may vary in amplitude and phase by changes of the Coriolus motion through shaft-position changes. Indeed, it is possible to avoid first harmonic in-plane motion in forward flight by locating the constant-speed shaft in the proper position—that position for which the Coriolus torques exactly compensate the air force torques.

It should be kept in mind that in all practical cases the periodic in-plane blade motion is quite small—of the order of $\frac{1}{2}$ to 2 degrees. Variations of the mean lag angle from one flight condition to another, however, may be much larger. From the full-power condition to the autorotation condition the lag angle may vary from 10 degrees to − 1 degree.

HIGHER HARMONIC IN-PLANE MOTION. In-plane motion, like flapping motion, contains higher harmonics as well as first harmonics. While these are usually small compared to the first harmonic motion (and accordingly were ignored in the discussions of the nature of the in-plane motion) they are important as sources of vibration. The mechanism by which higher harmonic motions produce rotor vibrations will be discussed in Chapter 12. The following paragraphs intend only to point out the physical reasons for these motions.

Second harmonic motions exist in the hovering case in proportion to the a_{1_s} flapping that is present. This may be understood physically by considering that the a_{1_s} motion results in the blade moving toward and away from the axis of rotation twice each revolution (Fig. 7–41).

The first harmonic motion has been shown in equation (43) to be proportional to the product $a_o a_{1_s}$. From equation (32) it is seen that if a_0 is zero, then only second harmonic in-plane motion exists, its amplitude being given by $\frac{1}{2}a_{1_s}{}^2$. Second harmonic motions also arise in forward flight in response to second harmonic air force inputs.

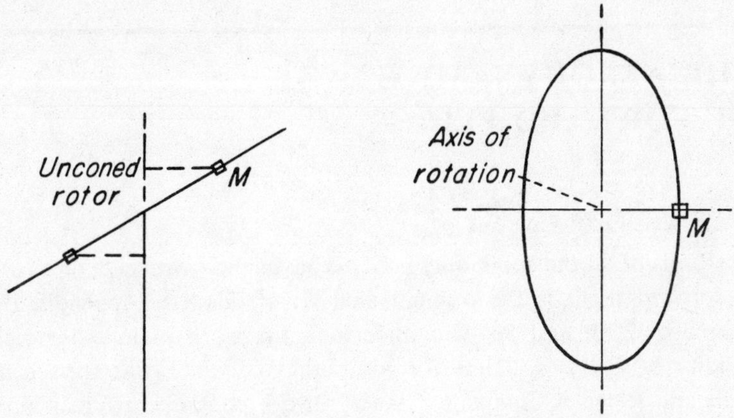

Fig. 7–41 Source of second harmonic in-plane motion in hovering.

Fourth harmonic motion can be shown to depend on second harmonic flapping as far as Coriolus inputs are concerned and on fourth harmonic air force variations. Higher harmonic in-plane motions are important with respect to fatigue stresses and rotor vibrations. They are, however, small enough to be safely neglected as far as their effects on the velocities and air forces encountered by the blade are concerned.

8

THE AERODYNAMICS OF FORWARD FLIGHT

The basis of all the commonly used aerodynamic treatments of a rotor in forward flight is the original analysis of Glauert (Appendix IIB, references 2, 3, and 5), who undertook the work to investigate the validity of Cierva's claims for his autogyro, which was then being built and tested in England. Glauert's work contained a great many simplifying assumptions which had to be adopted because of the mathematical complexity of the equations of forward flight. In its most general form, the analysis of the rotor in forward flight is extremely complex because of the number of degrees of blade freedom involved and the variations of velocities and forces over all parts of the disk. As time went on, and as the autogyro and helicopter received further study, certain of the simplifications contained in Glauert's work were investigated and replaced by more advanced treatments. Lock, Sissingh, and Wheatley, among others, were responsible for many basic advances in rotor theory. Still others (Bailey in particular) have simplified much of the previous work and put it in a form which could be used for practical engineering calculations without the adoption of new assumptions.

As a first step in the understanding of forward-flight rotor theory, the discussion will not start with the oversimplified Glauert theory, nor will it go to the other extreme and develop the latest work with all its refinements and ramifications. Instead, equations will be developed for some of the forces and torques acting on the rotor on the basis of certain simplifying assumptions which were found to be quite valid. Detailed mathematics will be avoided wherever possible inasmuch as an understanding of the problem is the objective. A certain amount of mathematical formulas which, though not difficult, are sometimes lengthy, cannot be avoided, however. Each assumption will be pointed out as it is reached, and its significance will be discussed. After developing the fundamental theory, refinements to it will be examined and discussed.

Definition of Reference Axes

The first step in an analysis of a rotor in forward flight is to define clearly the axes of reference; that is, the axes to which all blade motions

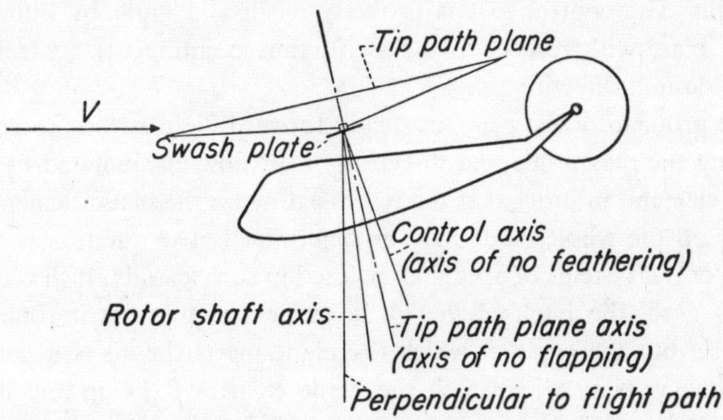

Fig. 8–1 Helicopter reference axes.

and air flow directions will be referenced. Various authors of basic rotor theory have used different systems of reference axes, and in order to avoid confusion in using the expressions in the literature, it is extremely important to understand clearly what axes are being employed.

Figure 8–1 shows the significant axes of a helicopter. These are:

(1) *The control axis or the axis of no feathering*. The control axis is the physical axis of a pure flapping rotor—a rotor with blades fixed in pitch but free to flap. (See Chapter 7.)

(2) *The tip-path-plane axis or axis of no flapping*. As pointed out in

Chapter 7, the blades change pitch periodically but do not flap with respect to the plane-of-the-tips axis. The control axis is the axis of no feathering, whereas the tip-path-plane axis is the axis of no flapping. The amount of feathering about the tip-path-plane axis is equal to the amount of flapping with respect to the control axis.

(3) *The shaft axis.* As shown in Chapter 7, the shaft axis of a conventional helicopter coincides with neither the control axis nor the tip-path-plane axis. The normal helicopter has blades which are free to flap with respect to the shaft and which in addition may be cyclically feathered with respect to the shaft. In flight, the shaft axis is located in space in accordance with the weight and moment characteristics of the fuselage. The fuselage behaves as a pendulum hung in a wind tunnel, swinging backward until the drag moment equals the weight moment. The control axis is properly located in flight by tilting the swash plate with respect to the shaft, thus pointing the control axis in the desired direction.

The problem of the rotor in steady forward flight is the problem of defining the magnitude and direction of air flow encountered by each blade element, in order that the forces acting on the blade element and finally on the whole blade may be determined. The question is, then, which of the systems of axes shall be used in such a study. If the control axis is used, the pitch of the blade element is fixed as it rotates in azimuth, but the angle at which the blade meets the air is dependent upon the velocity with which the blade element flaps up and down. If the tip-path-plane axis is used, the blade element encounters no angle-of-attack changes because of flapping velocity (except for harmonics higher than the first) but does experience a periodic change in pitch angle. If the shaft axis is used, the blade element experiences both flapping velocities and cyclic pitch changes. Note that the choice of reference axes in no way affects the validity of the analysis. They are all geometrically compatible.

For most purposes, using the shaft axis as a reference axis leads to unnecessary complications, since both flapping and feathering must be considered. It is then more convenient to use either the control axis or the tip-path-plane axis as the reference axis. *All analyses presented in this book will be referenced to the control axis—the axis of no feathering*. This choice is convenient because it is in agreement with the bulk of NACA rotating-wing publications.

Having established the axis of reference, the components that are chosen to resolve the resultant rotor force may be defined. It will be shown in the following pages and has been demonstrated by experiment that the resultant rotor vector remains approximately perpendicular to the plane of the tips even in forward flight. More accurately, for normal rotors, the resultant force is generally inclined slightly aft of the tip-path plane, usually less than a degree. The control axis tilts

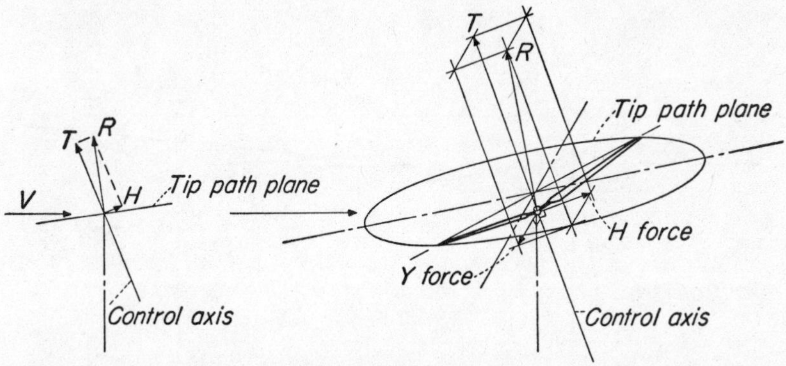

Fig. 8–2 Components of resultant rotor force.

backward (i.e. away from the direction of flight) from the axis of the tip-path plane in normal helicopter flight, the tilt increasing with forward speed. The angle is always small—less than 5 degrees for current helicopters. The components of rotor force with respect to the control axis are shown in Fig. 8–2. Thrust is taken as the component along the control axis; the H force is defined as the rearward-pointing component perpendicular to the control axis, and the Y force as the sideward-pointing component. Because the angle between the resultant rotor force and thrust is small, the magnitude of the thrust is taken as the magnitude of the resultant rotor force.

Sign Convention for Rotor Angle of Attack

The thrust vector of a helicopter rotor is inclined forward in order to secure a component of thrust to overcome the drag of the helicopter.

Consequently, there is a downflow through the tip-path plane because of the component of forward flight velocity parallel to the control axis. On the other hand, the tip-path plane of the autogyro rotor is inclined backward from the flight-path velocity because it is being dragged by the autogyro and an upflow is required to produce autorotative forces.

It is convenient to have one set of equations that applies for both the power-on and power-off (autorotative) conditions of flight. Rotor expressions will be developed on the basis of autogyro convention

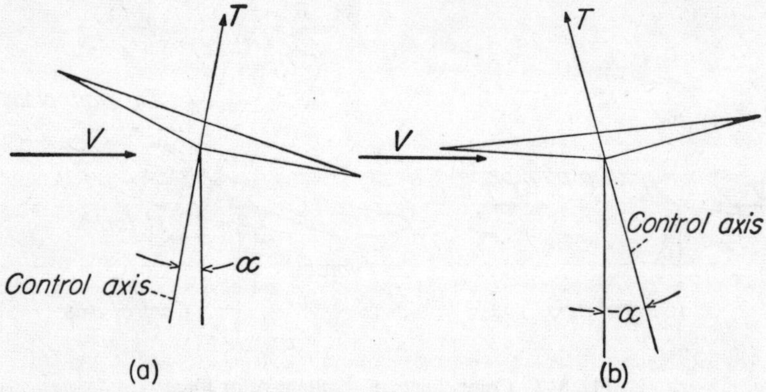

Fig. 8–3 Orientation of α and T in autorotation and in powered flight.
(a) Autorotation
(b) Power-on flight

because the most useful of existing theories were developed in that manner. The rotor angle of attack α, which is defined as the angle between the projection in the plane of symmetry of the control axis and a line perpendicular to the flight path, is positive when the axis is pointing rearward. For the power-off or autorotative condition, α is usually positive, whereas negative angles usually exist in the power-on or helicopter condition. (See Fig. 8–3.)

From the previous work in hovering, it is known that in order to develop expressions for rotor thrust and torque, the resultant velocity and angle of attack at each blade element must be known. Means for obtaining these items must therefore be considered before setting up force and torque expressions.

Rotor Induced Velocity

The total flow through the plane perpendicular to the control axis is composed of the component of forward-flight velocity $V \sin \alpha$, and the induced velocity v. It is convenient to express the flow through the plane in the form of a dimensionless parameter as follows:

$$\lambda = \frac{V \sin \alpha - v}{\Omega R} \tag{1}$$

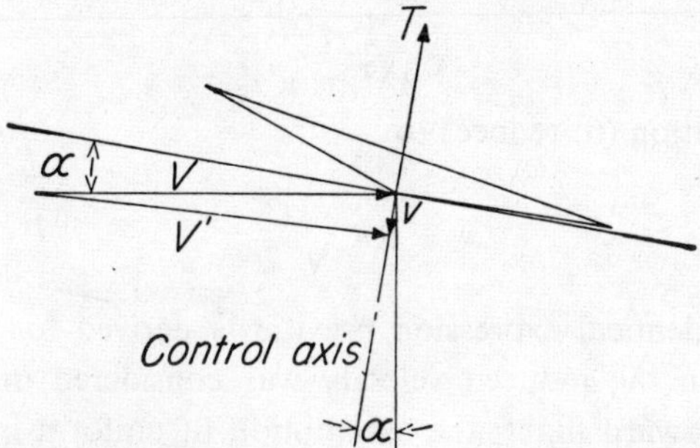

Fig. 8–4 Determination of resultant velocity V' at rotor.

In a similar manner, the component of forward-flight velocity in the plane perpendicular to the control axis ($V \cos \alpha$) is expressed in the form of a tip-speed ratio as:

$$\mu = \frac{V \cos \alpha}{\Omega R} \tag{2}$$

On the basis of momentum theory, the induced velocity in hovering was found from the expression

$$T = (\pi R^2 \rho v) 2v$$

where v, the induced velocity, comprises the total flow through the rotor in the static flight condition. In the same way, the induced velocity in forward flight may be found from the expression

$$T = (\pi R^2 \rho V') 2v$$

or

$$v = \frac{T}{2\pi R^2 \rho V'} \tag{3}$$

where V' is the resultant velocity at the rotor. From Fig. 8–4, it can be seen that V' is the vector sum of the translational and induced velocities. Then

$$V' = [(V \sin \alpha - v)^2 + (V \cos \alpha)^2]^{\frac{1}{2}} \tag{4}$$

When equations (1) and (2) are substituted into (4),

$$V' = \Omega R(\lambda^2 + \mu^2)^{\frac{1}{2}} \tag{5}$$

Expressing $T = C_T \pi R^2 \rho(\Omega R)^2$ and substituting it and equation (5) into equation (3), equation (3) becomes

$$v = \frac{\frac{1}{2}C_T \Omega R}{(\lambda^2 + \mu^2)^{\frac{1}{2}}} \tag{6}$$

If $V = 0$, equation (6) reduces to

$$v = \Omega R \sqrt{\frac{C_T}{2}}$$

which is the identical expression previously derived for the hovering condition when the induced velocity was considered uniform across the disk. In forward flight, the assumption of uniform inflow is more justified, inasmuch as the major part of the inflow is due to $V \sin \alpha$, which is, of course, uniformly distributed across the disk.

It can be shown that the induced velocity as expressed by equation (3) is equivalent to the induced velocity generated by a wing with assumed elliptical lift distribution (which results in uniform downwash).

An expression for the rotor angle of attack can be derived by means of equations (1), (2), and (6). Solving equations (1) and (2) for the sine and cosine of α.

$$\sin \alpha = \frac{\lambda \Omega R + v}{V}$$

$$\cos \alpha = \frac{\mu \Omega R}{V}$$

Substituting equation (6) into the above and dividing one expression by the other gives

$$\tan \alpha = \frac{\lambda}{\mu} + \frac{\frac{1}{2}C_T}{\mu(\mu^2 + \lambda^2)^{\frac{1}{2}}} \tag{7}$$

Or, considering α a small angle,

$$\alpha = \frac{\lambda}{\mu} + \frac{\frac{1}{2}C_T}{\mu(\mu^2 + \lambda^2)^{\frac{1}{2}}} \tag{8}$$

Blade Element Angle of Attack

Once the induced velocity has been calculated, the next step in the analysis is to determine the other velocities that contribute to the

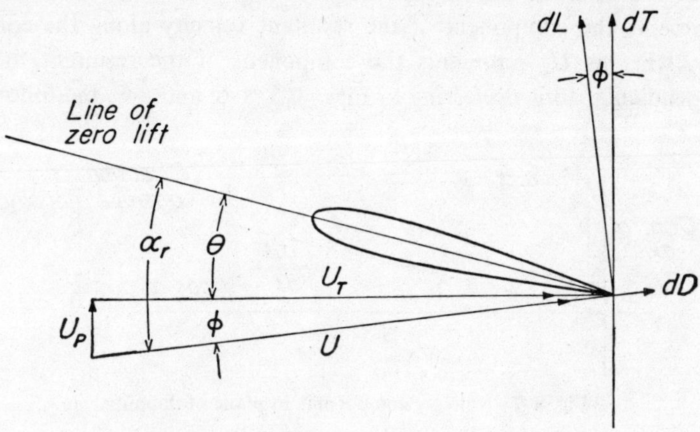

Fig. 8–5 Blade element in forward flight.

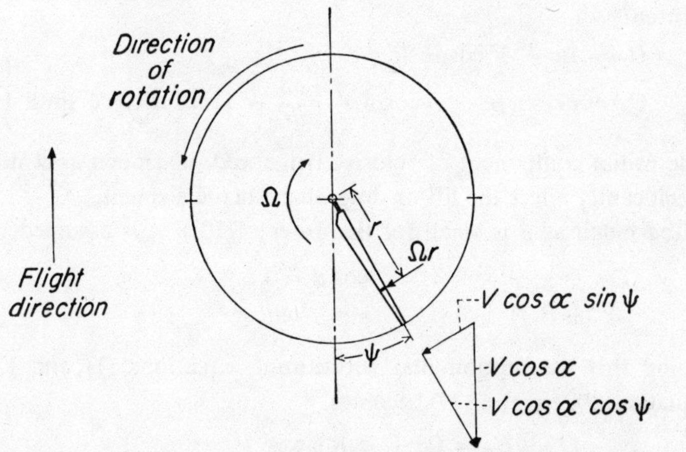

Fig. 8–6 Velocity components in plane perpendicular to control axis.

resultant velocity at the rotor blade elements. All of the contributing velocities are shown in Figs. 8–5, 8–6, and 8–7, referenced both to the plane perpendicular to the control axis and to the control axis for a rotor blade located at an azimuth angle ψ from its downwind position.

In addition to the induced flow, the three velocity sources are forward flight, blade rotation, and blade flapping.

The velocity picture at a blade element in forward flight, as shown in Fig. 8–5, is similar to that in hovering. The velocity vector U_P represents the component of the resultant velocity along the control axis, whereas U_T represents the component of the resultant that is perpendicular to it. Referring to Figs. 8–5, 8–6, and 8–7, the following

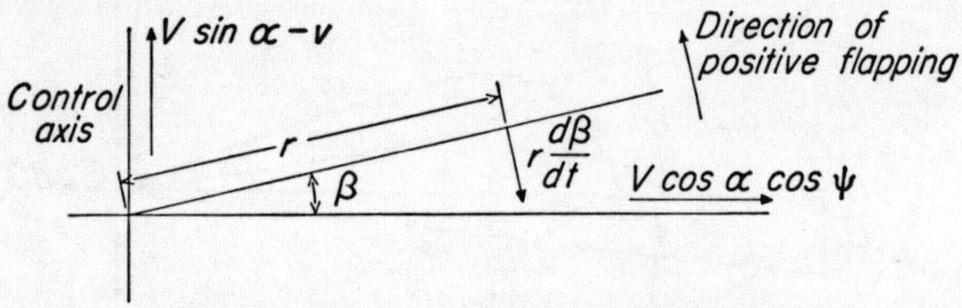

Fig. 8–7 Velocity components in plane of flapping.

expressions for the velocity components at a blade element may be written:

$$\left.\begin{aligned} U_T &= \Omega r + V\cos\alpha\sin\psi \\ U_P &= (V\sin\alpha - v)\cos\beta - r\frac{d\beta}{dt} - V\cos\alpha\cos\psi\sin\beta \end{aligned}\right\} \tag{9}$$

The radial component of velocity is ignored, inasmuch as it does not significantly affect the lift or drag of the blade element.

Inasmuch as β is small (of the order of 10°), it is assumed that

$$\cos\beta = 1$$
$$\sin\beta = \beta$$

Using this assumption and substituting equations (1) and (2) into equation (9), equation (9) becomes

$$\left.\begin{aligned} U_T &= \Omega r + \mu\Omega R\sin\psi \\ U_P &= \lambda\Omega R - r\frac{d\beta}{dt} - \mu\Omega R\beta\cos\psi \end{aligned}\right\} \tag{10}$$

Referring to Fig. 8–5, and noting that ϕ is small,

$$\alpha_r = \theta + \phi = \theta + \frac{U_P}{U_T} \tag{11}$$

Now that expressions for the angle of attack and resultant velocity at each blade element have been derived, expressions for the forces and torques acting on the rotor can be found by averaging the force or torque on each element around the disk, integrating these values along the blade, and finally multiplying by b, the number of blades. It should be noted, however, that equations expressing the flapping motion and the flapping velocity must be derived in terms of known parameters before all the rotor force and torque expressions can be properly evaluated. An expression for rotor thrust must first be found in order to derive the flapping constants.

Expression for Thrust

From the simple blade-element theory, the differential lift on a blade element of one blade can be written as

$$dL = \tfrac{1}{2}\rho U^2 c_l\, c\, dr \tag{12}$$

where U is the resultant velocity at the blade element, feet per second. The lift coefficient c_l of the blade element is assumed to be proportional to the angle of attack, an assumption that is valid for lift coefficients below the stall. Then,

$$c_l = a\alpha_r = a\left(\theta + \frac{U_P}{U_T}\right) \tag{13}$$

Inasmuch as ϕ is small,

$$\left.\begin{aligned} U &= U_T \\ dT &= dL \end{aligned}\right\} \tag{14}$$

Substituting equations (13) and (14) into equation (12), the differential thrust is

$$dT = \tfrac{1}{2}\rho a(\theta U_T{}^2 + U_P U_T)c\, dr \tag{15}$$

The total thrust produced by a rotor of b blades is then found by integrating the differential thrust around the azimuth and then along the blade span. This integration is expressed mathematically as follows:

$$T = \frac{b}{2\pi}\int_0^{2\pi}\int_0^R \frac{dT}{dr}\, dr\, d\psi \tag{16}$$

The integration of equation (16) can be simplified if the following integrations are used:

$$\left.\begin{aligned}\frac{1}{2\pi}\int_0^{2\pi}\sin\psi\,d\psi &= 0\\ \frac{1}{2\pi}\int_0^{2\pi}\cos\psi\,d\psi &= 0\\ \frac{1}{2\pi}\int_0^{2\pi}\sin^2\psi\,d\psi &= \frac{1}{2}\\ \frac{1}{2\pi}\int_0^{2\pi}\cos^2\psi\,d\psi &= \frac{1}{2}\\ \frac{1}{2\pi}\int_0^{2\pi}\sin\psi\cos\psi\,d\psi &= 0\\ \frac{1}{2\pi}\int_0^{2\pi}\sin^2\psi\cos^2\psi\,d\psi &= \frac{1}{8}\end{aligned}\right\} \quad (17)$$

Using equations (10) and (17), the average expressions for U_T^2 and U_PU_T around the azimuth become

$$\left.\begin{aligned}\bar{U}_T^2 &= (\Omega r)^2 + \frac{(\mu\Omega R)^2}{2}\\ \overline{U_TU_P} &= \Omega r(\lambda\Omega R) - \Omega r\left(r\frac{d\beta}{dt}\right)\end{aligned}\right\} \quad (18)$$

It will be demonstrated shortly [equation (23)] that the average value of $d\beta/dt$ around the azimuth is equal to zero, so that

$$\overline{U_TU_P} = \Omega^2\lambda Rr \quad (19)$$

Equation (16) then becomes

$$\begin{aligned}T &= \int_0^R b\,\frac{1}{2}\,\rho a\left[\theta\Omega^2\left(r^2 + \frac{\mu^2R^2}{2}\right) + \Omega^2\lambda Rr\right]c\,dr\\ &= \frac{1}{2}\,\rho abc\Omega^2R^3\left[\frac{\theta}{3} + \frac{\mu^2\theta}{2} + \frac{\lambda}{2}\right]\end{aligned} \quad (20)$$

Or, in coefficient form,

$$C_T = \frac{1}{2}\,a\sigma\left[\frac{\theta}{3} + \frac{1}{2}\,\mu^2\theta + \frac{1}{2}\,\lambda\right] \quad (21)$$

It should be noted that equation (21) is an expression for the thrust of an untwisted, untapered rotor, inasmuch as no variation of pitch angle or chord with radius was assumed before integrating.

Calculation of Flapping Coefficients

The flapping motion of the blade is repeated identically with each revolution and therefore can be expressed as a Fourier series in which the independent variable is ψ, the azimuth angle of the blade from its downwind position. The Fourier series expresses the acute angle β between the blade-span axis and the plane perpendicular to the control axis as a function of ψ, the form being

$$\beta = a_0 - a_1 \cos \psi - b_1 \sin \psi - a_2 \cos 2\psi - b_2 \sin 2\psi - a_3 \cos 3\psi \ldots \quad (22)$$

All harmonics above the second have been found experimentally to be small. For purposes of simplification in this development, only first harmonic flapping will be considered.

Flapping velocities and accelerations can now be computed by differentiating equation (22), remembering that $d\psi/dt = \Omega = \text{constant}$.

$$\left.\begin{aligned} \dot{\beta} &= \frac{d\beta}{dt} = \Omega(a_1 \sin \psi - b_1 \cos \psi) \\ \ddot{\beta} &= \frac{d^2\beta}{dt^2} = \Omega^2(a_1 \cos \psi + b_1 \sin \psi) \end{aligned}\right\} \quad (23)$$

Expressions for the blade flapping coefficients are derived from the fact that the resultant moment of all the forces acting on the blade about the flapping hinge is equal to zero. These forces are the blade thrust, weight, centrifugal, and inertia (resulting from blade flapping) forces, and are shown acting on a blade in Fig. 8–8. Taking moments about the flapping hinge.

$$M_T - M_W - M_{CF} - M_I = 0 \quad (24)$$

where M_T, M_W, M_{CF}, and M_I are the moments about the flapping hinge caused by the thrust, blade weight, centrifugal, and inertia forces, respectively. The sign of the inertia terms is negative because the inertia load acts downward when there is positive (upward) angular acceleration.

The weight, centrifugal force, and inertia moments can be expressed by the following equations:

$$M_W = \int_0^R mgr\,dr = mg\,\frac{R^2}{2} \tag{25}$$

$$M_{CF} = \int_0^R m\Omega^2 r(r\beta)\,dr = m\Omega^2\beta\,\frac{R^3}{3} \tag{26}$$

$$M_I = \int_0^R \left(m\,\frac{d^2\beta}{dt^2}\,r\right) r\,dr = m\,\frac{d^2\beta}{dt^2}\,\frac{R^3}{3} \tag{27}$$

As can be inferred from a consideration of the reference axes discussed at the beginning of the chapter, if the expressions for the bending

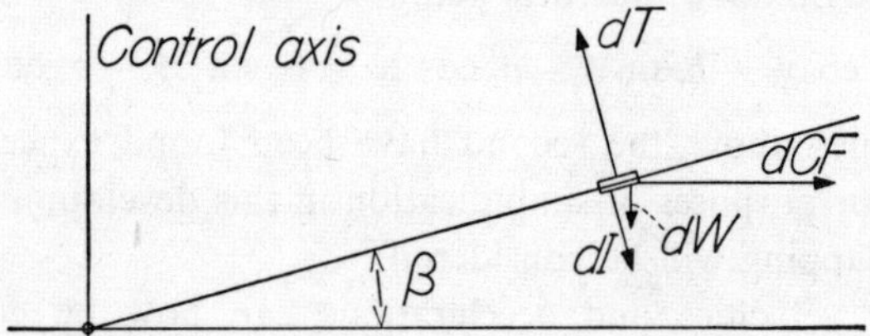

Fig. 8–8 Forces acting on a blade element.

moment caused by centrifugal and inertia forces in flapping are combined, the resultant moment has a value which is independent of the azimuth angle ψ. A proof of this follows.

Adding equations (26) and (27),

$$M_{CF} + M_I = m\,\frac{R^3}{3}\left(\Omega^2\beta + \frac{d^2\beta}{dt^2}\right) \tag{28}$$

Upon substitution of equations (22) and (23), equation (28) becomes

$$\begin{aligned} M_{CF} + M_I &= \frac{mR^3}{3}\,[\Omega^2(a_0 - a_1\cos\psi - b_1\sin\psi) \\ &\qquad + \Omega^2(a_1\cos\psi + b_1\sin\psi)] \\ &= m\Omega^2\,\frac{R^3}{3}\,a_0 \end{aligned} \tag{28a}$$

Because m was considered uniform, the moment of inertia I_1 of one blade about the flapping hinge is simply $mR^3/3$. Therefore,

$$M_{CF} + M_I = I_1\Omega^2 a_0 \tag{29}$$

It can now be seen from equations (24) and (29) that the thrust moment M_T is independent of the azimuth angle ψ inasmuch as $M_W + M_{CF} + M_I$ does not vary with ψ. This result is true only when first harmonic flapping motion is considered.

Substituting equation (29) into equation (24) and solving for a_0,

$$a_0 = \frac{M_T - M_W}{I_1\Omega^2} \tag{30}$$

The thrust moment can be derived as follows. The thrust moment at any azimuth position is

$$M_T = \int_0^R r\,\frac{dT}{dr}\,dr$$

Substituting equation (15) into the above, and dealing with but one blade,

$$M_T = \int_0^R \tfrac{1}{2}\rho ac(\theta U_T^2 + U_P U_T)r\,dr \tag{31}$$

Substituting for U_T and U_P from equation (10),

$$\begin{aligned} M_T = \tfrac{1}{2}\rho ac\Omega^2 \int_0^R & [(\theta r^3 + \tfrac{1}{2}\theta\mu^2R^2r + \lambda Rr^2) + \\ & \sin\psi(2\mu Rr^2\theta + \mu\lambda R^2r - a_1r^3 + \tfrac{1}{4}\mu^2R^2ra_1) + \\ & \cos\psi(r^3b_1 - \mu Rr^2a_0 + \tfrac{1}{4}\mu^2R^2rb_1)]dr \end{aligned} \tag{32}$$

It has already been shown that with the assumptions adopted, M_T is independent of ψ. The value of M_T can therefore be obtained by integrating only the first part of equation (32) (the part that is independent of sin ψ and cos ψ). It follows also that the coefficients of each harmonic term in equation (32) must be identically equal to zero. Therefore,

$$M_T = \frac{1}{2}\rho ac\Omega^2R^4\left[\frac{\theta}{4}(1 + \mu^2) + \frac{\lambda}{3}\right] \tag{33}$$

The coning angle a_0 may then be found by substituting equations (25) and (33) into equation (30). The flapping coefficients a_1 and b_1 are found by equating the coefficients of cos ψ and sin ψ equal to zero in equation (32). The expression for a_0 becomes

$$a_0 = \frac{\frac{1}{2}\rho ac\Omega^2R^4\left[\frac{\theta}{4}(1 + \mu^2) + \frac{\lambda}{3}\right] - M_W}{I_1\Omega^2} \tag{34}$$

Let

$$\gamma = \frac{c\rho a R^4}{I_1} \tag{35}$$

where γ is a nondimensional coefficient that represents the mass constant of the blade and expresses the relationship between the air forces and mass forces acting on the blade. This coefficient is often referred to as the Lock number. The Lock number, or mass factor, is a very important parameter in rotor behavior. It will occur frequently in blade motion and rotor stability analyses. An infinitely heavy blade has a Lock number of zero; normal blades have values ranging from 8 to 15.

Substituting equation (35) into (34),

$$a_0 = \frac{1}{2}\gamma\left[\frac{\theta}{4}(1+\mu^2)+\frac{\lambda}{3}\right] - \frac{M_W}{I_1\Omega^2} \tag{36}$$

and setting the coefficients of the cosine and sine terms in equation (32) to zero,

$$\left.\begin{aligned} a_1 &= \frac{\mu(\frac{8}{3}\theta + 2\lambda)}{1 - \frac{1}{2}\mu^2} \\ b_1 &= \frac{4\mu a_0}{3(1+\frac{1}{2}\mu^2)} \end{aligned}\right\} \tag{37}$$

The second term in the expression for a_0 represents the blade weight and is usually ignored as being small compared with the first term. Note that the coning angle is proportional to the Lock number, that a_1 is independent of the Lock number, and that b_1 is proportional to the coning angle (and hence also to the Lock number). The physical reasons for these relationships were pointed out in Chapter 7.

Expression for Torque

The torque dQ on a blade element (see Fig. 8–5) is

$$dQ = r(dD\cos\phi - dL\sin\phi) \tag{38}$$

Assume that the blade section profile-drag coefficient is constant along the blade and is equal to a mean value δ. Then,

$$dQ = \tfrac{1}{2}\rho U_T^2\delta c r\,dr - \tfrac{1}{2}\rho U_T^2\phi c_l c r\,dr \tag{39}$$

Q_0, the profile-drag torque, is then

$$Q_0 = \frac{b}{2\pi}\int_0^R \int_0^{2\pi} \frac{1}{2}\rho U_T^2 \delta cr \, d\psi \, dr \tag{40}$$

Substituting the average value of U_T^2 from equation (18) into the above, the profile-drag torque is

$$Q_0 = \tfrac{1}{2}\rho bc\delta\Omega^2 \int_0^R (r^3 + \tfrac{1}{2}\mu^2 R^2 r)dr = \tfrac{1}{2}\rho bc\delta\Omega^2 R^4[\tfrac{1}{4} + \tfrac{1}{4}\mu^2] \tag{41}$$

The induced torque Q_i is found by setting up the double integral of the second term of equation (39) as follows:

$$Q_i = \frac{b}{2\pi}\int_0^R \int_0^{2\pi} \frac{1}{2}\rho U_T^2 c_l cr\phi \, d\psi \, dr$$

which, after substituting equation (13) becomes

$$Q_i = \frac{b}{2\pi}\frac{1}{2}\rho ac \int_0^R \int_0^{2\pi} (\theta U_P U_T r + U_P^2 r) d\psi \, dr \tag{42}$$

Upon substituting the average values for $U_P U_T$ and U_P^2 over the disk, equation (42) becomes

$$\begin{aligned} Q_i = \tfrac{1}{2}\rho acb \int_0^R \Omega^2 [\theta\lambda R r^2 + \lambda^2 R^2 r + \tfrac{1}{2}r^3(a_1^2 + b_1^2) + \mu\lambda R^2 r a_1 \\ + \mu^2 R^2 r(\tfrac{1}{2}a_0^2 + \tfrac{3}{8}a_1^2 + \tfrac{1}{8}b_1^2) - \mu R r^2 a_0 b_1] \, dr \end{aligned} \tag{43}$$

Integrating the above and combining it with equation (41), the total torque expression is

$$\begin{aligned} Q = \frac{1}{2}\rho abc\Omega^2 R^4 \Bigg[\frac{\delta}{4a}(1 + \mu^2) - \frac{1}{3}\lambda\theta - \frac{1}{2}\lambda^2 - \frac{1}{8}(a_1^2 + b_1^2) \\ - \frac{1}{2}\mu^2\left(\frac{a_0^2}{2} + \frac{3}{8}a_1^2 + \frac{1}{8}b_1^2\right) - \frac{1}{2}\mu\lambda a_1 + \frac{1}{3}\mu a_0 b_1 \Bigg] \end{aligned} \tag{44}$$

or

$$\begin{aligned} C_Q = \frac{\sigma a}{2}\Bigg[\frac{\delta}{4a}(1 + \mu^2) - \frac{1}{3}\lambda\theta - \frac{1}{2}\lambda^2 - \frac{1}{8}(a_1^2 + b_1^2) \\ - \frac{1}{2}\mu^2\left(\frac{a_0^2}{2} + \frac{3}{8}a_1^2 + \frac{1}{8}b_1^2\right) - \frac{1}{2}\mu\lambda a_1 + \frac{1}{3}\mu a_0 b_1\Bigg] \end{aligned} \tag{45}$$

Rotor Drag—*H* Force

The component of the resultant rotor force in the plane perpendicular to the control axis (the *H* force) is composed of the components of the lift and profile-drag forces in that plane. The force is positive when

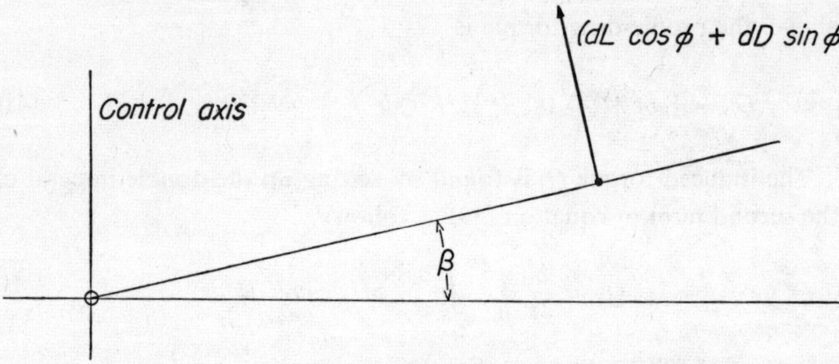

Fig. 8–9 Aerodynamic force components in plane of flapping.

it opposes the translational motion of the aircraft. The elemental expression for H may be obtained by reference to Figs. 8–5, 8–9, and 8–10.

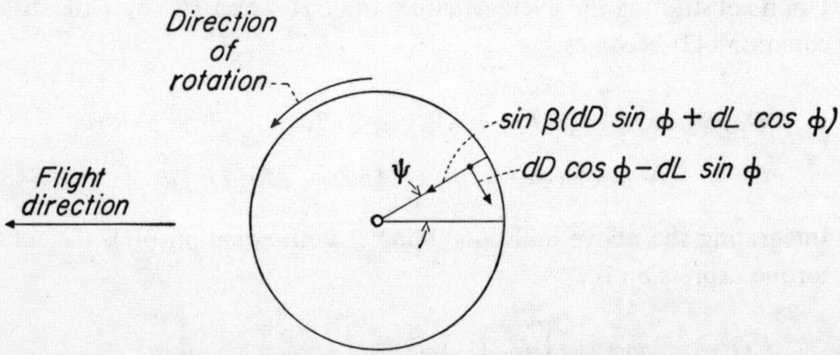

Fig. 8–10 Aerodynamic force components in plane perpendicular to control axis.

The projections of dL and dD acting in a plane perpendicular to the blade-span axis and taken parallel and perpendicular to the control axis are (see Fig. 8–5)

$$dL \cos \phi + dD \sin \phi \tag{46}$$

and

$$dD \cos \phi - dL \sin \phi \tag{46a}$$

and the components of these projections lying in the plane perpendicular to the control axis are (see Fig. 8–9)

$$(dL \cos \phi + dD \sin \phi) \sin \beta \tag{47}$$

and

$$dD \cos \phi - dL \sin \phi \tag{47a}$$

(Note that the latter expression is already in the desired plane.)

Taking components parallel to the velocity of translation of the projections as given by equations (47) and (47a) in the plane perpendicular to the control axis (see Fig. 8–10), the elemental H may be written as

$$dH = (dD \cos \phi - dL \sin \phi) \sin \psi - (dL \cos \phi + dD \sin \phi) \sin \beta \cos \psi \tag{48}$$

Using the assumption that ϕ and β are small angles, equation (48) becomes

$$dH = dD \sin \psi - dL(\beta \cos \psi + \phi \sin \psi) \tag{49}$$

The profile-drag part of H is

$$H_0 = \frac{b}{2\pi} \int_0^R \int_0^{2\pi} \frac{1}{2} \rho U_T^2 c\delta \sin \psi \, d\psi \, dr \tag{50}$$

Integrating around the disk, equation (50) reduces to

$$H_0 = \tfrac{1}{2}\rho bc\delta\mu\Omega^2 R \int_0^R r \, dr$$

which, when integrated spanwise, becomes

$$H_0 = \frac{\rho}{4} bc\delta\mu\Omega^2 R^3 \tag{51}$$

The "induced" component of rotor drag is

$$\begin{aligned} H_i &= \frac{b}{2\pi} \int_0^R \int_0^{2\pi} \frac{1}{2} \rho U_T^2 c_l c(\beta \cos \psi + \phi \sin \psi) \, d\psi \, dr \\ &= \frac{b}{2\pi} \int_0^R \int_0^{2\pi} \frac{1}{2} \rho ca[(\theta U_T^2 + U_P U_T)\beta \cos \psi \\ &\qquad + (\theta U_T U_P + U_P^2) \sin \psi] \, d\psi \, dr \end{aligned} \tag{52}$$

Substituting the series expression for β and integrating around the azimuth, H_i becomes

$$\begin{aligned} H_i = \tfrac{1}{2}\rho abc\Omega^2 \int_0^R & [\theta(\tfrac{1}{2}\mu\lambda R^2 - r^2 a_1) - \tfrac{3}{2}\lambda R r a_1 \\ & - \tfrac{1}{4}\mu R r a_1^2 + \tfrac{1}{2} r^2 a_0 b_1 - \tfrac{1}{2}\mu R r a_0^2] \, dr \end{aligned} \tag{53}$$

Integrating equation (53) spanwise, and combining it with equation (51) by means of equation (49), the total H force is found to be

$$H = \frac{1}{2}\rho abc\Omega^2R^3\left[\frac{\delta\mu}{2a} + \frac{1}{3}\theta a_1 - \frac{1}{2}\mu\lambda\theta + \frac{3}{4}\lambda a_1 + \frac{1}{4}\mu a_1^2 - \frac{1}{6}a_0b_1 + \frac{1}{4}\mu a_0^2\right] \tag{54}$$

Defining

$$C_H = \frac{H}{\pi R^2\rho(\Omega R)^2} \tag{55}$$

H in coefficient form is

$$C_H = \frac{\sigma a}{2}\left[\frac{\delta\mu}{2a} + \frac{1}{3}\theta a_1 - \frac{1}{2}\mu\lambda\theta + \frac{3}{4}\lambda a_1 + \frac{1}{4}\mu a_1^2 - \frac{1}{6}a_0b_1 + \frac{1}{4}\mu a_0^2\right] \tag{56}$$

Equations for the lateral force Y may be derived in a manner similar to the H derivation. The development can be made from the following elemental expression for Y (see Fig. 8–10).

$$Y = \frac{b}{2\pi}\int_0^R\int_0^{2\pi} - dD\cos\psi - dL(\beta\sin\psi - \phi\cos\psi)d\psi\, dr \tag{57}$$

Review of Assumptions

Rotor equations similar to those developed in the preceding sections can be applied to helicopter performance, stress, stability, and vibration analyses. In order to apply the equations intelligently for a particular analysis, however, the assumptions on which the equations are based must be thoroughly understood, inasmuch as an assumption that is valid for one type of investigation may lead to erroneous results in a different application. The uniform inflow assumption in forward flight, for example, is good in certain types of performance calculations but is not precise enough for most vibration analyses. In the same manner, the use of a tip-loss factor B may be sufficiently accurate for performance calculations but not for blade stress analysis.

It is therefore well to review at this time the assumptions on which the preceding analysis is based. If the use of an assumption is understood, the designer or research worker can decide for himself if a particular analysis is refined enough for his purpose.

The assumptions used are as follows:

(1) The blades are untwisted and untapered.

(2) The radial component of the resultant air velocity at each blade element may be neglected. (Radial flow may be expected to increase only slightly the total profile drag of the rotor and to have but little effect on the rotor thrust.)

(3) The induced velocity through the rotor disk is constant. (This assumption has been found reasonable for the calculated performance of a rotor operating at tip-speed ratios greater than about 0.10, but it does result in an error in computing the blade motion, particularly in b_1. In addition, certain types of vibration phenomena experienced by a helicopter in the transition region between hovering and forward flight (between zero and approximately 30 miles per hour) can be explained when the dissymmetry of inflow in this condition is calculated and used in the analysis.)

(4) The flapping angle β and the inflow angle of attack ϕ are small enough so that $\cos \beta$ and $\cos \phi = 1$, $\sin \beta = \beta$, and $\sin \phi = \phi$. (The assumption that ϕ is a small angle is satisfactory for most ranges of helicopter operation, but it may lead to significant errors in flight conditions where the inflow is large, such as in very steep rates of climb or descent, or if the rotor is used as a propeller as in some types of convertible aircraft.)

(5) In the Fourier expansion for the flapping angle, second and higher harmonics are negligible. (While this assumption is satisfactory for approximate rotor performance calculations and for estimates of rotor stability, higher harmonic motions, particularly the second and third harmonics, are a primary source of vibration and stress.)

(6) The effect of the reversed-flow region is negligible. (Reversed flow occurs at the inboard part of the retreating blades where the forward flight velocity is greater than the local rotational velocity, thus causing a flow over the blade section from the trailing to the leading edges. This region increases with tip-speed ratio.)

(7) The blades continue into the center of rotation and the flapping hinge is located on the axis of rotation. (Flapping-hinge offsets may be neglected safely in performance estimates but must be taken into account in stability and control analyses.)

(8) The profile-drag coefficient and lift-curve slope are assumed constant. (This ignores the fact that blade section angles of attack, and hence the profile-drag coefficient, vary over the rotor disk even when stall is not encountered. A more serious limitation is that blade stall is not taken into account. At high tip-speed ratios, when large parts of the retreating blades may be stalled, the blade profile drag increases rapidly while the lift falls off instead of rising further. The assumption will then result in very optimistic performance predictions.)

(9) The blades are infinitely rigid in all directions; i.e., the effect of torsional deflections and blade-bending deflections on the velocity and angle-of-attack distributions are neglected. (For normal blades, the effects of deformations on blade motion and rotor performance are small, but in vibration analyses and control-force studies deflections become important. Torsional deflections caused by displaced chordwise center-of-gravity positions are also quite important in stability calculations.)

(10) Tip losses are ignored. (This assumption results in somewhat optimistic predictions of rotor induced losses.)

Refined Treatments

The development of the basic rotor equations, as was presented herein, was repeated by many authors, each of whom extended the theory by removing one or more of the assumptions listed above. The contributions of two such authors will now be reviewed, inasmuch as their work is representative of refined treatments and is used as standard reference material. In the first paper, *NACA Rep. 487* (reference II–35, Appendix IIA), Wheatley extended the analyses of Glauert and Lock by removing some of the assumptions on which their studies were based. In the second paper, *NACA Rep. 716* (reference II–18, Appendix

IIA), Bailey further refined the theory and put it into a form that made it convenient for engineering calculations.

NACA Rep. 487

The following refinements were included in the theory of *NACA Rep. 487* by Wheatley; a few were considered by Glauert in a preliminary fashion in some of his early papers.

(1) The limitation of the analysis to blades of constant pitch was removed so that any linear pitch variation could be introduced. This was done by setting $\theta = \theta_0 + \theta_1(r/R)$ into the elemental expression for blade lift and drag, where θ_1 represents the difference between the root and tip pitch angles.

(2) A method of allowing approximately for tip losses in calculating thrust was introduced. The following expression for the tip loss factor B was suggested:

$$B = 1 - \frac{c_t}{2R} \tag{58}$$

where c_t is the blade tip chord. Sissingh, in reference II–21 (Appendix IIA) suggests:

$$B = 1 - \frac{\sqrt{2C_T}}{b} \tag{59}$$

Most analyses assume a value for the tip-loss factor $B = 0.97$. This factor is introduced into the equations by integrating the thrust and induced torque expressions from $r = 0$ to $r = 0.97R$. The profile-drag torque expression is integrated out to the blade tip inasmuch as drag exists even when there is a loss of thrust.

(3) The second harmonic terms in the blade flapping equation were retained, as well as all powers of μ up to and including the fourth.

(4) An approximate method of evaluating the influence of the reversed flow over the retreating blade was developed. This was done by integrating the rotor force and torque expressions in several parts. For example, the general thrust expression is

$$T = \frac{b}{2\pi}\int_0^{2\pi} d\psi \int_0^{BR} \frac{1}{2}\rho c U^2 c_l dr$$

Assuming that the lift-curve slope at 180-degree angle of attack is equal to that at 0 degrees and that linear twist is used, then

$$T = \frac{b}{2\pi}\int_0^{\pi} d\psi \int_0^{BR} \frac{1}{2}\rho ca U^2\left(\theta_0 + \theta_1\frac{r}{R} + \phi\right) dr$$
$$+ \frac{b}{2\pi}\int_{\pi}^{2\pi} d\psi \int_{-\mu R\sin\psi}^{BR} \frac{1}{2}\rho ac U^2\left(\theta_0 + \theta_1\frac{r}{R} + \phi\right) dr$$
$$+ \frac{b}{2\pi}\int_{\pi}^{2\pi} d\psi \int_0^{-\mu R\sin\psi} \frac{1}{2}\rho ac U^2\left(-\theta_0 - \theta_1\frac{r}{R} - \phi\right) dr \qquad (60)$$

From the aforementioned equation, it can be seen that the usual integration was performed over the "advancing" part of the disk (from $\psi = 0°$ to $180°$) and that the "retreating" part of the disk (from $\psi = 180°$ to $360°$) was integrated in two parts—one covering the reversed-velocity region which contributes negative lift in power-on flight, and the other the rest of the blade which acts in the normal manner.

(5) Another variation introduced by Wheatley was an energy method of evaluating the drag of the rotor. Instead of calculating an H force, Wheatley equated the sum of the energy dissipated in the generation of thrust and the losses arising from blade drag to the total energy loss of the rotor per second, DV, where D is the equivalent rotor drag. Thus

$$DV = vT + \frac{b}{2\pi}\int_0^{2\pi} d\psi \int_0^{R} \frac{1}{2}\rho c\delta U_T^3\, dr$$
$$- \frac{b}{\pi}\int_{\pi}^{2\pi} d\psi \int_0^{-\mu R\sin\psi} \frac{1}{2}\rho c\delta U_T^3\, dr \qquad (61)$$

With the aid of equation (6) and assuming that $L = T\cos\alpha$, the following expression for the rotor drag-lift ratio was obtained:

$$\left(\frac{D}{L}\right)_r = \left(\frac{D}{L}\right)_0 + \left(\frac{D}{L}\right)_i$$
$$= \frac{\sigma\delta(1 + 3\mu^2 + \frac{3}{8}\mu^4)}{8\mu C_T} + \frac{\frac{1}{2}C_T}{\mu(\mu^2 + \lambda^2)^{\frac{1}{2}}} \qquad (62)$$

Thus the power expended in moving the rotor through the air at a given airspeed may be obtained by multiplying the $(D/L)_r$ for that airspeed by LV.

(6) The effect of a varying induced flow on the rotor characteristics

was also studied because if the rotor behaves as a wing, it is reasonable to expect an increase in the magnitude of the induced velocity in passing from the leading to the trailing edges of the disk. Limited experimental data, as well as theoretical investigations, confirm this viewpoint. Wheatley studied the simplest case of a varying induced velocity; namely, one in which the induced velocity varied linearly with distance downstream from the leading edge, the average value over the disk being the same as obtained from uniform inflow considerations. Thus the assumption was made that there is superimposed upon the average induced velocity v an additional velocity v_1 expressed as

$$v_1 = Kv \frac{r}{R} \cos \psi \tag{63}$$

where K is the ratio between v_1 and v when $r = R$ and $\cos \psi = 1$. When an arbitrary value of $K = 0.5$ was used, it was found that fair agreement with experimental blade-motion data was obtained, but that the type of induced velocity assumed had only a secondary effect on the net rotor forces. Coleman, in reference II–14 (Appendix IIA), analytically determined values of K which are useful for calculating rotor blade-flapping motion and in explaining certain vibration phenomena that occur in the transition between hovering and forward flight.

NACA Rep. 716

Although Wheatley's analysis clarified and extended rotor theory beyond previous limits, the form in which the expressions were given was unsatisfactory for practical engineering calculations. Bailey found that considerable simplification could be effected if all expressions were reduced to three basic parameters that completely define the operating condition of any rotor; namely, the inflow velocity ratio, λ, the blade pitch, θ, and the tip-speed ratio, μ. Therefore, in *NACA Rep. 716* (see Appendix IIA, reference II–18) the theoretical expressions for thrust coefficient, flapping coefficients, torque coefficients, and the profile drag-lift ratio were reduced to simple functions of the inflow factor λ, the blade pitch angles, θ_0 and θ_1, and the tip-speed ratio, μ.

The coefficients of the λ and θ terms were expressed as functions of the tip-speed ratio μ, the mass constant γ, and the tip-loss factor B. Values of these coefficients, computed for $\gamma = 15$ and $B = 0.97$, were then tabulated for a series of specified values of μ. Because the departures of γ from 15 and B from 0.97 that are to be expected in modern rotor designs have a negligible effect on the values of these coefficients, the tabulated values may be used for calculating the aerodynamic characteristics of any conventional rotor. The coefficients involved in the formulas for some rotor characteristics, such as the coning angle a_0, the lateral flapping angle b_1, and the higher harmonic flapping terms, are essentially proportional to γ. For these expressions, the results can be ratioed to account for the actual mass constant of the blade in question.

In addition, the analysis of Wheatley was extended by approximating the relation between the section profile-drag coefficient c_{d_o}, and the angle of attack α_r, of a blade element by the same power series used in the hovering analysis of Chapter 4, namely,

$$c_{d_o} = \delta_0 + \delta_1\alpha_r + \delta_2\alpha_r^2$$

A convenient method was also developed in the report for assigning appropriate values to the coefficients δ_0, δ_1, and δ_2 for conventional airfoil sections at any Reynold's number.

The expressions for the flapping, thrust, torque, and profile drag-lift ratio coefficients are summarized below, together with the tables which provide numerical coefficients of the λ and θ terms. The numerical coefficients which depend on μ, B, and γ were evaluated from lengthy expressions which are presented in full in the appendix of reference II–18 (Appendix IIA).

FLAPPING COEFFICIENTS. The expressions for the coefficients of the constant term (coning angle) and of the first harmonic flapping terms are as follows:

$$\frac{a_0}{\gamma} = (t_{1,1})\lambda + (t_{1,2})\theta_0 + (t_{1,3})\theta_1 - \frac{1}{\gamma}\frac{M_W}{I_1\Omega^2} \tag{64}$$

$$a_1 = (t_{1,4})\lambda + (t_{1,5})\theta_0 + (t_{1,6})\theta_1 \tag{65}$$

$$\frac{b_1}{\gamma} = (t_{1,7})\lambda + (t_{1,8})\theta_0 + (t_{1,9})\theta_1 + (t_{1,10})\frac{1}{\gamma}\frac{M_W}{I_1\Omega^2} \tag{66}$$

TABLE 8-1

NUMERICAL VALUES OF THE COEFFICIENTS IN EQUATIONS (64), (65), AND (66)

($\gamma = 15$; $B = 0.97$)

Line	Coeff. of	$\mu = 0$	0.05	0.10	0.15	0.20	0.25	0.30	0.35	0.40	0.45	0.50
						a_0/γ in Equation (64)						
1	λ	0.15211	0.15212	0.15215	0.15224	0.15240	0.15266	0.15303	0.15353	0.15417	0.15495	0.15588
2	θ_0	0.11066	0.11096	0.11183	0.11329	0.11530	0.11786	0.12097	0.12448	0.12847	0.13286	0.13760
3	θ_1	0.08587	0.08606	0.08663	0.08758	0.08889	0.09056	0.09257	0.09492	0.09759	0.10054	0.10378
						a_1 in Equation (65)						
4	λ	0	0.1064	0.2134	0.3215	0.4315	0.5438	0.6592	0.7781	0.9011	1.0290	1.1622
5	θ_0	0	0.1377	0.2767	0.4184	0.5641	0.7154	0.8737	1.0404	1.2172	1.4055	1.6071
6	θ_1	0	0.1002	0.2013	0.3043	0.4102	0.5199	0.6343	0.7545	0.8813	1.0159	1.1589
						b_1/γ in Equation (66)						
7	λ	0	0.01045	0.02086	0.03122	0.04151	0.05174	0.06191	0.07205	0.08220	0.09241	0.10273
8	θ_0	0	0.00762	0.01534	0.02326	0.03148	0.04011	0.04926	0.05904	0.06955	0.08093	0.09329
9	θ_1	0	0.00591	0.01188	0.01798	0.02426	0.03080	0.03765	0.04490	0.05261	0.06087	0.06974
10	$\frac{1}{\gamma}\frac{M_W}{I 1 \Omega^2}$	0	−0.0686	−0.1367	−0.2038	−0.2693	−0.3328	−0.3939	−0.4522	−0.5071	−0.5585	−0.6058

TABLE 8—2

NUMERICAL VALUES OF THE COEFFICIENTS IN EQUATIONS (67) AND (68)

($B = 0.97$)

Line	Coeff. of	$\gamma = 0$	2	4	6	8	10	12	14	15	16	18	20
							a_2/μ^2 in Equation (67)						
1	λ	0	0.2136	0.4167	0.6025	0.7694	0.9195	1.0566	1.1844	1.2459	1.3061	1.4242	1.5401
2	θ_0	0	0.1981	0.3833	0.5476	0.6890	0.8100	0.9152	1.0092	1.0532	1.0958	1.1780	1.2575
3	θ_1	0	0.1504	0.2912	0.4165	0.5246	0.6176	0.6988	0.7717	0.8059	0.8391	0.9031	0.9652
							b_2/μ^2 in Equation (68)						
4	λ	0	−0.0130	−0.0488	−0.0997	−0.1573	−0.2145	−0.2675	−0.3142	−0.3351	−0.3544	−0.3884	−0.4171
5	θ_0	0	−0.0157	−0.0591	−0.1209	−0.1907	−0.2601	−0.3243	−0.3810	−0.4063	−0.4297	−0.4710	−0.5058
6	θ_1	0	−0.0117	−0.0440	−0.0901	−0.1423	−0.1938	−0.2416	−0.2838	−0.3027	−0.3201	−0.3509	−0.3768

It will be noted that the subscripts of the symbols $(t_{1,1})$, $(t_{1,2})$, etc., used in the equations were chosen to indicate the table and the line in that table where the appropriate numerical values are to be found. Thus the $(t_{1,n})$ coefficients are listed in Table 8–1. This procedure is used subsequently throughout.

The second harmonic flapping terms are given below. The coefficients of λ, θ_0 and θ_1 in the following equations are not independent of γ but are independent of μ. Values of these coefficients for specified values of γ are given in Table 8–2.

$$\frac{a_2}{\mu^2} = (t_{2,1})\lambda + (t_{2,2})\theta_0 + (t_{2,3})\theta_1 \tag{67}$$

$$\frac{b_2}{\mu^2} = (t_{2,4})\lambda + (t_{2,5})\theta_0 + (t_{2,6})\theta_1 \tag{68}$$

THRUST COEFFICIENT.

$$\frac{2C_T}{\sigma a} = (t_{3,1})\lambda + (t_{3,2})\theta_0 + (t_{3,3})\theta_1 \tag{69}$$

Numerical values of the coefficients of λ, θ_0, and θ_1 in equation (69) are listed in Table 8–3.

TABLE 8—3

NUMERICAL VALUES OF THE COEFFICIENTS IN EQUATION (69) FOR $2C_T/\sigma a$

($\gamma = 15$; $B = 0.97$)

Line	Coeff. of	$\mu = 0$	0.05	0.10	0.15	0.20	0.25	0.30	0.35	0.40	0.45	0.50
1	λ	0.4704	0.4711	0.4730	0.4762	0.4807	0.4868	0.4944	0.5038	0.5152	0.5286	0.5445
2	θ_0	0.3042	0.3054	0.3090	0.3148	0.3229	0.3333	0.3460	0.3612	0.3790	0.3996	0.4231
3	θ_1	0.2213	0.2219	0.2237	0.2267	0.2310	0.2366	0.2437	0.2523	0.2627	0.2749	0.2892

CONTROL AXIS ANGLE OF ATTACK. After the thrust coefficient C_T has been determined, the control axis angle of attack α, which is the angle between the plane perpendicular to the control axis and the flight path, can be obtained from the previously derived equation (7); namely,

$$\tan \alpha = \frac{\lambda}{\mu} + \frac{C_T}{2\mu(\mu^2 + \lambda^2)^{\frac{1}{2}}} \tag{70}$$

ACCELERATING TORQUE. In the previous sections, it was found convenient to divide the aerodynamic torque into two parts: one dependent on the components of the lift vectors of the blade elements parallel to the plane perpendicular to the control axis, and the other dependent on the components of the profile-drag vectors parallel to the same plane. In the case of the autogyro, the torque arising from the inclination of the lift vectors tends to accelerate the rotor and was therefore designated the accelerating torque. This designation will be retained even though, in the case of the helicopter, the inclination of the elemental lift vectors usually tend to decelerate the rotor. The expression for the accelerating torque coefficient is:

$$\frac{2C_{Q_a}}{\sigma a} = (t_{4,1})\lambda^2 + (t_{4,2})\lambda\theta_0 + (t_{4,3})\lambda\theta_1 + (t_{4,4})\theta_0^2 + (t_{4,5})\theta_0\theta_1 + (t_{4,6})\theta_1^2 \quad (71)$$

Numerical values of the coefficients of λ, θ_0, and θ_1 in equation (71) are listed in Table 8–4.

TABLE 8—4

NUMERICAL VALUES OF THE COEFFICIENTS IN EQUATION (71) FOR $2C_{Q_a}/\sigma a$

($\gamma = 15$; $B = 0.97$)

Line	Coeff. of	$\mu = 0$	0.05	0.10	0.15	0.20	0.25	0.30	0.35	0.40	0.45	0.50
1	λ^2	0.4704	0.4739	0.4844	0.5018	0.5265	0.5585	0.5982	0.6457	0.7016	0.7662	0.8399
2	$\lambda\theta_0$	0.3042	0.3112	0.3324	0.3685	0.4205	0.4900	0.5787	0.6891	0.8238	0.9859	1.1789
3	$\lambda\theta_1$	0.2213	0.2264	0.2419	0.2681	0.3059	0.3562	0.4205	0.5003	0.5975	0.7144	0.8536
4	θ_0^2	0	0.0023	0.0093	0.0216	0.0399	0.0656	0.1000	0.1450	0.2029	0.2762	0.3676
5	$\theta_0\theta_1$	0	0.0033	0.0136	0.0316	0.0584	0.0958	0.1460	0.2116	0.2959	0.4026	0.5357
6	θ_1^2	0	0.0012	0.0050	0.0115	0.0213	0.0350	0.0533	0.0772	0.1079	0.1467	0.1951

It might be noted that the accelerating torque of a rotor with linearly twisted blades is, for practical purposes, equivalent to that of a rotor with untwisted blades if the pitch of the blade elements at 75 per cent of the effective radius *BR* is identical in the two rotors. Considerable time can therefore be saved in calculating the accelerating torque with linearly twisted blades by first determining the blade pitch at 0.75 *BR*

TABLE 8—5

NUMERICAL VALUES OF THE COEFFICIENTS IN EQUATION (72) FOR $2C_{Qd}/\sigma$

($\gamma = 15$; $B = 0.97$)

Line	Coeff. of	$\mu = 0$	0.05	0.10	0.15	0.20	0.25	0.30	0.35	0.40	0.45	0.50
1	δ_0	0.2500	0.2506	0.2525	0.2556	0.2600	0.2655	0.2722	0.2802	0.2892	0.2993	0.3105
2	$\delta_1\lambda$	0.3333	0.3333	0.3333	0.3333	0.3333	0.3332	0.3330	0.3327	0.3323	0.3316	0.3307
3	$\delta_1\theta_0$	0.2500	0.2506	0.2525	0.2556	0.2599	0.2654	0.2721	0.2799	0.2887	0.2985	0.3093
4	$\delta_1\theta_1$	0.2000	0.2004	0.2017	0.2037	0.2066	0.2103	0.2147	0.2198	0.2257	0.2322	0.2393
5	$\delta_2\lambda^2$	0.5000	0.5038	0.5153	0.5345	0.5607	0.5971	0.6411	0.6940	0.7563	0.8287	0.9116
6	$\delta_2\lambda\theta_0$	0.6667	0.6743	0.6976	0.7373	0.7946	0.8713	0.9694	1.0918	1.2413	1.4217	1.6367
7	$\delta_2\lambda\theta_1$	0.5000	0.5056	0.5225	0.5513	0.5927	0.6480	0.7187	0.8066	0.9139	1.0432	1.1974
8	$\delta_2\theta_0^2$	0.2500	0.2532	0.2630	0.2798	0.3047	0.3389	0.3839	0.4419	0.5150	0.6061	0.7181
9	$\delta_2\theta_0\theta_1$	0.4000	0.4046	0.4186	0.4429	0.4787	0.5280	0.5930	0.6768	0.7827	0.9147	1.0773
10	$\delta_2\theta_1^2$	0.1667	0.1684	0.1735	0.1824	0.1955	0.2135	0.2372	0.2676	0.3060	0.3537	0.4124

TABLE 8—6

NUMERICAL VALUES OF THE COEFFICIENTS IN EQUATION (73) FOR $\mu \frac{2C_T}{\sigma a}\left(\frac{D}{L}\right)_0$

($\gamma = 15$; $B = 0.97$)

Line	Coeff. of	$\mu = 0$	0.05	0.10	0.15	0.20	0.25	0.30	0.35	0.40	0.45	0.50
1	δ_0/a	0.2500	0.2519	0.2575	0.2669	0.2802	0.2973	0.3183	0.3433	0.3724	0.4057	0.4434
2	$\delta_1\lambda/a$	0.3333	0.3337	0.3348	0.3367	0.3394	0.3430	0.3475	0.3532	0.3601	0.3683	0.3781
3	$\delta_1\theta_0/a$	0.2500	0.2507	0.2529	0.2566	0.2618	0.2685	0.2767	0.2866	0.2982	0.3115	0.3266
4	$\delta_1\theta_1/a$	0.2000	0.2004	0.2017	0.2038	0.2067	0.2106	0.2153	0.2210	0.2277	0.2353	0.2441
5	$\delta_2\lambda^2/a$	0.5000	0.5024	0.5097	0.5220	0.5395	0.5626	0.5918	0.6276	0.6707	0.7217	0.7816
6	$\delta_2\lambda\theta_0/a$	0.6667	0.6716	0.6863	0.7111	0.7467	0.7940	0.8545	0.9298	1.0219	1.1333	1.2666
7	$\delta_2\lambda\theta_1/a$	0.5000	0.5030	0.5121	0.5277	0.5502	0.5804	0.6194	0.6684	0.7286	0.8019	0.8900
8	$\delta_2\theta^2/a$	0.2500	0.2522	0.2587	0.2700	0.2867	0.3095	0.3395	0.3779	0.4263	0.4864	0.5602
9	$\delta_2\theta_0\theta_1/a$	0.4000	0.4029	0.4116	0.4266	0.4487	0.4788	0.5182	0.5685	0.6316	0.7096	0.8051
10	$\delta_2\theta_1{}^2/a$	0.1667	0.1677	0.1709	0.1765	0.1845	0.1955	0.2097	0.2278	0.2502	0.2779	0.3114

and then calculating the torque of a rotor with untwisted blades having this pitch.

DECELERATING TORQUE. The expression for the decelerating torque coefficient attributable to profile drag is

$$\frac{2C_{Q_d}}{\sigma} = \delta_0(t_{5,1}) + \delta_1[(t_{5,2})\lambda + (t_{5,3})\theta_0 + (t_{5,4})\theta_1] + \delta_2[(t_{5,5})\lambda^2 + (t_{5,6})\lambda\theta_0 + (t_{5,7})\lambda\theta_1 + (t_{5,8})\theta_0^{\,2} + (t_{5,9})\theta_0\theta_1 + (t_{5,10})\theta_1^{\,2}] \tag{72}$$

Numerical values of the coefficients in the terms of equation (72) are given in Table 8–5.

PROFILE DRAG-LIFT RATIO. The rotor profile drag-lift ratio is expressed as

$$\mu\frac{2C_T}{\sigma a}\left(\frac{D}{L}\right)_0 = \frac{\delta_0}{a}(t_{6,1}) + \frac{\delta_1}{a}[(t_{6,2})\lambda + (t_{6,3})\theta_0 + (t_{6,4})\theta_1] + \frac{\delta_2}{a}[(t_{6,5})\lambda^2 + (t_{6,6})\lambda\theta_0 + (t_{6,7})\lambda\theta_1 + (t_{6,8})\theta_0^{\,2} + (t_{6,9})\theta_0\theta_1 + (t_{6,10})\theta_1^{\,2}] \tag{73}$$

Numerical values of the t coefficients in equation (73) are given in Table 8–6.

In addition to Table 8–6, the following formula [derived from equation (69)] and Table 8–7 are helpful in evaluating $(D/L)_0$ from equation (73).

$$\mu\frac{2C_T}{\sigma a} = (t_{7,1})\lambda + (t_{7,2})\theta_0 + (t_{7,3})\theta_1 \tag{74}$$

TABLE 8—7

NUMERICAL VALUES OF THE COEFFICIENTS IN EQUATION (74) FOR $\mu\dfrac{2C_T}{\sigma a}$

($\gamma = 15$; $B = 0.97$)

Line	Coeff. of	$\mu = 0$	0.05	0.10	0.15	0.20	0.25	0.30	0.35	0.40	0.45	0.50
1	λ	0	0.02355	0.04700	0.07141	0.09609	0.12152	0.14789	0.17538	0.20418	0.23448	0.26648
2	θ_0	0	0.01527	0.03089	0.04720	0.06450	0.08308	0.10322	0.12515	0.14911	0.17530	0.20390
3	θ_1	0	0.01110	0.02237	0.03399	0.04615	0.05901	0.07275	0.08755	0.10358	0.12103	0.14006

SUMMARY REMARKS. It should be emphasized that although the numerical coefficients listed in the preceding tables were computed with $\gamma = 15$, negligible errors will result for any other value of γ between 0 and 25, except, of course, for the expressions for a_0, b_1, a_2, and b_2. Also, it should be remembered that the analysis is strictly valid only for constant-chord blades having linear or no twist.

The equations and tables presented above are a convenient means of rapidly and accurately calculating the blade motion, control-axis angle of attack, thrust and power characteristics of a rotor in forward flight. The application of all the equations and the problems of calculating rotor characteristics is known as the "performance problem" and is fully discussed in the next chapter. The details of using the equations and tables, however, will be illustrated by the following problem (from reference II–18 of Appendix IIA).

PROBLEM. Calculate the performance characteristics of a rotor operating in the autorotative state at a tip-speed ratio of 0.35 and at a pitch angle of 4°, and having untwisted ($\theta_1 = 0$) blades that can be represented by the following profile drag equation:

$$c_{d_0} = 0.0087 - 0.0216\alpha_r + 0.400\alpha_r^2$$

SOLUTION. Inasmuch as the rotor condition is completely represented by μ, θ_0, θ_1, and λ, of which θ_0, θ_1, and μ are already known, the first step is to calculate λ. The magnitude of λ is determined by the resultant torque on the rotor and can, therefore, be calculated by means of the torque equations (71) and (72). In the present example, the resultant torque is zero, for the rotor is in the autorotative condition. Hence,

$$\frac{2C_{Q_a}}{\sigma} = \frac{2C_{Q_d}}{\sigma}$$

At $\mu = 0.35$, from equations (71) and (72) and Tables 8–4 and 8–5, the following relationship exists:

$$\begin{aligned} a(0.646\lambda^2 + 0.689\lambda\theta_0 + 0.500\lambda\theta_1 + 0.145\theta_0^2 + 0.212\theta_0\theta_1 + 0.077\theta_1^2) \\ = 0.280\delta_0 + \delta_1(0.333\lambda + 0.280\theta_0 + 0.220\theta_1) + \delta_2(0.694\lambda^2 \\ + 1.092\lambda\theta_0 + 0.807\theta_1\lambda + 0.443\theta_0^2 + 0.676\theta_0\theta_1 + 0.268\theta_1^2) \end{aligned}$$

Substitution of the appropriate values of δ_0, δ_1, δ_2, θ_0, θ_1, and a into the preceding equation gives a quadratic equation with λ as the unknown variable. Solution of this quadratic gives two values of λ. The smaller (algebraic) value corresponds to operation at a negative angle of attack and can be ignored. The larger value, which corresponds to operation at a positive angle of attack, is $\lambda = -0.0050$.

With λ, θ_0, and θ_1 known, the flapping angles can be calculated by means of Tables 8–1 and 8–2. Thus

$$a_0 = 15[(0.1536)(-0.0050) + (0.1244)(0.0698)] - \frac{M_W}{I_1\Omega^2}$$
$$= 0.1187 - \frac{M_W}{I_1\Omega^2}$$

Using a typical value of $\frac{M_W}{I_1\Omega^2} = 0.006$, the first harmonic flapping coefficients are computed to be

$$a_0 = 0.1127 \text{ radians}$$
$$a_1 = (0.777)(-0.0050) + (1.041)(0.0698)$$
$$= 0.0687 \text{ radians}$$
$$b_1 = 15[(0.0721)(-0.0050) + (0.0591)(0.0698)] - (0.452)(0.006)$$
$$= 0.0536 \text{ radians}$$

In a similar manner, the second harmonic flapping coefficients are found to be

$$a_2 = 0.0082 \text{ radians}$$
$$b_2 = -0.0033 \text{ radians}$$

It will be noted that a value of $\gamma = 15$ was used for this example in applying Table 8–2 to the calculations.

The thrust coefficient is obtained from equation (69) and Table 8–3.

$$\frac{2C_T}{\sigma a} = (0.504)(-0.0050) + (0.361)(0.0698)$$
$$= 0.0227$$

The profile drag-lift ratio of the rotor is obtained by using equation (73), Table 8–6, and the known value of $\mu(2C_T/\sigma a)$ as obtained from Table 8–7. (The values of $\mu(2C_T/\sigma a)$ in Table 8–7 are consistent with assumptions regarding omission of powers of μ greater than μ^4. Values differ slightly from those obtained by taking $\mu \times (2C_T/\sigma a)$ from Table 8–3.)

$$\left(\frac{D}{L}\right)_o [(0.1753)(-0.0050) + (0.1252)(0.0698)]$$

$$= \left(\frac{0.0087}{5.73}\right)(0.343) + \left(\frac{-0.0216}{5.73}\right)[(0.353)(-0.0050) + (0.286)(0.0698)] + \left(\frac{0.400}{5.73}\right)[(0.627)(-0.0050)^2 + (0.930)(-0.0050)(0.0698) + (0.378)(0.0698)^2]$$

which gives

$$\left(\frac{D}{L}\right)_o = 0.0711$$

In order to find the complete rotor drag-lift ratio $\left(\frac{D}{L}\right)_r$, the induced contribution must be determined. From equation (62), it can be seen that

$$\left(\frac{D}{L}\right)_i = \frac{C_T}{2\mu(\mu^2 + \lambda^2)^{1/2}} \tag{75}$$

Now obtaining $2C_T/\sigma a$ from equation (69) and Table 8–3, and taking $\sigma = 0.060$, C_T is calculated as

$$C_T = 0.0227 \times \frac{0.060 \times 5.73}{2} = 0.00390$$

and

$$\left(\frac{D}{L}\right)_i = \frac{0.0039}{2(0.35)[(0.35)^2 + (-0.0050)^2]^{1/2}} = 0.0159$$

Hence,

$$\left(\frac{D}{L}\right)_r = \left(\frac{D}{L}\right)_o + \left(\frac{D}{L}\right)_i = 0.0711 + 0.0159 = 0.0870$$

Limits of Validity of Theory

Another contribution to rotor theory contained in *NACA Rep. 716* was the presentation of a convenient means of estimating the limiting conditions of operation for an autorotating rotor beyond which the theory is invalid because of blade stall. The method, which was later extended to include power-on flight, will be included in the discussion of stall limits in Chapter 10. It is considered desirable, however, to make the following remarks at this time.

The series used to approximate the profile-drag coefficient of the blade elements begins to underestimate seriously the drag coefficient at angles of attack near and beyond the stall. It is impossible, moreover, to limit the application of the theory to flight conditions in which the stall angle is never exceeded by any blade element. For moderate values of thrust and tip-speed ratio, however, these high values of the angle of attack are either confined to parts of the rotor disk in which the square of the velocity of the air relative to the blade element is quite low or to very small areas. Under such conditions the total contribution of these blade elements to the rotor thrust, torque, and flapping is very small, and the error in their estimation is negligible.

As the thrust coefficient or tip-speed ratio of the rotor is increased, the high angle-of-attack region spreads and the accuracy of the theory is correspondingly reduced. For the power-on flight condition, high angles of attack first appear near the tip of the retreating blades, whereas high angles first appear near the root end of the retreating blades when the rotor is autorotating. The difference in the two power conditions is due to the fact that the resultant flow is down through the rotor disk in power-on flight and up in autorotative flight.

Satisfactory limits to the use of the theory and approximately to the practical operating condition of the rotor in powered flight are the conditions at which the tip of the retreating blade reaches its stalling angle of attack. For the autorotative case, limits to the theory would consist of the conditions at which the velocity of the blade elements of the stalled inboard sections reached high enough values so that the contributions of these elements to the total thrust and torque of the rotor become significant.

Once a condition of operation is reached where significant stall is imminent, further increases in speed or blade loading result in a very rapid spreading of the stall over the rotor disk. Because of the very high profile-drag losses associated with rotor stall, optimum rotor performance is reached at the flight condition at which tip stall first occurs. Accordingly, it may be stated that the conditions for thrust coefficient and tip-speed ratio for optimum performance are only slightly different from the limits of accuracy of the theory. The conditions for optimum performance will be discussed in detail in Chapter 10.

9

FORWARD-FLIGHT PERFORMANCE

The problem of computing helicopter performance in forward flight is complicated by the many variables involved, and by the length and complication of the equations that define the rotor characteristics. For these reasons, any "exact" performance method (i.e., a method that uses the most refined theory available, or the minimum number of approximations) necessarily involves the use of tables and charts in order to facilitate the work. The performance charts may be based upon a method that utilizes the fact that (1) a power balance exists for a helicopter in steady flight (i.e., the power expended at the main rotor shaft must equal the sum of all the power losses expended by the rotor and the fuselage), or (2) the resultant force on the helicopter in steady flight must be equal to zero. Both methods should yield the same performance if they are both calculated from the same group of rotor equations. Method (1) is known as an *energy* method, although the name has also been applied to simplified methods that employ approximations to the rotor losses; whereas method (2) might be called a *balance of force* method.

The energy method of performance calculation that will be described in this chapter was developed and used extensively by the *NACA*. The method is among the most accurate available and involves less time than many so-called "rapid" or "simplified" methods of performance calculations which yield only approximate results.

Basic Performance Equation

Helicopter performance analyses, like those applied to the airplane, are based essentially on curves of power required against airspeed for various flight conditions. In order to obtain curves of this type, it is necessary to consider the various sources of power expended by a helicopter in steady flight. These power-absorbing elements are:

(1) Rotor

(a) Induced power loss

(b) Blade profile-drag loss

(2) Parasite drag of the fuselage, rotor hub, and tail rotor (if any)

(3) The power necessary to change the potential energy of the helicopter at a given rate of speed in the climb or glide condition.

The total power required is equal to the sum of all these items.

These power expenditures may be expressed by the following relationship

$$HP_{\text{total}} = HP_o + HP_i + HP_p + HP_c \tag{1}$$

Each individual power loss may be expressed as the energy dissipated per second by an equivalent drag force moving at the translational velocity of the aircraft. Thus, if P represents the *total equivalent drag force* (not power) and D_o, D_i, D_p, and D_c the equivalent drag forces corresponding to each of the sources of power expenditure, then

$$\left.\begin{aligned} D_oV &= HP_o \\ D_iV &= HP_i \\ D_pV &= HP_p \\ D_cV &= HP_c \\ PV &= HP_{\text{total}} \end{aligned}\right\} \tag{2}$$

If equation (2) is substituted into equation (1), then

$$P = D_o + D_i + D_p + D_c \tag{3}$$

Equation (3) may be made nondimensional by dividing through by the rotor lift L. Thus,

$$\frac{P}{L} = \left(\frac{D}{L}\right)_o + \left(\frac{D}{L}\right)_i + \left(\frac{D}{L}\right)_p + \left(\frac{D}{L}\right)_c \tag{4}$$

The first two terms of the right side of equation (4) are sometimes combined by the relation

$$\left(\frac{D}{L}\right)_r = \left(\frac{D}{L}\right)_o + \left(\frac{D}{L}\right)_i \tag{5}$$

wherein $(D/L)_r$ is termed the rotor drag-lift ratio.

The symbol P/L represents the total rotor-shaft power input and is analogous to the drag-lift ratio of an airplane. It is proportional to the drag that would absorb the same power at the velocity along the flight path as the power being supplied through the rotor shaft. Thus, P/L may be defined as:

$$\frac{P}{L} = \frac{\text{shaft power}}{VL} = \frac{Q\Omega}{VL} \tag{6}$$

In a similar manner, $(D/L)_o$, $(D/L)_i$, $(D/L)_p$, and $(D/L)_c$ represent the power required to overcome the rotor profile drag, the rotor induced drag, the parasite drag of the helicopter without the rotor, and the climb power. Thus, in order to compute the power required by a helicopter to fly along a flight path at a given speed, it is only necessary to compute the individual drag-lift ratios in equation (4), add them to find P/L, and then determine the shaft power required from equation (6). The next step, therefore, will be to evaluate each power-absorbing item as expressed by its drag-lift ratio.

Calculation of Drag-Lift Ratios

INDUCED DRAG-LIFT RATIO. In equation (75) of the preceding chapter, the induced drag-lift ratio was written as

$$\left(\frac{D}{L}\right)_i = \frac{C_T}{2\mu(\mu^2 + \lambda^2)^{\frac{1}{2}}} \tag{7}$$

With the assumption that $L = T \cos \alpha$, equation (7) becomes

$$\left(\frac{D}{L}\right)_i = \frac{C_L}{4}\left[\frac{\mu}{\cos^3 \alpha(\mu^2 + \lambda^2)^{\frac{1}{2}}}\right] \tag{8}$$

At tip-speed ratios above 0.10 (approximately), the bracketed expression in equation (8) may be considered equal to unity, thereby allowing the induced drag-lift ratio to be calculated from the rotor lift coefficient alone. The reasoning behind the limitation in the use of the simpler expression may be understood from the following.

Consider the rotor of a helicopter as a fixed airplane wing. Then, if it is assumed that the downwash at the rotor is distributed uniformly across the rotor span, according to simple wing theory the amount of air influenced by the rotor per second may be considered equal to a circle of radius R multiplied by the velocity of flight. The lift or thrust developed by the rotor is then, from momentum considerations

$$L = \pi R^2 \rho V(2v) \tag{9}$$

(Note that equation (9) is equal to equation (3) of Chapter 8 if it is assumed that the rotor angle of attack α in Fig. 8–4 is zero and that the contribution of the induced flow v to the total flow through the rotor is negligible.)

Inasmuch as the induced drag D_i is equal to the component of the lift force on the velocity V, it follows that

$$\frac{D_i}{L} = \left(\frac{D}{L}\right)_i = \frac{v}{V} \tag{10}$$

Combining equations (9) and (10),

$$\left(\frac{D}{L}\right)_i = \frac{L}{2\pi R^2 \rho V^2}$$

or

$$\left(\frac{D}{L}\right)_i = \frac{C_L}{4} \tag{11}$$

The induced drag-lift ratio can be simply calculated, therefore, as $C_L/4$ at all speeds except near hovering or at large rotor angles of attack. At large angles or at very low speeds (between zero and approximately 30 miles per hour) wherein the induced velocity is large compared with the forward-flight speed, $(D/L)_i$ should be calculated from equation (7) or (8).

PARASITE DRAG-LIFT RATIO. The parasite drag force of the fuselage, rotor hub, and all of the nonlifting components of the helicopter may be expressed as

$$D_p = C_{D_p} \tfrac{1}{2} \rho V^2 \pi R^2 \tag{12}$$

where C_{D_p} is the parasite drag coefficient. It is usually more convenient to combine the drag coefficient and the area term in equation (12) into a single parameter called the equivalent flat-plate drag area. If such a parameter is denoted by the symbol f, then

$$D_p = f\tfrac{1}{2}\rho V^2 \tag{13}$$

Dividing equation (13) by rotor lift, the parasite drag-lift ratio is obtained as

$$\left(\frac{D}{L}\right)_p = \frac{f\frac{1}{2}\rho V^2}{L} = \frac{1}{C_L}\frac{f}{\pi R^2} \tag{14}$$

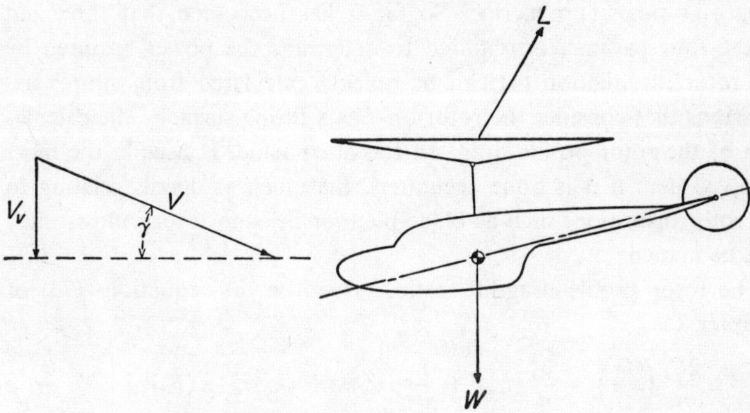

Fig. 9–1 Helicopter in climb.

CLIMB DRAG-LIFT RATIO. Consider a helicopter climbing at an angle γ as in Fig. 9–1. If the vertical rate of climb of the helicopter is V_v, then the power that must be supplied by the engine to the rotor to maintain this rate of climb is equal to WV_v. The equivalent drag force that would absorb this power at the helicopter climb speed V can therefore be implicitly expressed as

$$D_c V = WV_v$$

or

$$D_c = W\frac{V_v}{V} \tag{15}$$

Replacing W by $L/\cos\gamma$, and V_v/V by $\sin\gamma$, equation (15) becomes

$$\left(\frac{D}{L}\right)_c = \tan\gamma \tag{16}$$

For small angles of climb, the climb drag-lift ratio may be simply taken as equal to the rate of climb divided by the velocity along the flight path, so that for most calculations

$$\left(\frac{D}{L}\right)_c = \frac{V_v}{V} \tag{17}$$

When the helicopter is descending, gravity reduces the amount of power required from the engine to maintain flight and, therefore, $(D/L)_c$ is negative in glides.

PROFILE DRAG-LIFT RATIO. So far it has been seen that three out of the four parameters required to determine the power required by the rotor in equation (4) can be quickly calculated from simple expressions that consider the rotor only as a lifting surface. The calculation of the rotor profile drag, on the other hand, is usually the most involved item if it is done accurately, inasmuch as details relating to the rotor operation, such as blade-pitch angles and rotor inflow, must first be known.

The rotor profile drag-lift ratio, as given by equation (73) of Chapter 8 is

$$\begin{aligned}\mu \frac{2C_T}{\sigma a}\left(\frac{D}{L}\right)_o = \frac{\delta_0}{a}(t_{6,1}) &+ \frac{\delta_1}{a}[(t_{6,2})\lambda + (t_{6,3})\theta_0 + (t_{6,4})\theta_1] \\ &+ \frac{\delta_2}{a}[(t_{6,5})\lambda^2 + (t_{6,6})\lambda\theta_0 + (t_{6,7})\lambda\theta_1 \\ &+ (t_{6,8})\theta_0^2 + (t_{6,9})\theta_0\theta_1 + (t_{6,10})\theta_1^2]\end{aligned} \tag{18}$$

wherein the profile drag of the rotor-blade elements is expressed by the following equation:

$$c_{d_o} = \delta_0 + \delta_1\alpha_r + \delta_2\alpha_r^2 \tag{19}$$

For a given machine, flying at a known speed and altitude, all of the quantities in equation (18) are known except λ and θ. In order to determine λ and θ, use is made of the thrust and torque equations developed in the preceding chapter. It can be seen from equations (69), (71), and (72) of that chapter that C_T and C_Q are functions of λ, θ, and μ, or expressed symbolically

$$C_T = f(\lambda, \theta, \mu)$$
$$C_Q = F(\lambda, \theta, \mu)$$

In order to solve the above for λ and θ, μ, C_T and C_Q must be known.

Although μ and C_T are known quantities, C_Q, which represents the total torque or power absorbed at the rotor shaft, is not known until the profile-drag contribution is determined. Thus, in order to calculate $(D/L)_o$, various values of C_Q must be assumed until an assumed torque value is equal to the sum of the individual contributions to the total torque. This procedure may be summarized as follows:

(1) Assume a value of C_Q and solve for λ and θ from the thrust and torque equations.

(2) Using the values of λ and θ calculated from step (1), calculate $(D/L)_o$ from equation (18).

(3) Calculate P/L by means of equation (4).

(4) Convert P/L to Q and thence to C_Q by means of equation (6) and the definition of C_Q.

(5) Compare the assumed value of C_Q with that calculated from step (4). If a difference exists, repeat steps (1) through (4) until the assumed value of C_Q is equal to the calculated value.

It should be understood that the necessity for a trial-and-error process is not the result of using one particular method of calculation but must be carried on, in one form or another, with any method. In calculating $(D/L)_o$, however, laborious computations and the use of lengthy equations can be eliminated through the use of charts presented in reference II–16 (Appendix IIA), which will now be discussed.

Profile Drag-Lift Ratio Charts

METHOD OF CALCULATION. The charts shown in Figs. 9–2 and 9–3 give rotor profile drag-lift ratios for a helicopter rotor operating in forward flight and having hinged, rectangular, untwisted blades. Charts are given for a range of power input (i.e., various P/L values) covering glides, level flight, and moderate rates of climb. Each chart of Fig. 9–2 (from reference II–16 of Appendix IIA) expresses the relation between the lift and the profile-drag characteristics of the rotor for various combinations of pitch angle, tip-speed ratio, and solidity for a particular value of shaft power input as represented by the parameter P/L. The charts of Fig. 9–2 are extended to $\mu = 0.10$ in Fig. 9–3, which gives curves of $(D/L)_o$ for all P/L values at $\mu = 0.10$.

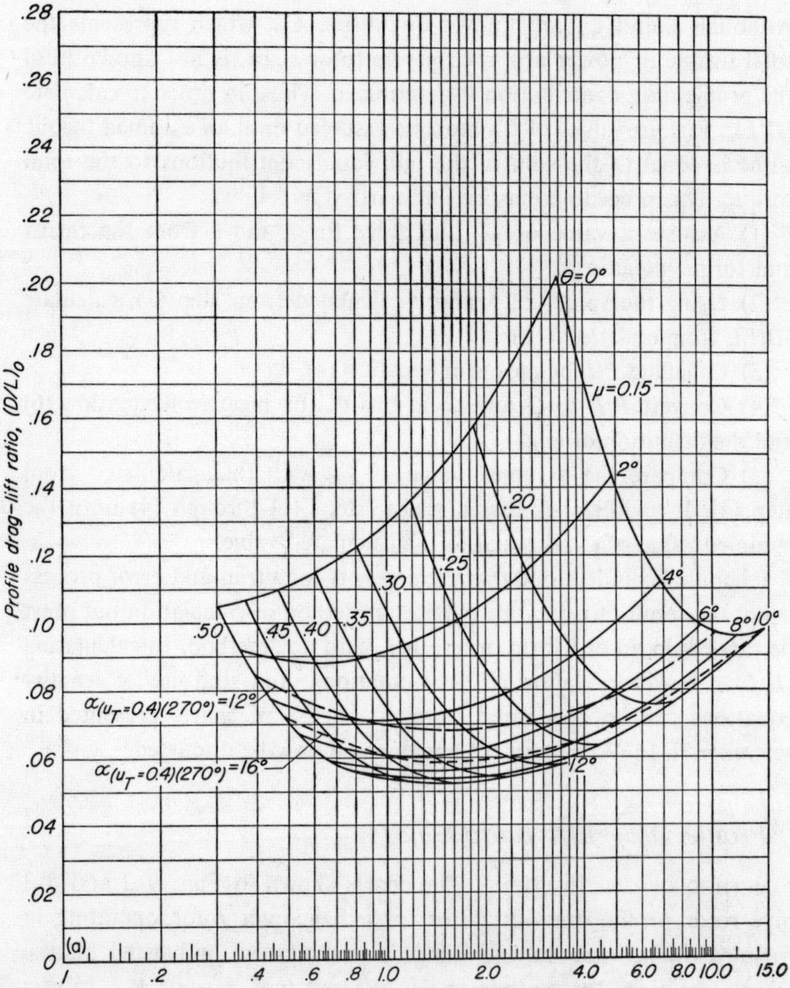

Fig. 9–2 Performance charts for computation of profile drag-lift ratios. (a) $P/L = 0$

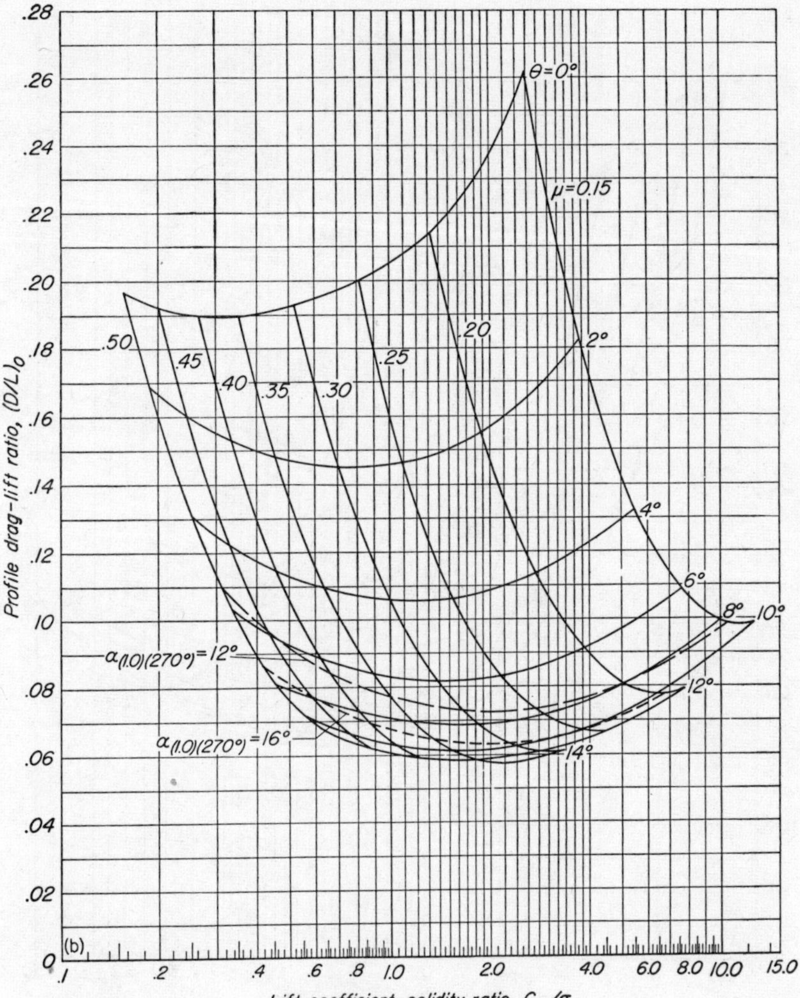
θ=0°
μ=0.15
.20
.25
.30
.35
.40
.45
.50
2°
4°
6°
8°
10°
12°
14°
α(1.0)(270°)=12°
α(1.0)(270°)=16°
(b)
Profile drag-lift ratio, (D/L)o
Lift coefficient-solidity ratio, CL/σ

(b) $P/L = 0.1$

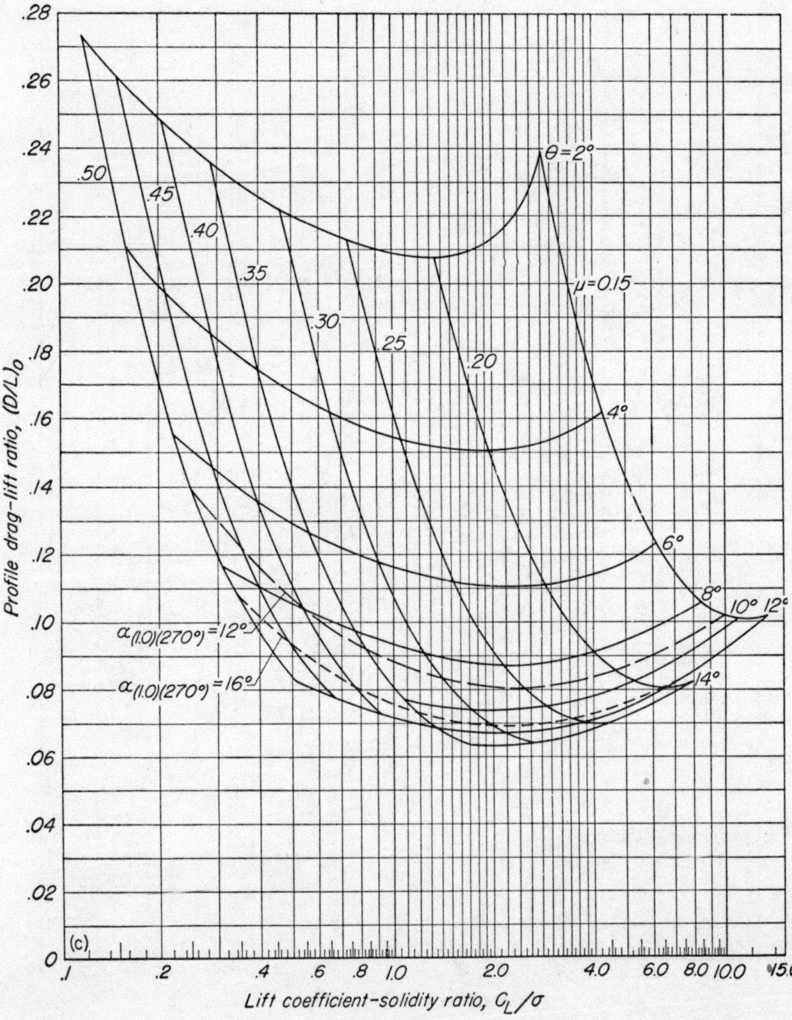

.28
.26
.24
.22
.20
.18
.16
.14
.12
.10
.08
.06
.04
.02
0
.50
.45
.40
.35
.30
.25
.20
μ=0.15
θ=2°
4°
6°
8°
10°
12°
14°
$\alpha_{(1.0)(270°)}$=12°
$\alpha_{(1.0)(270°)}$=16°
(c)
.1
.2
.4
.6
.8
1.0
2.0
4.0
6.0
8.0
10.0
15.0
Profile drag-lift ratio, $(D/L)_o$
Lift coefficient-solidity ratio, C_L/σ

(c) $P/L = 0.2$

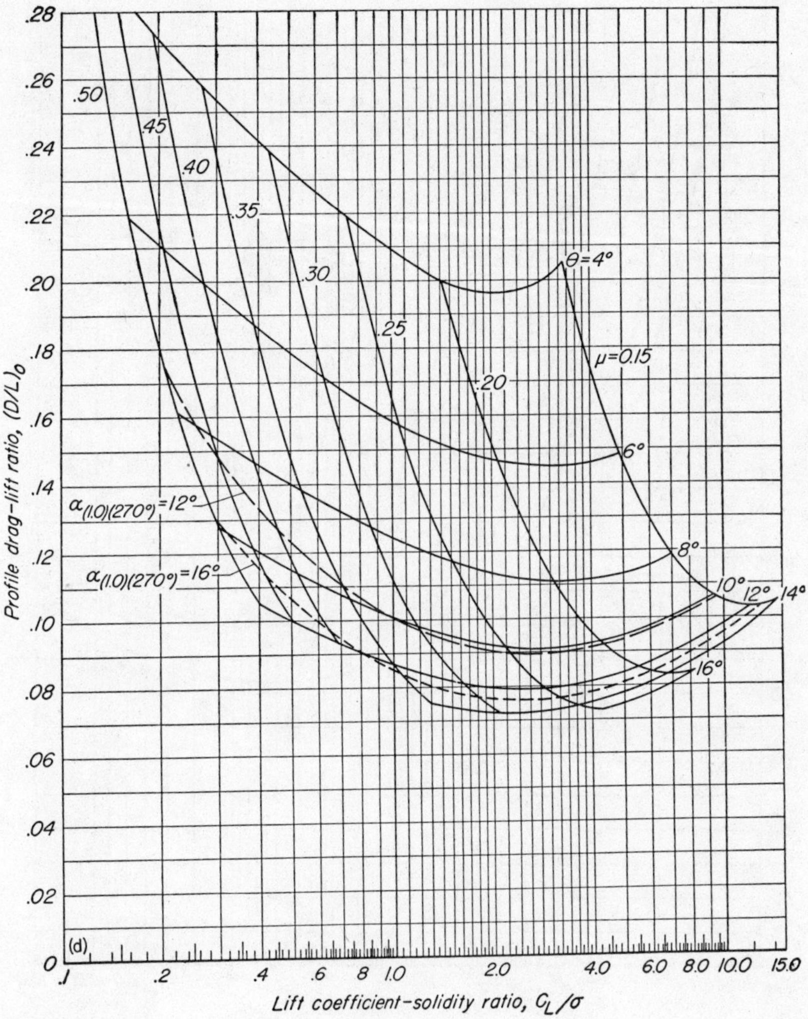

Profile drag-lift ratio, $(D/L)_0$
Lift coefficient-solidity ratio, C_L/σ
$\theta=4°$
6°
8°
10°
12°
14°
16°
$\mu=0.15$
.20
.25
.30
.35
.40
.45
.50
$\alpha_{(1.0)(270°)}=12°$
$\alpha_{(1.0)(270°)}=16°$
(d)

(d) $P/L = 0.3$

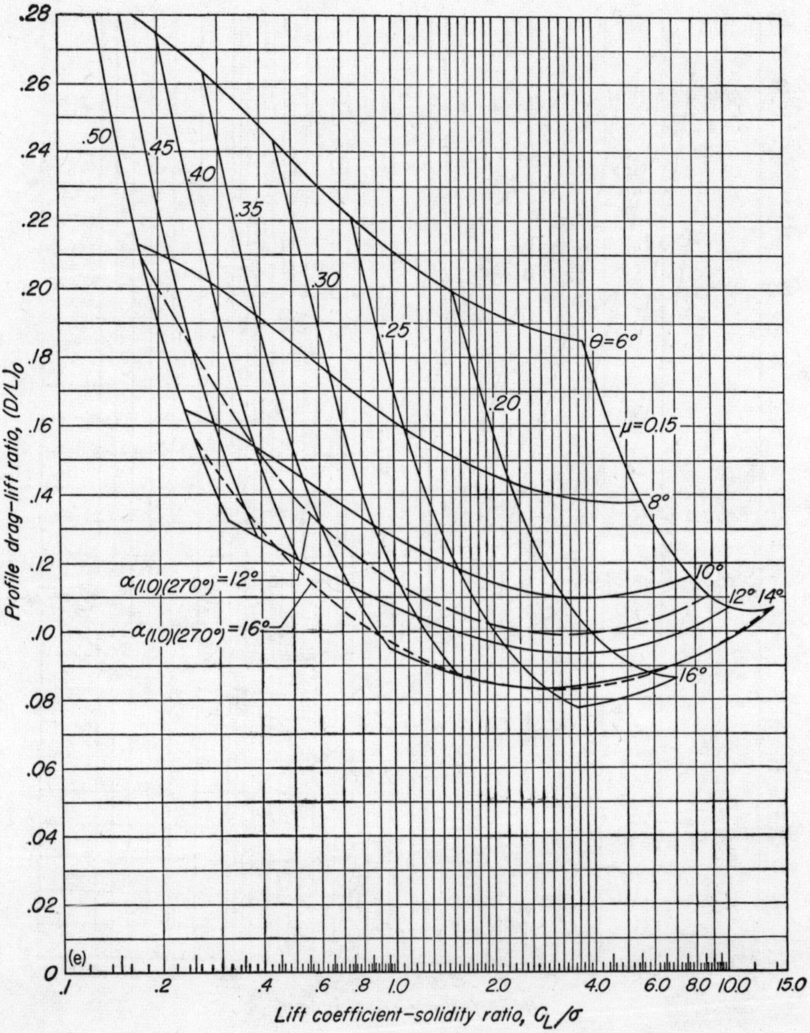

Profile drag-lift ratio, $(D/L)_0$
Lift coefficient-solidity ratio, C_L/σ
.50
.45
.40
.35
.30
.25
.20
$\theta=6°$
$\mu=0.15$
8°
10°
12°
14°
16°
$\alpha_{(1.0)(270°)}=12°$
$\alpha_{(1.0)(270°)}=16°$
(e)

(e) $P/L = 0.4$

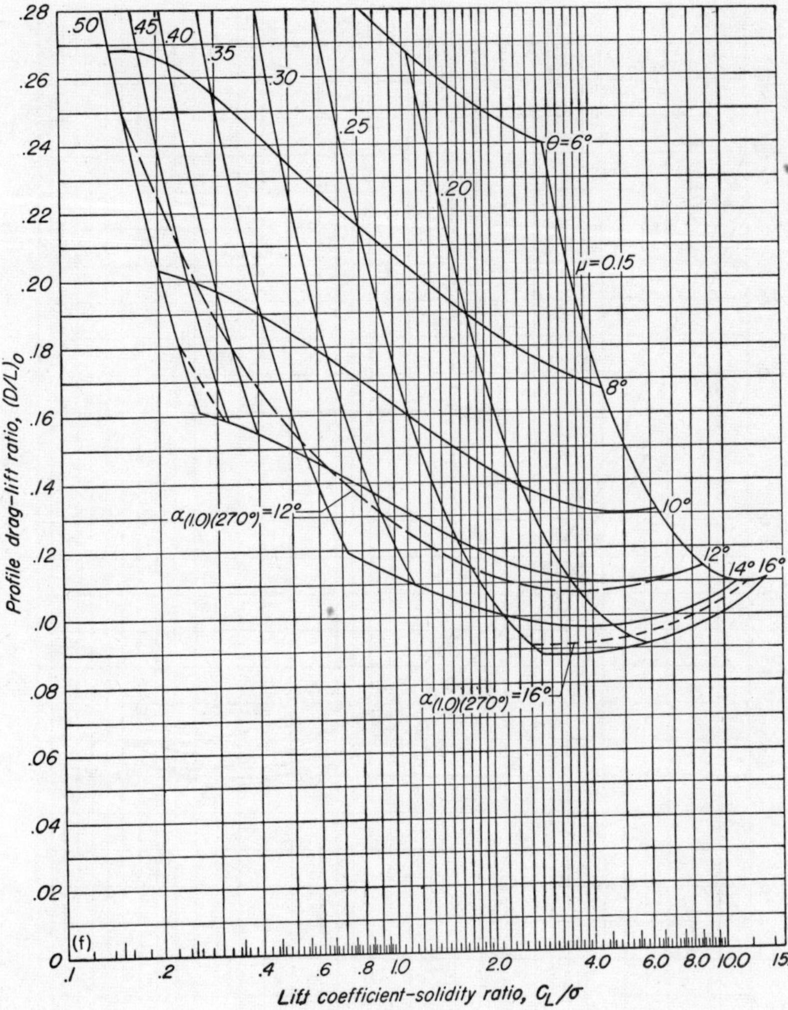
.28
.26
.24
.22
.20
.18
.16
.14
.12
.10
.08
.06
.04
.02
0
.50
.45
.40
.35
.30
.25
.20
θ=6°
μ=0.15
8°
10°
12°
14°
16°
$\alpha_{(1.0)(270°)}$=12°
$\alpha_{(1.0)(270°)}$=16°
(f)
.1
.2
.4
.6
.8
1.0
2.0
4.0
6.0
8.0
10.0
15.0
Profile drag-lift ratio, $(D/L)_0$
Lift coefficient-solidity ratio, C_L/σ

(f) $P/L = 0.5$

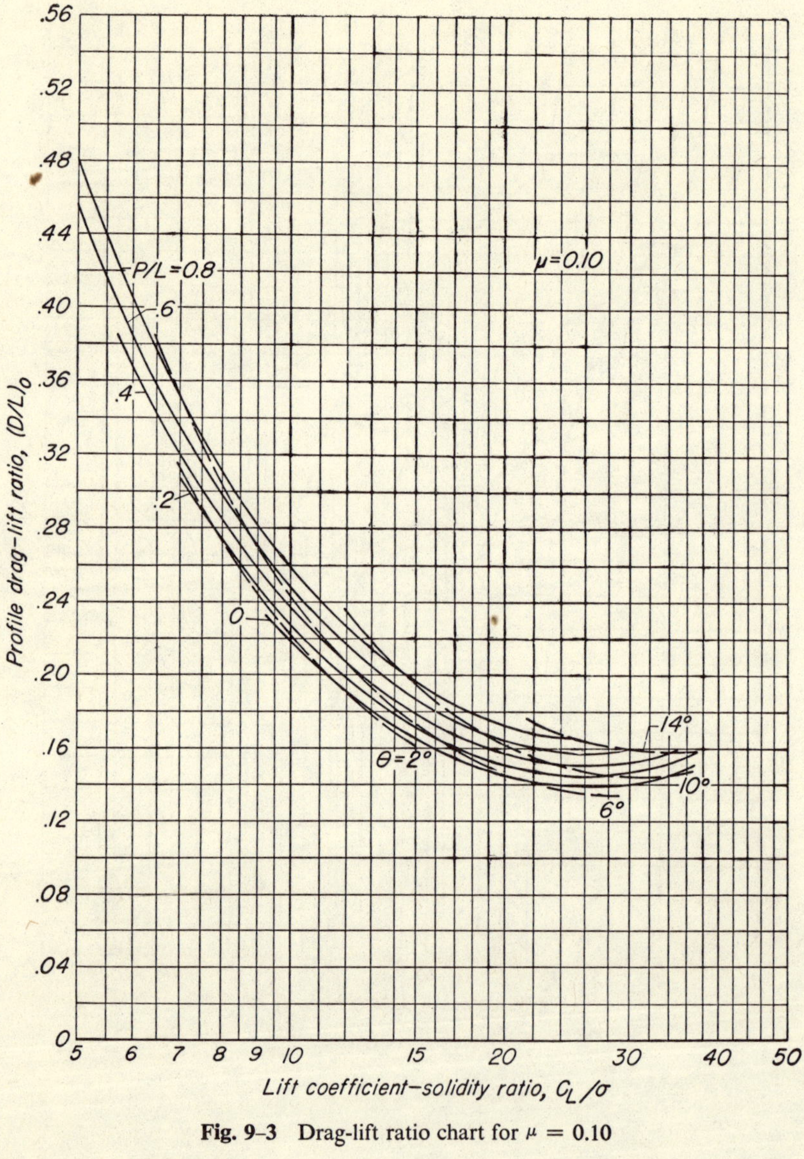

Fig. 9–3 Drag-lift ratio chart for $\mu = 0.10$

The profile drag-lift ratio charts were constructed by using the following relationships as well as the equations and tables given in the preceding chapter.

From equation (6) and assuming that $L = T \cos \alpha$:

$$\frac{P}{L} = \frac{C_Q}{C_T \mu} \qquad (20)$$

Also, assuming that $\cos \alpha = 1$,

$$W = C_L \tfrac{1}{2} \rho V^2 \pi R^2 = C_T \pi R^2 \rho (\Omega R)^2$$

or

$$\frac{C_L}{\sigma} = \frac{2}{\mu^2} \frac{C_T}{\sigma} \qquad (21)$$

The calculations were made by fixing a value of P/L for each chart, and then varying C_L/σ for a range of μ's. For each value of C_L/σ and μ, corresponding values of $2C_T/\sigma a$ and $2C_Q/\sigma$ were calculated from equations (20) and (21). With thrust, torque, and tip-speed ratio known, the pitch angle θ and the inflow ratio λ were then determined. With the three fundamental variables, λ, θ, and μ known, the profile drag-lift ratio corresponding to the fixed conditions was then calculated by means of equation (18). Because a knowledge of the pitch angle is useful, the values of θ that were determined in the course of the calculations were cross-plotted on the charts.

RANGE OF APPLICATION. Inasmuch as the charts are based on the theory of reference II–18 (Appendix IIA), they are subject to the same limitations arising from the assumptions used in developing that theory. The range of usefulness of the charts as limited by the most pertinent assumptions will now be pointed out.

The sample rotor for which the charts were constructed has a blade mass factor $\gamma = 15$, but the charts are considered applicable to rotors having values of γ ranging from 0 to 25. This range includes any conventional rotor.

Although the charts were derived for rotors having rectangular blades, rotor performance as predicted by the charts agreed very well with data obtained from flight tests of blades having as much as 3 : 1 taper ratio, provided that the rotor solidity was calculated with an equivalent weighted chord as described in Chapter 4.

Although the rotor theory derived in reference II–18 (Appendix IIA) was set up to include blades having any amount of linear twist, the charts of Fig. 9–2 and Fig. 9–3 were constructed for blades having zero twist. Strictly speaking, therefore, different sets of charts should be constructed for different amounts of blade twist. The charts may be used for blades of conventional twist, however, by remembering that, in general, blades having $-8°$ twist may be considered to have of the order of 5 per cent less profile drag than untwisted blades on the basis of calculations made with the same basic theory.

The same three-term drag curve used in the hovering analysis of Chapter 4, namely,

$$c_{d_o} = 0.0087 - 0.0216\alpha_r + 0.400\alpha_r^2 \tag{22}$$

was used in preparing the charts. This drag curve may be considered as representative of practical construction blades of conventional airfoil section having fairly accurate leading-edge profiles and rigid surfaces. The charts may be applied, however, to rough or poorly built blades of conventional section merely by multiplying the profile drag-lift ratio obtained from the charts by a constant "roughness" factor equal to the ratio of the average of the ordinates of the drag curve of the actual blade to the average of the ordinates of the drag curve used in the charts. If the two drag curves do not have similar shapes, the determination of this factor should take into account the relative importance of different angles of attack; a basis for doing this is provided later in the discussion of "weighting curves." The use of the "roughness" factor will be illustrated in a later section of this chapter.

It should be realized that the drag polar on which the charts are based becomes invalid at angles of attack above the stall and that the theory would become correspondingly optimistic for conditions of operation at which stall is present. Lines representing the limits beyond which the theory becomes increasingly optimistic (because it neglects stall) are therefore placed on the drag-lift ratio charts. The lines are drawn for conditions at which the calculated tip angle of attack of the retreating blade is 12° and 16°. The two values are given on the supposition that the stall angles of conventional sections lie somewhere between 12° and 16°. In addition to pointing out the conditions beyond which the theory is optimistic, it will be shown in the next chapter that the limit lines also serve to indicate the regions of optimum performance. (The reason for choosing the angle of attack of the retreating blade as an indication of blade stall, as well as the effects of blade stall on power required and helicopter top speed, are also discussed in the following chapter.)

SAMPLE CALCULATIONS. In order to illustrate the manner in which the charts may be used to estimate the power required by a helicopter under a given set of conditions, the following example (from reference II–16 of Appendix IIA) is included:

Assume a helicopter with the following characteristics to be operating in level flight at sea level at a forward speed of 80 feet per second, and at a tip-speed ratio of 0.20:

Disk loading, pounds/square foot	2.50
Gross weight, pounds	3140
Blade plan form	Rectangular
Blade twist	None
Solidity	0.07
Parasite-drag area, square feet	15.0

From these values, $C_L = 0.329$ and $C_L/\sigma = 4.70$. In order to obtain a first approximation to $(D/L)_o$, P/L is assumed to be 0.2. Figure 9–2c then gives a value of 0.086 for $(D/L)_o$ at the intersection of the curve for $\mu = 0.2$ with the line for $C_L/\sigma = 4.70$. The value of $(D/L)_i$, assumed equal to $C_L/4$, is computed to be 0.082. The parasite drag-lift ratio is calculated as

$$\left(\frac{D}{L}\right)_p = \frac{\frac{1}{2} \times 0.002378 \times 15 \times \overline{80}^2}{3140} = 0.036$$

Because level flight is being considered, $(D/L)_c$ is zero, and the value of P/L from equation (3) is

$$\frac{P}{L} = 0.086 + 0.082 + 0.036 + 0$$
$$= 0.204$$

As a second approximation, the value of $(D/L)_0$ may be obtained by interpolation between the charts for $P/L = 0.20$ and for $P/L = 0.30$.

The value so obtained is, within the limits of accuracy in reading the charts, equal to the original value of 0.086, and no further approximation is necessary. (After some experience has been gained with the method, it will be found that even the second approximation can frequently be omitted, and that in making the first approximation it will often be sufficiently accurate to use the chart having the nearest P/L value, rather than interpolating between charts.)

The total rotor-shaft power required for the specified condition may now be calculated as

$$\frac{0.204 \times 3140 \times 80}{550} = 93.2 \text{ horsepower}$$

The power absorbed by the tail rotor may also be calculated by means of the charts. Before proceeding with the calculations it should be understood that the power absorbed by the tail rotor depends on the fixed relation of its control axis to the flight path. If the control axis is fixed backwards from the flight path, there will be an "upflow" through the tail rotor disk, corresponding to an autorotative condition, and little or no power is expended at the tail rotor shaft. The force necessary to drag the tail rotor through the air, however, must be supplied by the main rotor, which overcomes the tail rotor drag as a parasite-drag force. As the tail rotor control axis is inclined forward, it absorbs more and more power through its own shaft, acting as a helicopter rotor by pulling itself through the air. At the same time, the main rotor expends less power in pulling the tail rotor along. Thus, there is a division of power between the main and the tail rotors, depending on the inclination of the tail rotor control axis.

Proceeding with the calculations, assume that the tail rotor is traveling at $\mu = 0.2$, has rectangular, nontwisted blades of solidity equal to 0.10 and radius equal to 4 feet. Let its control axis be set at zero degrees and be situated at a distance of 25 feet from the main rotor shaft.

The angular velocity of the main rotor is then

$$\Omega_{mr} \cong \frac{V}{\mu R} = \frac{80}{0.2 \times 20} = 20 \text{ radians/second}$$

Also

$$\Omega_{tr} = \frac{80}{0.2 \times 4} = 100 \text{ radians/second}$$

The thrust of the tail rotor is therefore

$$T_{tr} = \frac{\text{hp}_{mr} \times 550}{\Omega_{mr} \times 25 \text{ ft}} = \frac{93.2 \times 550}{20 \times 25} = 102.7 \text{ pounds}$$

Then

$$C_{T_{tr}} = \frac{102.7}{\pi \times 4^2 \times 0.002378 \times (100 \times 4)^2} = 0.00536$$

The inflow velocity ratio may be calculated from equation (8) of Chapter 8, which, for low values of λ, becomes

$$\alpha = \frac{\lambda}{\mu} + \frac{C_T}{2\mu^2} \tag{23}$$

Setting α equal to zero and solving for λ,

$$\lambda = -\frac{C_T}{2\mu} = \frac{-0.00536}{2 \times 0.2} = -0.0134$$

Substituting this value of λ into equation (69) of the preceding chapter yields a blade pitch angle θ of 4.47°. Then, using the relation given by equation (21),

$$\frac{C_L}{\sigma} = \frac{2C_T}{\sigma\mu^2} = \frac{2 \times 0.00536}{0.10 \times 0.004} = 2.68$$

The value of P/L corresponding to the specified combination of values μ, θ, and C_L/σ is found by interpolation between the $P/L = 0.10$ and $P/L = 0.20$ charts to be 0.138.

Then

$$\text{hp}_{tr} = \frac{P}{L} \times \frac{TV}{550} = 2.1 \text{ horsepower}$$

Inasmuch as the tail rotor may also be producing a drag, the total power charged to torque counteraction (distributed between the main and tail rotors) should be calculated from the sum of the values of $(D/L)_o$ and $(D/L)_i$ rather than from P/L. The value of $(D/L)_o$ at $P/L = 0.138$ is found by interpolation between the charts to be 0.120. Also,

$$\left(\frac{D}{L}\right)_i = \frac{C_L}{\sigma}\frac{\sigma}{4} = 0.067$$

The total $(D/L)_{tr}$ is 0.187, corresponding to 2.8 horsepower.

The difference between the two values of power (2.8 and 2.1) results from the drag on the tail rotor and hence must be supplied through the main rotor shaft as an additional parasite drag. The revised value of the main rotor-shaft power is then 93.2 + 0.7 = 93.9 horsepower.

EFFECTS OF OPERATING CONDITION ON PROFILE DRAG. It is necessary from a design point of view to be able to predict the conditions of operation at which the rotor will perform most efficiently. For a helicopter of given disk loading and parasite drag, both the induced and parasite losses are fixed at a particular speed. Rotor profile-drag losses, however, which form a significant part of the total rotor losses in forward flight, are dependent on variables that are more directly under the designer's control. These variables include blade-pitch angles, rotor thrust or lift coefficient, and rotor solidity. The effects of these quantities can be readily seen by an inspection of the charts of Fig. 9–2.

An inspection of the charts reveals that minimum profile drag-lift ratios are obtained at any tip-speed ratio at the highest pitch angles, or at the highest values of rotor mean lift coefficients (as represented by C_L/σ) permitted by the high section angles of attack encountered on the retreating side of the rotor disk. In other words, optimum performance is obtained at a given tip-speed ratio by operating as close to the stall limit lines as possible. This conclusion applies to all flight conditions, for it will be remembered that most efficient hovering performance is also obtained when the rotor is operating at section angles of attack close to the stall. (The consequences of operating at conditions beyond the stall will be shown in the next chapter.) The charts also show that the optimum tip-speed ratio for helicopters of conventional design is approximately 0.25.

Climb Performance Calculations

The problem of computing climb performance can be handled quite simply by means of the performance charts. In relation to climb, the solution to either one of two alternate problems is required.

(1) Compute the rate of climb at a given airspeed for a given available power.

(2) Compute the power required to climb at a given rate of climb and at a given airspeed.

The information may be desired for various gross weights, altitudes, rotor speeds, etc.

The procedure for calculating the rate of climb (or descent) for a given available power would be as follows:

(1) Calculate P/L from the available power and given gross weight (assuming that $W = L$).

(2) Calculate C_L/σ, $(D/L)_i$, and $(D/L)_p$ from the given weight, airspeed, altitude, and rotor dimensions.

(3) Using the calculated values of P/L, C_L/σ, and the given value of μ, determine $(D/L)_o$ from the performance charts.

(4) Calculate $(D/L)_c$ from the general performance equation [equation (4)].

(5) Solve for the rate of climb by means of equation (17).

If the angle of climb γ is very large, the above process may be repeated with $L = W \cos \gamma$ instead of the assumption that $L = W$. Step (1) may be omitted for the power-off condition, for then $P/L = 0$.

The procedure as applied to the sample helicopter discussed in the preceding section would be as follows if 140 horsepower were assumed available:

(1) $\dfrac{P}{L} = \dfrac{140 \times 550}{80 \times 3140} = 0.306$

(2) $C_L/\sigma = 4.70$; $(D/L)_i = 0.082$; $(D/L)_p = 0.036$

(3) Using the $P/L = 0.30$ chart (Fig. 9–2d), the intersection of $C_L/\sigma = 4.70$ and $\mu = 0.20$ yields a value of $(D/L)_0 = 0.089$

(4) $(D/L)_c = 0.306 - 0.089 - 0.082 - 0.036 = 0.099$

(5) $V_v = 0.099 \times 80$ fps $= 7.9$ fps $= 475$ feet per minute

Inasmuch as the climb angle γ is small (i.e., about 6°), the calculations need not be repeated with a new value of L.

A plot of rate of climb against airspeed for a typical helicopter is given in Fig. 9–4.

The power required to climb at a given rate at a particular airspeed can be solved in a manner similar to finding the level-flight power of a sample helicopter, except that a known value of $(D/L)_c$ is inserted into the general performance equation before P/L is calculated.

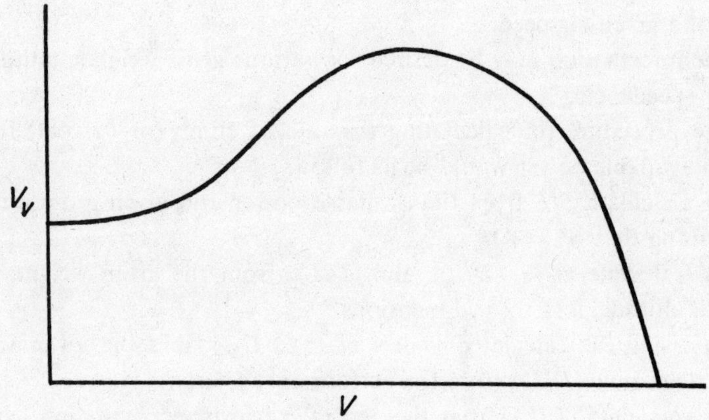

Fig. 9–4 Rate-of-climb curve for typical helicopter.

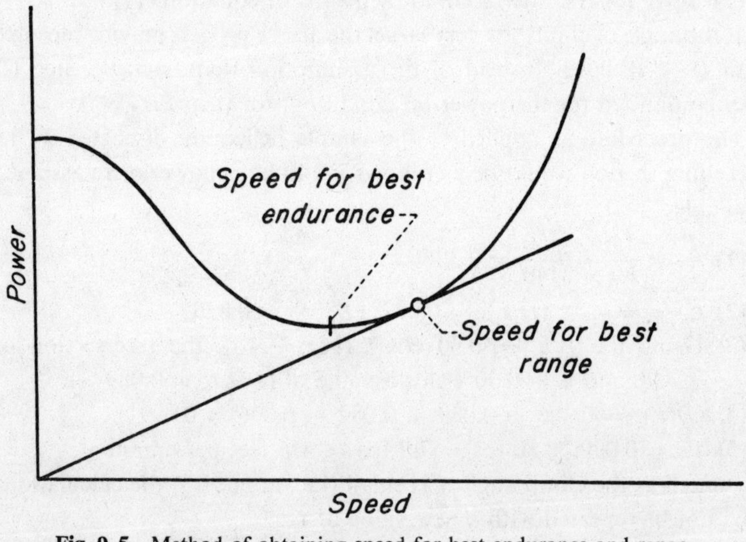

Fig. 9–5 Method of obtaining speed for best endurance and range.

Inasmuch as the climb angle is known at the start, the rotor lift can be calculated directly as $W \cos \gamma$.

Range and Endurance Calculations

While it would be possible to formulate an analytical expression for the speed for best range and speed for best endurance, it is usually simpler to calculate the level-flight power-required curve and pick off these values. As shown in Fig. 9–5, the speed for best range is found at the point at which the power-required curve is tangent to a line drawn through the origin, for at this point the ratio of speed to power (and, therefore, of distance to fuel) is greatest.

For best endurance, the flight should be made at the speed for minimum power.

Experimental Data and Comparison with Theory

As was true in the case of hovering, a valid comparison of helicopter performance with predicted values in the forward-flight condition was not possible for many years because of the absence of good experimental data. In this connection, two important quantities which have to be measured rather precisely are the horsepower delivered to the main rotor shaft and the true speed at which the helicopter is traveling. In addition, the parasite drag of the fuselage, tail rotor, and all components of the helicopter other than the lifting rotor must be accurately known if the performance of the rotor alone, rather than the over-all performance of the helicopter, is to be determined from tests of the entire aircraft. By means of careful instrumentation, the NACA was able to obtain accurate performance data from flight and full-scale wind-tunnel tests on several sets of rotors, with which to compare the general rotor theory. (See Appendix IIA, references II–3, II–4, II–5, and II–6.)

A comparison of the measured and calculated rotor performance of a fabric-covered set of rotor blades is shown in Fig. 9–6 (from reference II–4 of Appendix IIA). The data, which were obtained from flight tests, are presented in the general, nondimensional form of rotor drag-lift ratio, $(D/L)_r$, plotted against tip-speed ratio, μ, and are grouped according to thrust coefficients. In order to indicate the data for which blade stall was present, all points having a calculated angle of attack at the tip of the retreating blade greater than 12° are shaded (i.e., ▲). (Tuft observations on the rotor, as well as wind-tunnel blade-section data, indicated that 12° closely represented the stalling angle

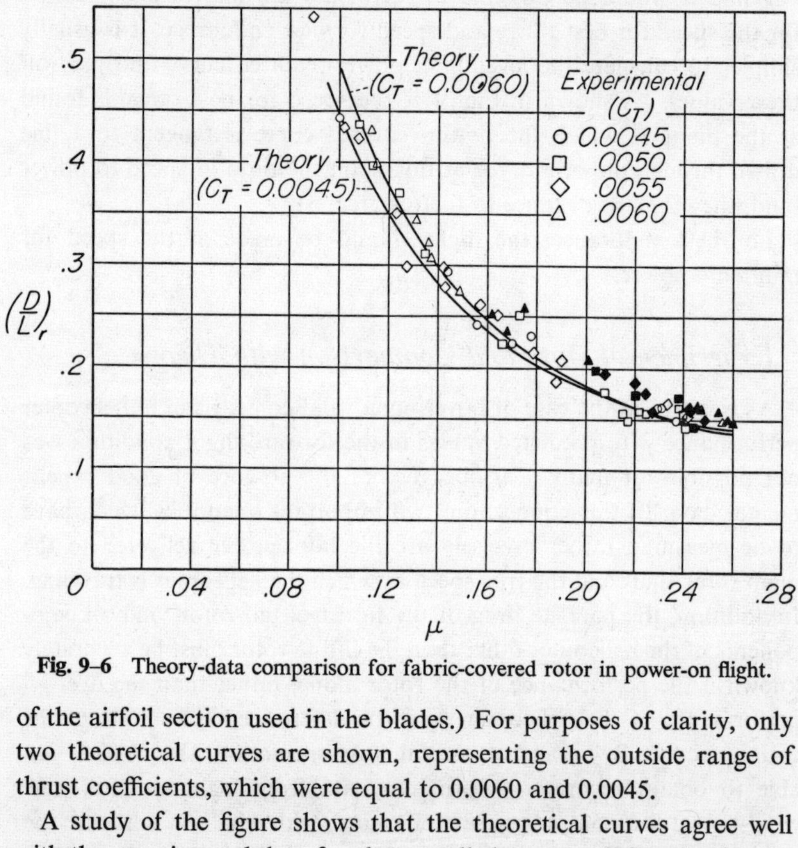

Fig. 9–6 Theory-data comparison for fabric-covered rotor in power-on flight.

of the airfoil section used in the blades.) For purposes of clarity, only two theoretical curves are shown, representing the outside range of thrust coefficients, which were equal to 0.0060 and 0.0045.

A study of the figure shows that the theoretical curves agree well with the experimental data for the unstalled rotor and that the theory, which omits any allowance for blade stall, underestimates the rotor losses as the stall becomes more severe at the higher tip-speed ratios.

The theoretical curves also provide an example of how the performance charts of Figs. 9–2 and 9–3 can be applied to blades having different drag characteristics other than the drag curve on which the charts are based. As previously mentioned, the charts are based on a drag curve which is representative of well-built, plywood-covered blades. In order to apply the charts to the relatively rough, deformable, fabric-covered blades of the rotor represented by the performance curves of Fig. 9–6,

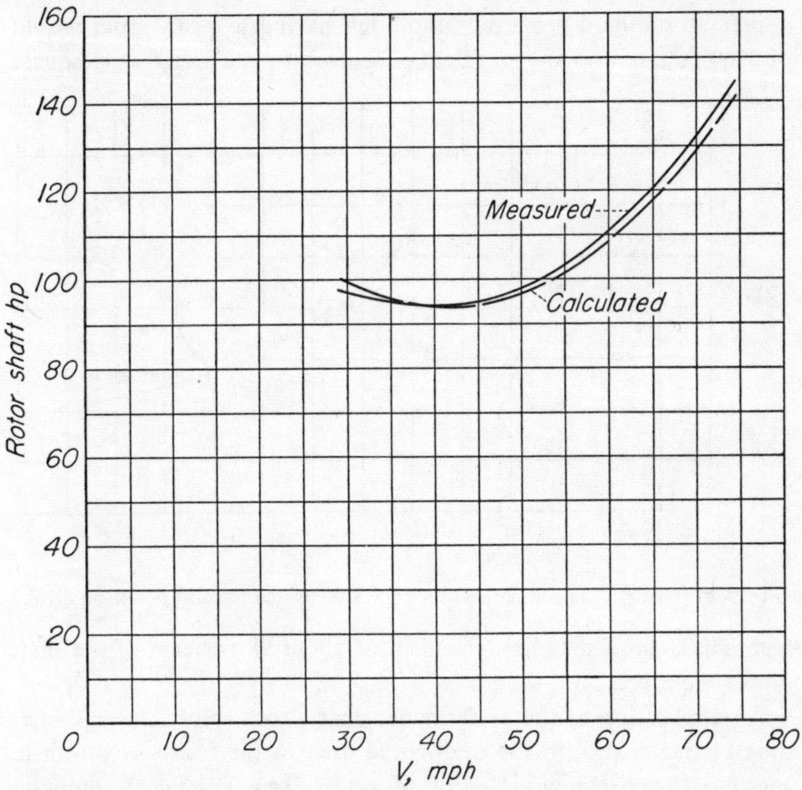

Fig. 9–7 Comparison of measured and calculated power required in power-on flight (from Appendix IIB, reference 20).

the profile drag-lift ratios obtained from the charts were increased by 28 per cent. The use of this factor was equivalent to increasing the drag of the basic airfoil section in the perfectly smooth condition by 50 per cent, inasmuch as the drag polar of the charts already included a 17 per cent roughness allowance (i.e., $1.17 \times 1.28 = 1.50$). The resulting drag polar obtained by applying the 50 per cent roughness allowance checked with wind-tunnel measurements of practical construction, fabric-covered blade specimens.

The nondimensional theory-data comparison of Fig. 9–6 is shown in the form of the familiar plot of rotor-shaft power against velocity in Fig. 9–7, which was calculated to yield the performance of a helicopter at standard sea-level conditions having a mean gross weight of 2560 pounds and an equivalent parasite flat-plate area $f = 15$ square

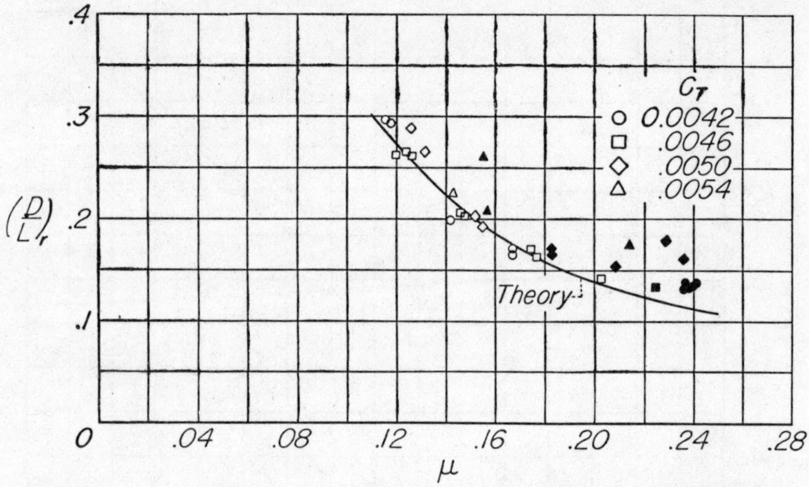

Fig. 9–8 Theory-data comparison for plywood-covered rotor in power-on flight.

feet. The figure shows an agreement of about 98 per cent of the shaft power required.

An application of the performance charts to a rotor having similar drag characteristics to the drag curve used in the charts is shown in Fig. 9–8 (from reference II–4 of Appendix IIA). The figure compares the nondimensional performance of a relatively smooth, plywood-covered set of rotor blades having $-8°$ twist, as measured in flight, with a single theoretical curve representative of the average thrust coefficient at which the data were taken. Good agreement between the theory and the unstalled (unshaded) points is indicated. When individual, theoretical drag-lift ratios were calculated at the thrust coefficient corresponding to each of the data points, instead of comparing the data with a single curve computed for an average thrust coefficient, agreement within 1 to 2 per cent was shown for the unstalled flight conditions. (See Fig. 10–6.)

A check of the theory in autorotative flight is given in Figs. 9–9 and 9–10 (from reference II–4, Appendix IIA), which compare predicted and flight-measured rotor drag-lift ratios for the fabric and

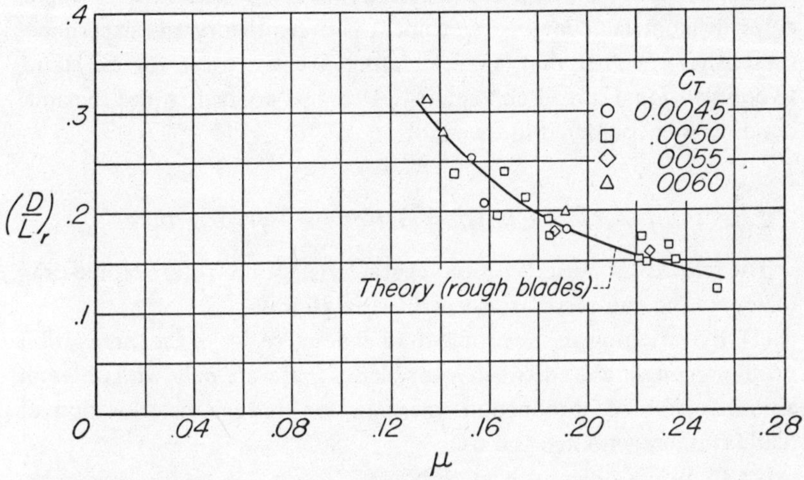

Fig. 9–9 Theory-data comparison for fabric-covered rotor in autorotation.

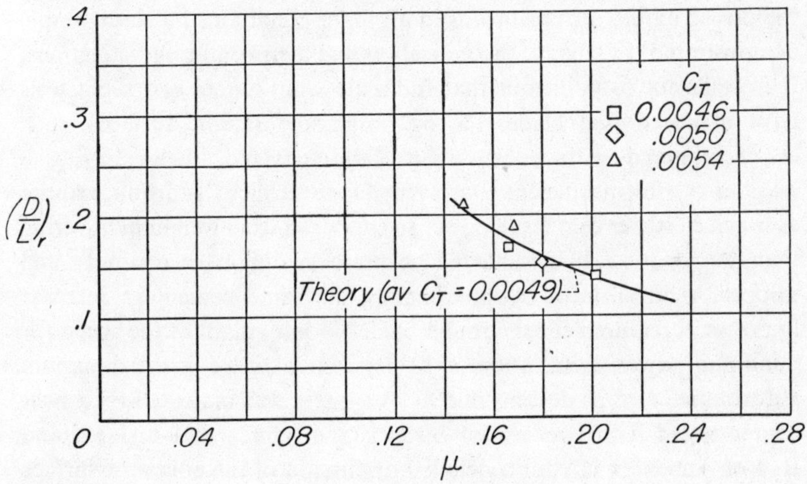

Fig. 9–10 Theory-data comparison for plywood-covered rotor in autorotation.

plywood-covered rotors in power-off flight. It can be seen that the theoretical performance curves, computed with the aid of the $P/L = 0$ chart, serve as a good fairing for the data obtained with both rotors.

In general, it may be said that the comparison between theoretical analyses and reliable flight data indicate that rotor theory may be used with confidence for predicting the steady-flight characteristics of single-rotor helicopters. The good agreement between theory and experiment establishes the fact that, in calculating average rotor forces, static, two-dimensional airfoil characteristics can be applied to the dynamic conditions encountered in the rotor.

Effects of Airfoil Characteristics on Performance

The effects of airfoil-section characteristics on rotor profile-drag losses can be considered as occurring in two ways:

(1) By variations in the profile-drag characteristics of the same airfoil profile, as might exist between rotor blades that were built with different amounts of blade production tolerance, or through deterioration of blade surfaces with age and use.

(2) By use of different blade profiles.

The importance of smooth, nondeformable blade surfaces in reducing the power required by the rotor in all flight conditions has been amply demonstrated by many theoretical and experimental investigations. It is sufficient to point out that full-scale wind-tunnel and flight tests have indicated that blades having rough, deformable surfaces, such as encountered with poorly built, fabric-covered blades having a number of ribs insufficient to prevent blade surface distortion, require as much as 10 per cent more power to support a fixed weight in hovering than do identical blades having accurate leading-edge contours and smooth, rigid surfaces. Approximately the same percentage increase in power is required for the rough blades in level flight at the speed for minimum power. Also, about a 10 per cent increase in the minimum autorotative rate of descent due to roughness was indicated by a helicopter tested with two sets of blades. (See references I–8, I–12, and II–4 of Appendix IIA for a detailed discussion of the effects of surface roughness.)

Theoretical investigations also have indicated that significant savings in profile-drag power may be realized by the use of airfoil sections designed especially for rotors as distinguished from wings or propellers. Although most rotor blades at the present time are composed of conventional wing sections, attention is being given to the development of special airfoil sections. In addition to a high stall angle and a high critical Mach number, the desirable aerodynamic characteristics of airfoil sections suitable for use as rotor-blade sections are:

(1) Nearly zero pitching moment
(2) Low drag throughout the range of low and moderate lifts
(3) Moderate drag at high lifts

Most of the NACA low-drag airfoils that have been developed for use in wings and control surfaces have too high a pitching-moment coefficient to warrant consideration for use with current helicopter designs. (High pitching-moment coefficients lead to undesirable periodic stick forces, to vibrations, and to undesirable control-position gradients, all brought about by periodic blade twist.) Although this objection is removed with the low-drag symmetrical sections, these sections are not applicable because half of the low drag "bucket," or, in other words, half of the limited range of lift coefficients in which the drag reductions are achieved is below zero lift whereas the faster moving portions of the helicopter blade are nearly always operating at positive lift coefficients.

In order to place the low drag "bucket" in a useful range of lift coefficients and still retain zero or almost zero moment coefficients, a number of special airfoils have been derived by the NACA (references IV–2, IV–4, and IV–7 of Appendix IIA). A comparison of two such airfoils with the conventional and commonly used NACA 23012 section is given in Fig. 9–11. It can be seen that the two *H* (helicopter) sections have lower drag over the lower range of lift coefficients, but that they show an earlier and more violent drag rise at the higher angles than the 23012 profile. The 8–*H*–12 section, which was derived later than the 3–*H*–13.5 section, is seen to have better drag characteristics at the higher angles than the 3–*H*–13.5.

The section characteristics shown in Fig. 9–11 indicate that if rotor operation could be restricted to relatively low lift coefficients, the

8–H–12 section would be superior to the 23012. However, a plot of c_{d_0} against α_r does not supply sufficient information to reach a conclusion in this regard in forward flight, inasmuch as the section angle of attack normally varies from low values on the advancing side to high values on the retreating side of the disk for the same flight conditions,

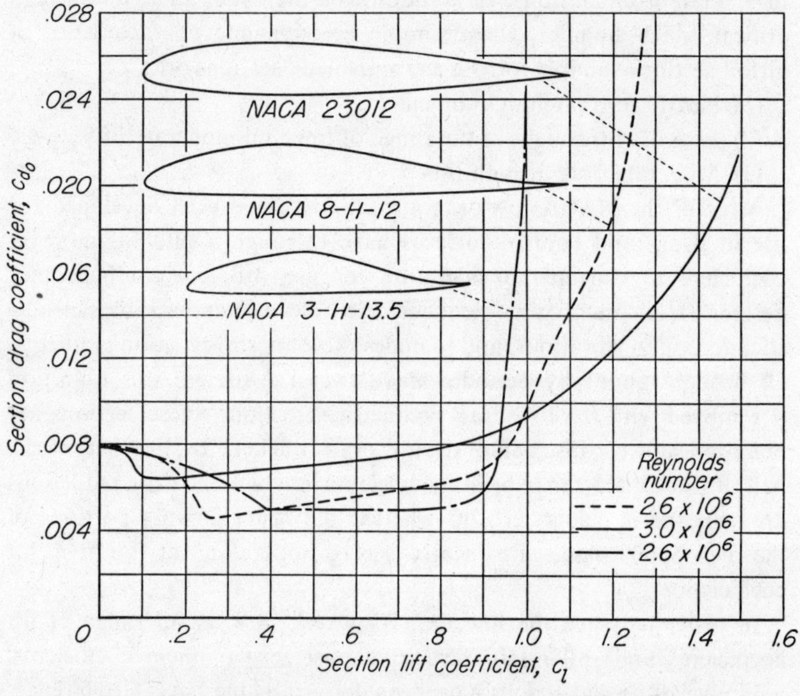

Fig. 9–11 Aerodynamic characteristics of conventional and helicopter airfoil sections.

and because a given increment in drag coefficient has a smaller effect on the power absorbed at the low velocity retreating parts of the disk than at the high velocity advancing side. This idea is illustrated in Fig. 9–12 (from reference I–12 of Appendix IIA), which shows the angle-of-attack distribution and the distribution of power loss per unit value of profile-drag coefficient for a typical rotor in cruising flight.

In order to permit more rapid quantitative judgment of such factors, the data may be combined for the two sets of contours into a single curve showing the relative importance of different parts of the curve of airfoil-section profile-drag coefficient against section angle of attack.

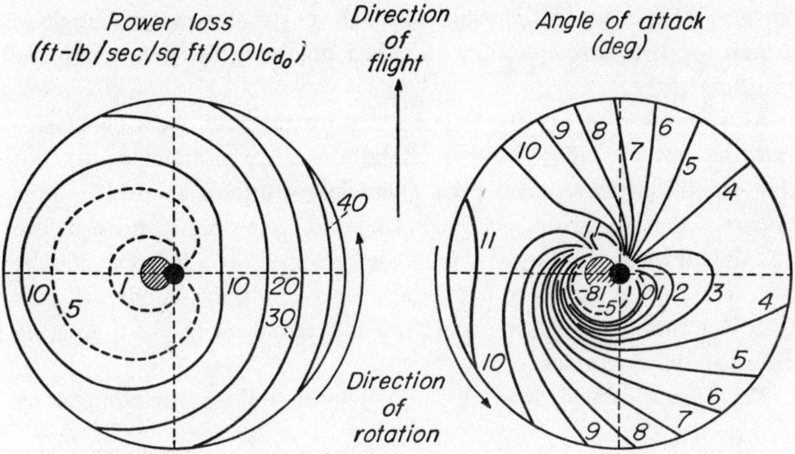

Fig. 9–12 Power loss and angle-of-attack contours: (V = 55 miles per hour; μ = 0.2; W/S = 2.5 pounds per square foot).

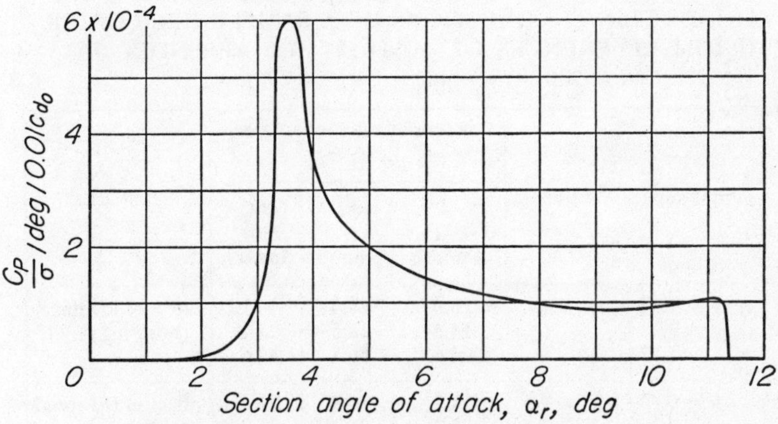

Fig. 9–13 Weighting curve for rotor of Fig. 9–12.

Such a curve, called a *weighting curve* (developed and given in reference I–12 of Appendix IIA), is shown in Fig. 9–13 for the rotor of Fig. 9–12. The ordinates of the curve represent the power consumed in overcoming the profile drag of all the blade elements operating at a particular angle of attack for a unit value of c_{d_0}. The total profile drag power absorbed by the rotor may be calculated simply by multiplying the ordinates of the weighting curve by the ordinates of the curve of c_{d_0} against angle of attack for the airfoil section and then obtaining the area under the resulting curve.

An inspection of Fig. 9–13 shows that for the case under consideration the greatest losses occur over the low angle-of-attack range, but that significant losses also exist at angles of attack up to 12°. It is apparent that an airfoil section that has no large drag increase before 12° should be used if large power losses are to be avoided. The NACA 3–*H*–13.5 would therefore not be appropriate for this condition. It is thus seen that the weighting curve is a convenient means for evaluating different airfoil profiles.

The effects of disk loading and tip-speed ratio as determined by

TABLE 9—1

COMPARISON OF ROTOR-BLADE PROFILE-DRAG LOSS OF THE NACA 3–*H*–13.5, 8–*H*–12, AND 23012 AIRFOIL SECTIONS FOR VARIOUS FLIGHT CONDITIONS OF A SAMPLE HELICOPTER (FROM REFERENCES I–12 AND IV–4 OF APPENDIX IIA)
(R = 20 ft, ΩR = 400 fps, σ = 0.07, f = 15)

	Operating Conditions		Rotor-Blade Profile-Drag Loss, *HP*			Remarks
			NACA 3–*H*–13.5 Smooth	NACA 8–*H*–12 Smooth	NACA 23012 Smooth	
1	W/S = 1.55	μ = 0	16.0	14.4	20.1	Effect of loading (hovering flight)
2	3.33	0	14.5	18.5	24.1	
3	5.42	0	204.6	56.8	42.6	
4	μ = 0	W/S = 2.5	14.2	16.3	21.7	Effect of tip-speed ratio
5	0.2	2.5	23.2	21.2	25.7	
6	0.3	2.5	54.5	36.7	31.0	
7	W/S = 1.9	μ = 0.2	18.2	17.5	23.5	Effect of loading (forward flight)
5	2.5	0.2	23.2	21.2	25.7	
8	3.1	0.2	54.3	28.6	29.2	

weighting curves are illustrated in Table 9–1 for three airfoil profiles. The table shows that for the low loading conditions and all but the high tip-speed ratio condition, the two *H* profiles have about the same losses, and that they are more efficient by as much as 30 per cent than the 23012 profile. For the high loading and the high tip-speed ratio conditions when stall is present, the 3–*H*–13.5 section is the worst because of its early stall characteristics, whereas the 8–*H*–12 and the 23012 profiles have similar power losses.

It may be concluded, therefore, that the 8–*H*–12 section shows promise of being superior to the conventional sections from an aerodynamic standpoint. Full-scale tests of practical construction blades incorporating the NACA 8–*H*–12 section are needed, however, to determine the true worth of the airfoil under actual operating conditions.

10

THE PREDICTION AND EFFECTS OF ROTOR BLADE STALL

This chapter deals with rotor blade stall as it affects the limiting conditions of operation of the helicopter. Whereas the stall of an airplane wing limits the low speed characteristics of the airplane, stall on a helicopter blade limits the high speed possibilities of the helicopter. This paradox is understandable when one considers that the retreating blade of the helicopter rotor (that is, the blade going with the wind in forward flight) encounters lower and lower velocities as the forward speed is increased. The retreating blade, however, must produce its portion of the lift; therefore, as the velocity decreases with forward speed, the blade angle of attack must be increased. It follows that at some forward speed the retreating blade will stall.

Growth of Blade Stall

It is clear, then, that stall must always occur on the retreating side of the rotor disk. In forward flight the angle-of-attack distribution along the blade is far from uniform, so that it must be expected that some portion of the blade will stall before the rest. It was shown in Chapter 7 (see Fig. 7–22) that the angle-of-attack distribution along a blade is primarily a function of the magnitude and direction of the rotor inflow. As the flow down through the rotor increased (helicopter condition), the angles of attack were increased at the tip with respect to those at the root. As the flow up through the disk increased (autogyro state), the load shifted toward the blade root. It has also been shown that in helicopter flight in the higher speed range, downflow through the disk does increase with speed, for the rotor must be tilted farther into the wind to provide thrust to overcome the fuselage drag. Therefore, in

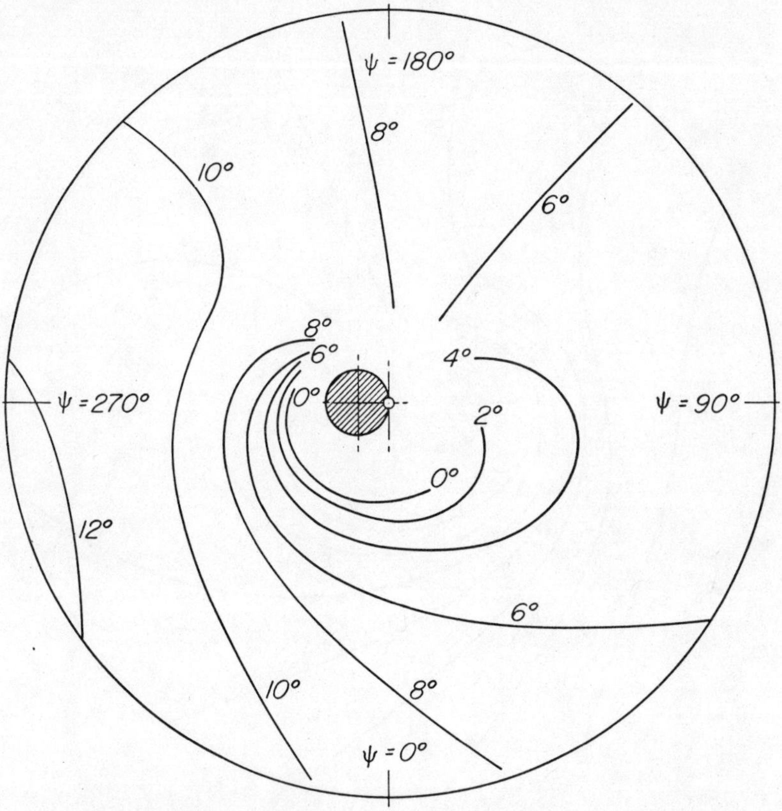

Fig. 10–1 Angle-of-attack-contour plot: V = 40 miles per hour; ΩR = 408 feet per second; (μ = 0.15; C_T/σ = 0.11; f = 23 square feet).

the helicopter, angles of attack on the retreating blade increase with forward speed and the highest blade angles of attack are at the blade tip.

In the helicopter, stall begins at the tip of the retreating blade, spreading inboard as forward speed is increased. In the autogyro, stall begins at the root of the retreating blade, spreading outboard as speed is increased.

The nature of the angle-of-attack distribution around a rotor disk operating in the helicopter condition is shown in the contour diagrams

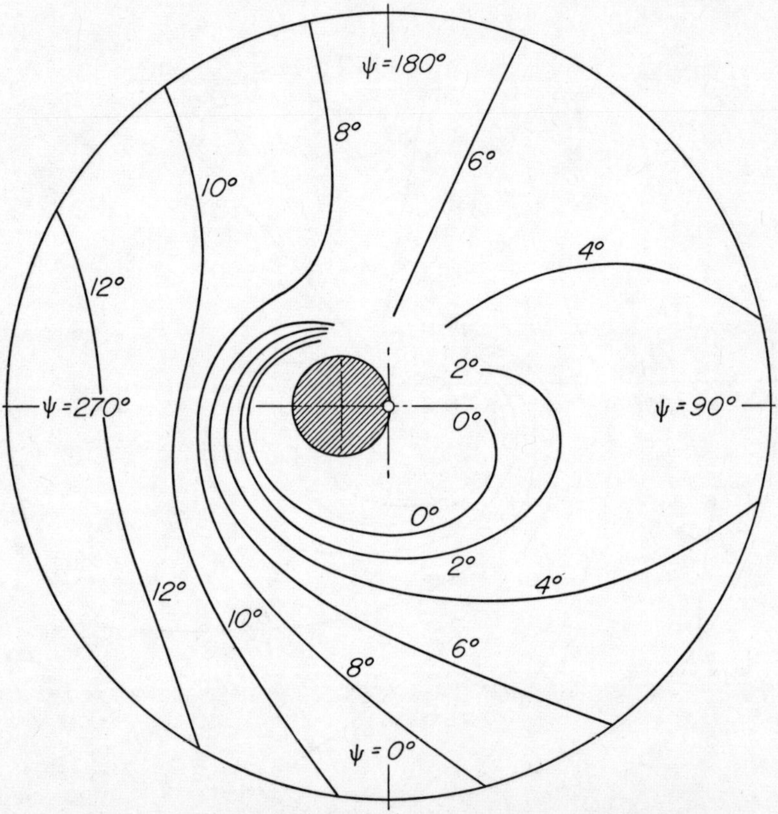

Fig. 10–2 Angle-of-attack-contour plot: V = 70 miles per hour; ΩR = 448 feet per second; (μ = 0.23; C_T/σ = 0.09; f = 23 square feet).

of Figs. 10–1, 10–2, and 10–3. These diagrams represent plan views of a typical rotor in three progressively more severe flight conditions, the direction of rotation and flight direction being defined by the azimuth angles. These cases were calculated to correspond with flight test conditions which will be discussed later. The air speeds and rotor speeds given in the figure legends correspond to the measured conditions. The contour lines are lines of constant angle of attack. The small shaded circles represent the reversed velocity region. The distribution of angle of attack along a blade at any azimuth position is found by

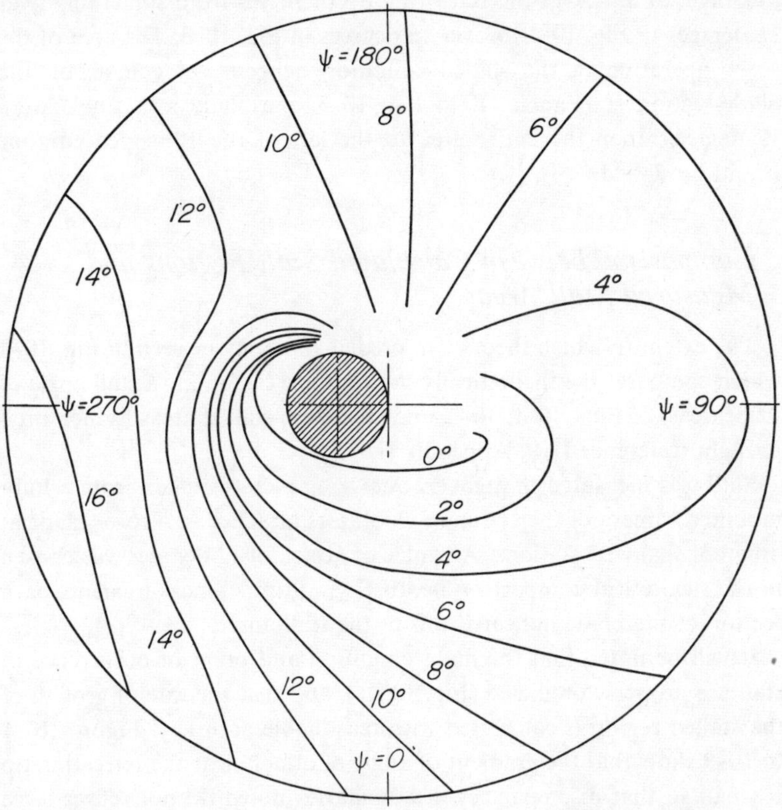

Fig. 10–3 Angle-of-attack-contour plot: V = 70 miles per hour; ΩR = 408 feet per second; (μ = 0.25; C_T/σ = 0.11; f = 23 square feet).

noting what contour lines cut a radial line drawn at the azimuth in question.

It will be noted that the retreating tip encounters the highest angles of attack, and that the gradient of angle of attack is steep. The advancing blade, on the other hand, has relatively uniform, low angles of attack.

Note that the rate of change of angle of attack with azimuth at the blade tip is relatively smooth and sinusoidal in nature, while farther inboard —say at the 0.4 radius—the rate of change of angle of attack is very rapid on the retreating side. Comparing the three figures, it is seen that the angle of attack at the retreating tip increases from something over 12 degrees in Fig. 10–1 to over 16 degrees in Fig. 10–3. The area of the rotor operating in the stalled condition depends, of course, on the blade-section stall angle. If in Fig. 10–3, the blade stall angle were 12 degrees, then the entire area to the left of the 12-degree contour would be stalled.

Comparison between Calculated Stall Regions and Measured Stall Area

The extent to which theory can predict stall may be seen in Fig. 10–4 which compares the theoretically stalled areas (assuming a stall angle of 12 degrees) of Figs. 10–1, 10–2, and 10–3 with stalled areas as measured in flight (reference II–9, Appendix IIA).

Stall was measured in flight by analyzing pictures taken from a hub-mounted camera of the action of wool tufts attached to a rotor blade at different spanwise stations. An angle of attack of 12 degrees was chosen in the theoretical comparison because wind-tunnel measurements on a section of the blade indicated this as the stall angle.

It will be noted that the flight condition and point of occurrence of stall are properly predicted (Fig. 10–4a), and that the rate of growth of the stalled region is calculated with reasonable accuracy. Figures 10–1 to 10–3 show that the gradient of angle of attack near the retreating tip is small so that discrepancies in boundaries noted do not reflect large differences in angle of attack. (By the same reasoning the contours make it clear that small reductions in the stalling angle of a blade can result in large increases in stalled area.)

Pilot reactions to the flight conditions represented in Fig. 10–4 are quite interesting. The condition shown in Fig. 10–4c (where a large amount of stall is present) represented the most extreme condition that the pilot was able to maintain long enough to take a record. For this helicopter, therefore, the condition with this large stalled area (extending about one-quarter of the rotor circumference) appeared to represent the limiting condition of operation.

In the condition of moderate stall, the pilot, although uncomfortable, was able to control the helicopter satisfactorily and take records. No effects that would be associated with stall were noted in the first marginal stall case. It would seem then that although the effects of a

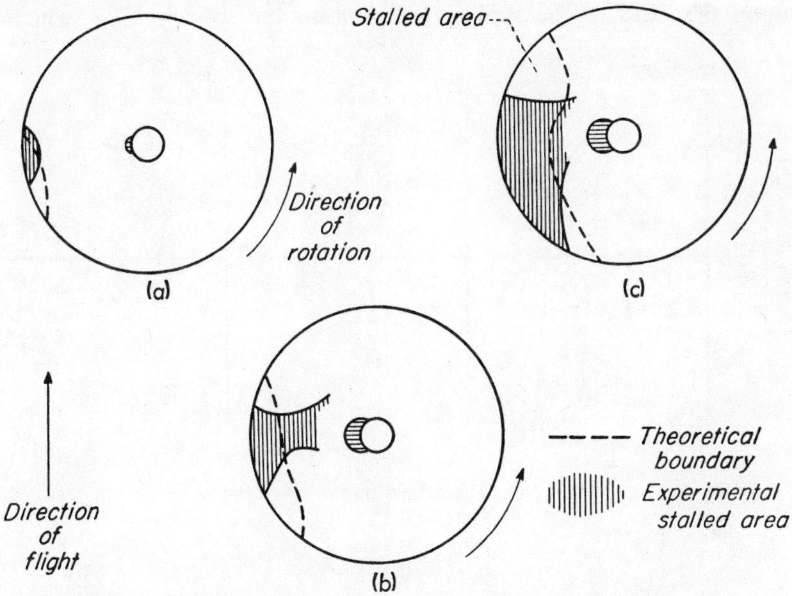

Fig. 10–4 Comparison between theoretical and experimental stalled areas.
(a) V = 40 miles per hour, ΩR = 408 feet per second
(b) V = 70 miles per hour, ΩR = 448 feet per second
(c) V = 70 miles per hour, ΩR = 408 feet per second

small amount of stall are tolerable, operation with large amounts of stall is prohibitive.

The theory-data comparison of Fig. 10–4 has shown that stall was observed and predicted to begin at the retreating blade tip. The calculated angle of attack at this point is therefore an important criterion of rotor condition of operation. By expressing stall in terms of a single parameter in this way, it is possible to illustrate graphically the effect on stall of the variables defining the condition of operation.

Factors That Affect Blade Stall

The factors that affect blade stall may be conveniently summarized and examined on the charts of Fig. 10–5 (reference II–9, Appendix IIA). The angle of attack at the retreating tip of the blade depends upon three variables. It depends first upon the tip-speed ratio μ; that is, the ratio of forward speed to rotational speed. It also depends upon the ratio of the thrust coefficient to the solidity C_T/σ (which

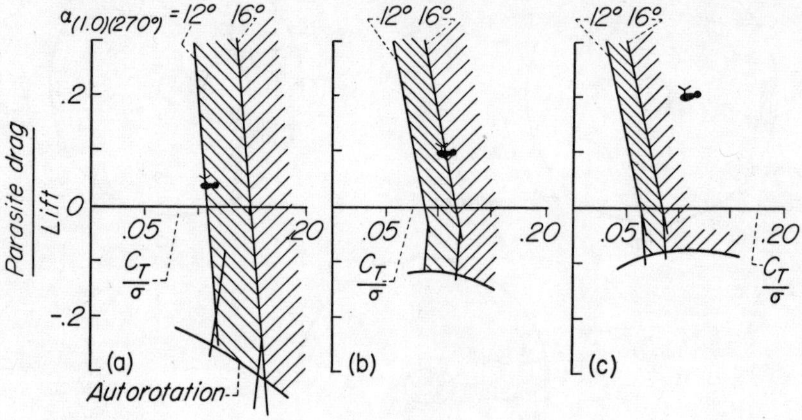

Fig. 10–5 Theoretical stalling-angle plots.
(a) $\mu = 0.15$
(b) $\mu = 0.25$
(c) $\mu = 0.35$

is a measure of the mean blade lift coefficient), for as the angles of attack all around the disk increase the angle of the retreating tip also increases. Finally, the angle of attack at the retreating tip depends upon the ratio of parasite drag to lift; that is, the degree to which the thrust vector must tilt forward to overcome the drag of the fuselage. If a propeller is installed on the fuselage, it is possible in effect to reduce the parasite drag which the rotor must overcome. In fact, with increased propeller power, negative values of parasite drag can be produced. Then, although for positive values of parasite drag the rotor is dragging the fuselage, for negative values the propeller and fuselage are dragging the rotor. The limiting condition is that of the autogyro, for which no power is applied to the rotor. Thus, increasing the ratio of parasite drag to lift (Fig. 10–5) is equivalent to applying more and more power to the rotor. Values of the ratio of parasite drag to lift for present-day helicopters at cruising speeds are of the order of 0.1.

It should be realized that climbing at a given rate of speed will affect the tip angle in a manner similar to the effects of increased parasite drag-lift ratio, inasmuch as both parameters tend to increase the rotor inflow and consequently the angles of attack. Therefore, the ordinates of the plots of Fig. 10–5 may be considered as $(D/L)_p + (D/L)_c$ instead of parasite drag/lift [or $(D/L)_p$] alone.

Once the tip-speed ratio μ is fixed, the angle of attack at the retreating tip is determined for combinations of the ratio of parasite drag to lift and mean blade lift coefficient. The line labeled $\alpha_{(1.0)(270°)} = 12°$ in Fig. 10–5 represents combinations of these quantities for which the calculated angle of attack at the retreating tip (at $\psi = 270°$) is 12°. Similarly, the line labeled $\alpha_{(1.0)(270°)} = 16°$ represents, for a fixed value of μ, combinations of parasite-drag and mean lift coefficient for which the retreating tip angle of attack is 16°. Because the stall begins inboard in the autorotation or near-autorotation conditions, the type of inboard limit utilized in reference II–18 of Appendix IIA and briefly discussed in Chapter 8 has been used for conditions where this limit is more stringent than the tip angle-of-attack limitations. The condition represented by the short lines crossing the curves labeled "autorotation" in Fig. 10–5 is one wherein the blade angle of attack shown has been reached at 270 degrees azimuth at a radius such that the tangential velocity is equal to four-tenths the rotational tip speed. The 12-degree and 16-degree lines represent the range of angle of attack in which conventional blade airfoils would be expected to stall and have been included in NACA theoretical papers as probable limiting conditions of the validity of the theory (see Fig. 9–2).

Definition of Limiting Conditions of Operation

It is of interest to spot the flight test data points of Figs. 10–1, 10–2, and 10–3 on the theoretical plots of Fig. 10–5. The condition for which extreme stall was observed in flight (Fig. 10–4c) was at a forward speed of 70 miles per hour and a rotor speed of 205 revolutions per minute, which corresponds to a tip-speed ratio μ of approximately 0.25. This condition applies therefore to the plot of Fig. 10–5b, and the values of the ratio of parasite drag to lift and C_T/σ locate the point as shown. Because this point represents the extreme amount of tip stall operationally tolerable, it would appear that, for this rotor, the 16° tip angle-of-attack line may be taken as the limit of practical conditions of operation.

If in a similar manner the point at which stall was just beginning to occur is plotted (for a forward speed of 40 miles per hour and a rotor rotational speed of 205 revolutions per minute, giving a tip-speed ratio of approximately 0.15 and locating the point on the plot of Fig. 10–5a), it is found that the coordinate values are such as to place the point on the 12-degree angle-of-attack line. This result indicates again the agreement with theory shown earlier for this marginal stall case. It will be noted that, for the helicopter tested, the operational limit occurred when the calculated angle of attack at the tip of the retreating blade exceeded the stalling angle of the airfoil section by approximately 4 degrees.

The 12-degree and 16-degree angle-of-attack lines may be considered boundaries of three regions, the first representing conditions for which no stall will be encountered, the second (hatched region in Fig. 10–5) representing conditions for which a moderate amount of stall is present, and third (cross-hatched region in Fig. 10–5) representing conditions for which stall is so severe as to prohibit operation.

If, for example, an attempt were made to fly this helicopter at 100 miles per hour, and hence at a tip-speed ratio of approximately 0.35 and still use the low rotor speed used in the first two cases, the operating condition would be far beyond the 16-degree angle-of-attack line that represents the maximum tolerable amount of stall. This point is shown in Fig. 10–5c.

Power Losses Due to Blade Stall

It is important to determine the power losses (and therefore rotor efficiency losses) resulting from blade stall in order to know the extent to which rotor efficiency is reduced before the operating limitations due to vibration and loss of control are reached. In order to determine the effect of stall on power required, profile-power losses as measured on a typical helicopter have been compared with calculated losses under the assumption of no blade stall in reference II–7 of Appendix IIA. This comparison is given in Fig. 10–6.

Performance data obtained with the blades used in these tests indicate that the parabolic blade-section drag polar discussed in previous chapters represents the blade profile-drag characteristics in unstalled

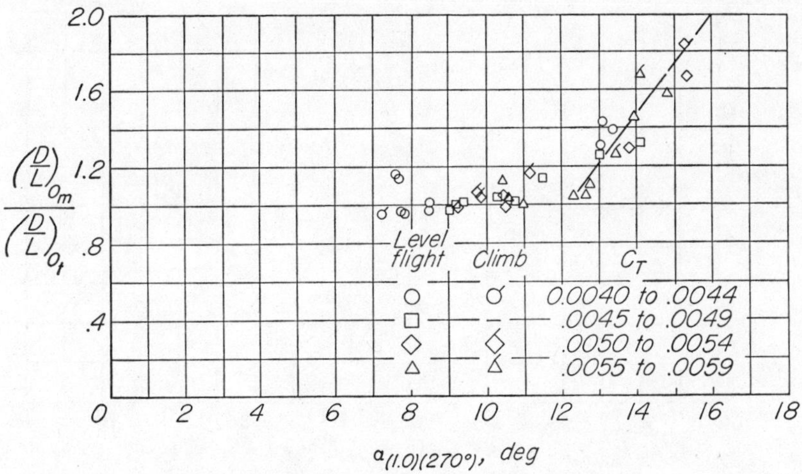

Fig. 10–6 Effect of rotor stall on power required.

conditions with good accuracy. The parameter which is used to indicate the stall losses in Fig. 10–6 is the calculated angle of attack at the retreating blade tip. For each data point the ratio of the measured profile power to the theoretical profile power (each expressed as a drag-lift ratio) is plotted against the calculated angle of attack at the retreating tip. Data for different thrust coefficients and different flight conditions are included.

It will be noted from Fig. 10–6 that measurements and calculations of profile losses are in good agreement (the ratio of actual to calculated power is approximately 1.0) up to the point where the calculated angle of attack at the retreating tip is approximately 12 degrees. For conditions in which the calculated angle of attack at the retreating tip was greater than 12 degrees, measured profile losses were progressively greater than predicted values which did not account for the increase in blade drag with stall. For conditions in which the calculated angle of attack at the retreating tip was 16 degrees, the measured profile losses were approximately double the predicted values.

These measurements show that power losses due to blade stall begin as stall is first encountered, and that profile losses are approximately doubled by the time the calculated angle of attack at the retreating tip

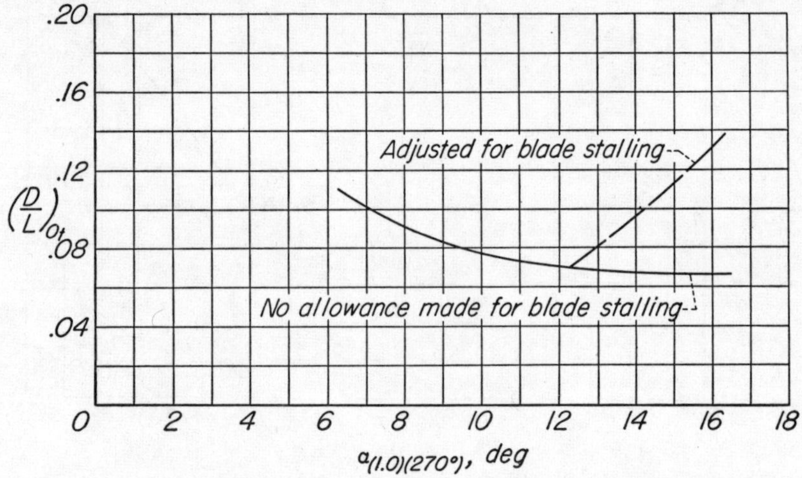

Fig. 10–7 Theoretical profile-drag losses adjusted for rotor stall.

exceeds the stall angle by about 4 degrees. The limiting flight conditions, from a vibration and control standpoint, therefore correspond to conditions of excessive power losses due to stall. For conventional helicopters, this extreme stall condition may result in a 20 per cent to 30 per cent increase in horsepower required.

Figure 10–7 (from reference II–7 of Appendix IIA) shows how the theoretical variation of a rotor drag-lift ratio with tip angle of attack for a tip-speed ratio μ of 0.25 would be modified by an allowance for stalling losses based on Fig. 10–6. The choice of $\mu = 0.25$ is pertinent because theory shows it to be nearly optimum as regards profile drag-lift ratio. The losses due to stall (dashed line in Fig. 10–7) are seen to outweigh greatly the gains otherwise anticipated from use of higher blade-section angles of attack. Further, since the optimum profile drag-lift ratio is shown to occur approximately as stall sets in, the lines of constant tip angle of attack, which were put on the theoretical drag-lift ratio charts of Fig. 9–2 as limiting the application of the theory, are seen to be applicable as lines of optimum performance as well.

Means for Delaying Blade Stall in Helicopter Design

While blade stall always imposes a barrier to high speed and high altitude operation of the helicopter, several means are available to extend the range of operating conditions free of stall. Some involve compromises in rotor efficiency in low-speed flight conditions; some involve only care on the part of the designer and, perhaps, compromises in cost of production. The important factors are discussed below.

IMPROVEMENT IN SECTION STALL CHARACTERISTICS. Very significant gains are available if the stalling characteristics of present-day rotor blades are improved. Two means are available for increasing the stalling angle: (1) irregularities in the section that bring on premature stall should be avoided and (2) airfoil profiles having higher stalling angles may be used insofar as is possible without producing drag increases at low angles and without producing large pitching moments. Then, too, the successful application of various high lift devices that would substantially increase the section stall angle without prohibitive drag increases in the high-velocity, low-angle-of-attack regions of the disk would offer large possibilities for increasing the top speed of helicopters.

REDUCTION IN FUSELAGE DRAG. A second method of avoiding stall from which only benefits accrue is cleaning up the fuselage, inasmuch as a reduction in parasite drag reduces the forward vector tilt, the inflow, and thus the angles of attack on the outer portion of the blade. (See discussion of Fig. 10–5.)

REDUCTION IN BLADE LOADING. Reductions in the design mean lift coefficient C_T/σ is a powerful means of avoiding blade stall (see Fig. 10–5) but this, unfortunately, usually involves losses in rotor efficiency in hovering and low-speed flight as already discussed. If a gear shift is used to provide lower blade loadings at high speed while retaining the efficient high loadings at low speed, its practicability becomes a question of weighing initial cost and increased empty weight against its advantages. Probably a more practical solution to the variable rotor speed problem is to use part of the weight which would be used in the two-speed transmission installation to "beef up" the engine bearings in order to allow engine operation at higher rotor speeds. The result would be an engine with a considerable range of rotational speeds, allowing the pilot to operate at lower pitch angles and higher speeds in high-speed or

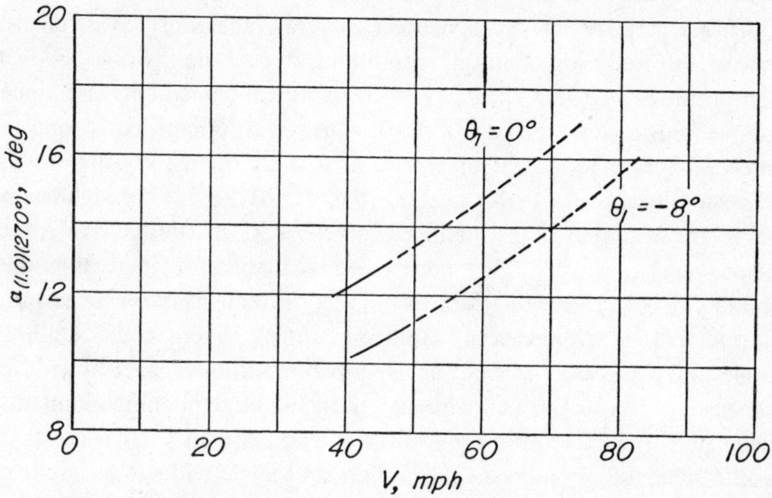

Fig. 10–8 Effect of blade twist on helicopter top speed.

high-altitude flight. This method has the significant advantage of using the increased weight to increase the reliability of the machine in normal operation as compared to decreasing the reliability by adding additional wearing parts to the system as does the two-speed transmission.

Another means of varying blade loading with flight condition which is extremely effective for high-speed flight involves a fixed lifting surface which unloads the rotor as forward speed is increased. Speeds in excess of 200 miles per hour should be possible with rotor combinations involving a fixed wing which unloads the rotor at high speed.

BLADE TWIST. Theory indicates, and flight measurements have shown, that blade twist is effective in delaying stall. Twisting the blade so as to lower the pitch at the tip with respect to the root pitch tends to distribute the lift more evenly along the blade and therefore avoids the high angle-of-attack region at the tip. Although the lower tip angles are obtained at the expense of somewhat higher angles inboard, the highest angles would still occur at the blade tip for normal amounts of twist (in the neighborhood of $-8°$).

The degree to which twist would be expected to delay the occurrence of high tip angles is illustrated in Fig. 10–8 for a typical two-place

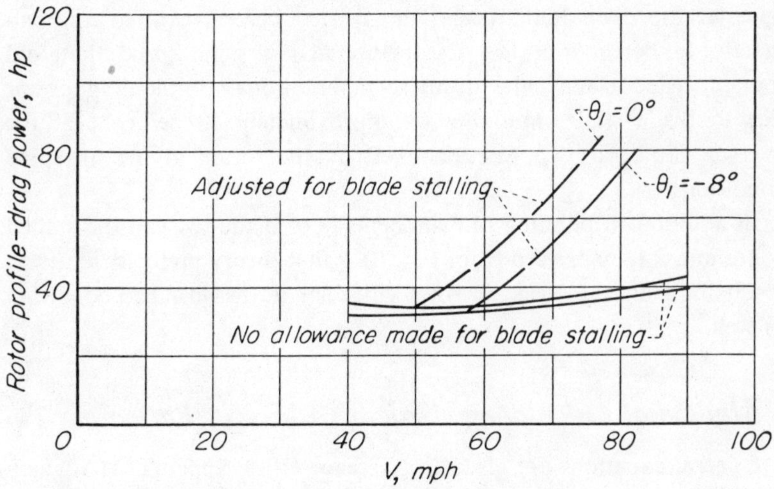

Fig. 10–9 Effect of blade twist on power required.

helicopter at typical operating conditions. The figure (from reference II–10 of Appendix IIA) shows that, at the same airspeed, the calculated tip angles of attack of the blades having $-8°$ twist are about 2.5° less than those of the untwisted blades over the speed range shown.

The effects of twist on rotor stall were experimentally determined by flight testing two sets of rotors that were identical except for the fact that one set had untwisted blades, whereas the other set had $-8°$ twist (reference II–3, Appendix IIA). The results of the tests, showing the effectiveness of twist in extending the speed range of the helicopter by delaying blade stall and in reducing the profile-drag power losses due to stall, are illustrated in Fig. 10–9. The figure gives the variation of profile-drag power with speed for the test helicopter at a typical operating condition. The solid-line curves in the figure represent the calculated profile-drag power with no allowance for stall, whereas the dash-line curves represent the theoretical power plus an experimental correction for blade stall. These experimental corrections were determined separately for the two rotors.

The results shown in Fig. 10–9 indicate that stall losses began at a speed 7 miles per hour (about 10 per cent) higher with the twisted blades than with the untwisted blades. The figure also shows that, once stall had developed on both rotors, the twisted blades required approximately 15 horsepower less to operate at the same speed than did the untwisted blades, the decrease in additional profile-drag power due to blade stall amounting to approximately 40 per cent of the average profile-drag power absorbed by the rotors in the unstalled conditions.

In addition to pointing out the benefits of blade twist in the stalled condition, it may be seen from Fig. 10–9 that theory predicts that twist results in a slight increase in rotor efficiency in the unstalled condition as well.

Method of Calculating Angle of Attack at Retreating Tip

General equations are given in reference II–18 Appendix IIA which define the angle of attack at any point on the blade at $\psi = 270$ degrees in terms of the variables λ, θ, μ, and θ_1 (the amount of linear blade twist). Calculations according to these equations have been included as the "limit lines" on the NACA profile-drag charts (Fig. 9–2), which show the combinations of the variables P/L, C_L/σ, and μ for which angles of 12 and 16 degrees are encountered at the retreating tip for rectangular, untwisted blades. These, in turn, have been presented in a form more convenient for engineering calculations in Figs. 10–10a and 10–10b. Figure 10–10a gives combinations of the variables C_T/σ, P/L, and μ which result in 12-degree angle of attack at the retreating tip. Figure 10–10b defines the 16-degree angle-of-attack conditions.

In calculating the speed at which stall is encountered for a given helicopter the following procedure is suggested:

(1) For the helicopter in question—for a given rotor diameter,

solidity, weight, altitude, and tip speed—calculate C_T/σ and the $\mu - P/L$ values (taken from the calculated power-required curve) for several forward speeds.

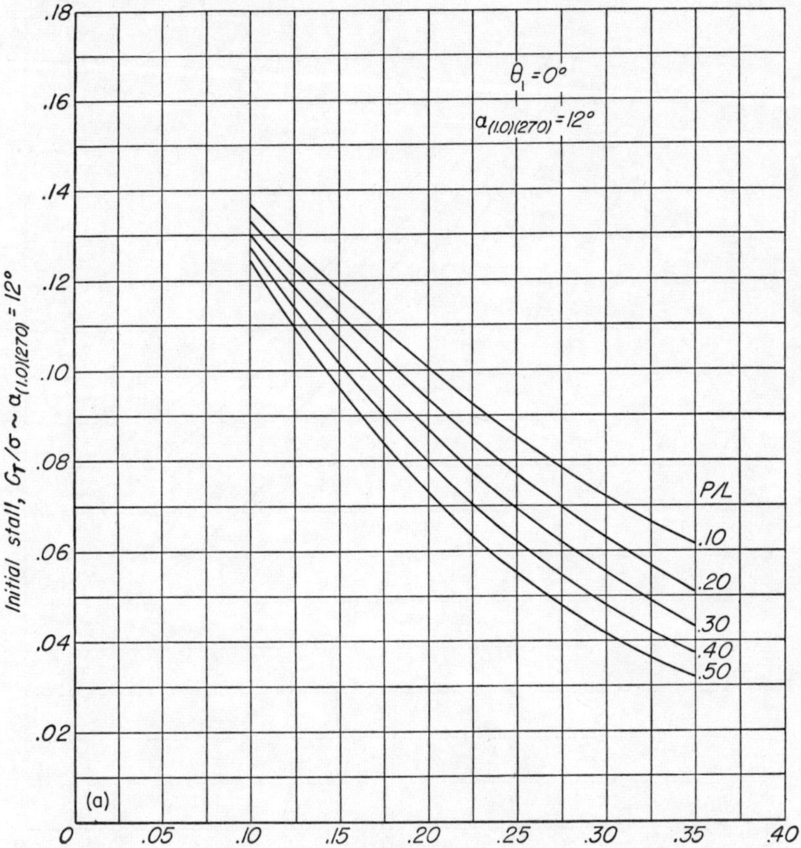

Fig. 10–10 Charts for determining tip angle of attack of retreating blade. (a) $\alpha_{(1.0)(270°)} = 12°$

(2) Determine C_T/σ for 12 degrees for each combination of μ and P/L from Fig. 10–10a (for three μ values, for example).

(3) Graphically interpolate to find the μ value corresponding to 12 degrees at the design C_T/σ value.

(4) Repeat above for 16 degrees using Fig. 10–10b. Conditions for intermediate angles may be determined with reasonable accuracy by linear interpolation between the 12-degree and 16-degree states.

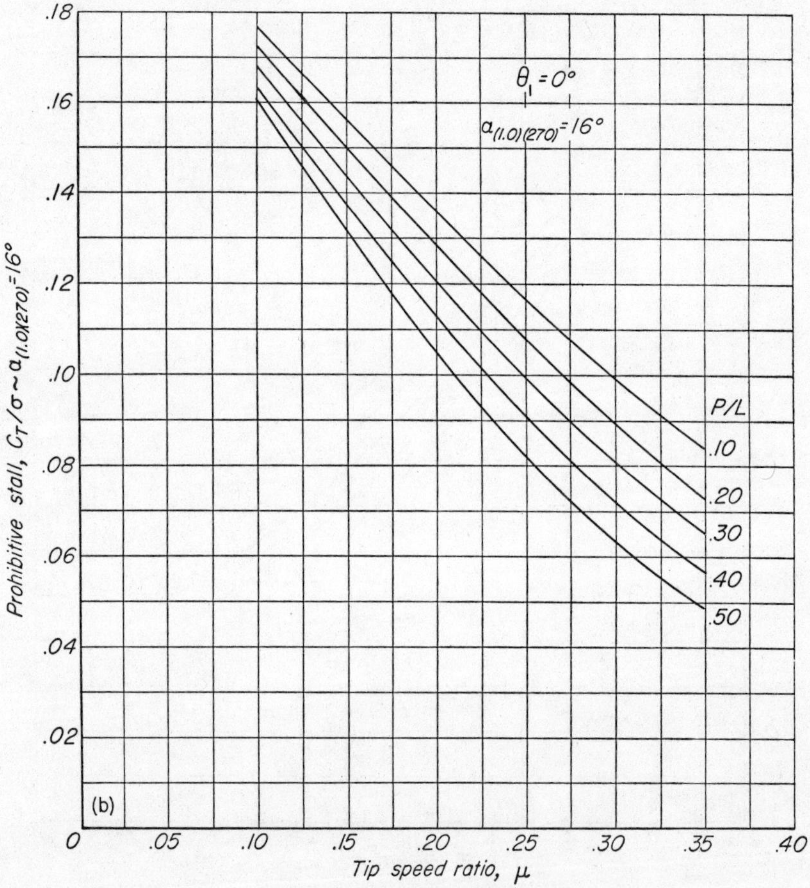

(b) $\alpha_{(1.0)(270°)} = 16°$

Method of Taking into Account Stall Losses in Performance Calculations

Using the experimentally determined fact that the actual profile power is about twice the calculated value when the calculated tip angle of attack exceeds the stall angle by 4 degrees, the following procedure allows estimation of power losses for a given helicopter.

(1) From the level flight power-required curve, determine the velocity at which tip angles of 12 and 16 degrees occur, according to the method previously outlined. (This method is especially applicable when the power-required curve is available from measurements. If the curve is being calculated by means of Fig. 9–2, it may be convenient to read off the stall angles from the performance charts at the same time. Also, the stall angle depends, of course, on the blade section and blade construction used. The flight tests previously reported indicate, however, that for blades of normal construction these values are reasonable.)

(2) Add an increment of power at the speed where the tip angle of 16 degrees was encountered equal to the calculated profile power of that point without stall.

(3) Fair a curve between the $\alpha_r = 12°$ point and $\alpha_r = 16°$ point so as to maintain a uniform rate of divergence from the original curve.

(4) Recalculate the $\alpha_r = 16°$ condition for the new power-required curve.

This condition will now occur at a somewhat lower speed because of the increased P/L value of the new curve. For this new point, the stall power increment may be recalculated as above and applied to the original curve. This recalculation provides some refinement in that in the theory-data comparison of reference II–7 (Appendix IIA), the calculations were based on the measured power-required curve which included the stall effects.

For high-speed and high-altitude performance calculations, where significant amounts of stall may be encountered, the above power corrections appreciably influence performance and should not be neglected.

11

AN INTRODUCTION TO HELICOPTER STABILITY[1]

The aim of stability and control theory is to enable an aircraft to be designed with satisfactory flying qualities. The flying, or handling, qualities of an aircraft have been defined as the stability and control characteristics that have an important bearing on the safety of flight and on the pilot's impressions of the ease of flying and maneuvering an aircraft. An introductory survey of helicopter stability would therefore most profitably concern itself with those factors that directly affect helicopter flying qualities.

NACA flight-test experience with various makes and types of helicopters (published in references VI–3 and VI–6 of Appendix IIA) has indicated that there are certain primary stability and control requirements which must be satisfied before the helicopter can be said to have satisfactory handling qualities. These requirements may, in some cases, depend solely on the mechanical design of the helicopter, control-friction limits being an example. In most other cases, the requirements are primarily aerodynamic in nature and depend on the response of the rotor and fuselage to control motions or atmospheric disturbances. It is the object of the present chapter to explain the sources of the most important of those aerodynamic stability and control characteristics that are known to affect helicopter flying qualities.

The subject will be introduced by a discussion of the means of heli-

[1] The material in this chapter first appeared in references VI–4 (Appendix IIA) and 22 (Appendix IIB).

copter control and of rotor damping caused by a pitching or rolling velocity; these two factors determining a very important handling characteristic—control sensitivity. Two additional important stability characteristics that will be discussed are static stability with speed and static stability with angle of attack, the latter being objectionably deficient in many present-day helicopters. The need for considering both these aspects of static stability will be brought out.

The dynamic stability of the helicopter in both hovering and forward flight will also be considered in this chapter. It will be shown that the dynamic behavior of the helicopter is to a large extent determined by the stability parameters previously mentioned.

Because the mathematical theory of helicopter stability is a complex subject to the non-specialist in stability theory, and because many of its conclusions have not yet been soundly established, it is considered desirable to separate the fundamental ideas of helicopter stability from the complicated mathematics surrounding them and to explain the fundamentals in a rather general fashion and in physical terms. One means of accomplishing this is to correlate the stability concepts of the helicopter with those of the more familiar and thus more easily understood airplane.

Only the single-rotor helicopter with fully articulated blades and a conventional control system is considered in this chapter as it is the fundamental configuration.

Symbols

The following list contains symbols that are used in this chapter and which are not defined, or are defined differently, in Appendix I. The sign conventions following the symbols are also pertinent to the stability discussion.

W gross weight of helicopter or airplane, pounds

L airplane or helicopter lift, pounds

V true airspeed of helicopter or airplane along flight path, feet per second

b airplane wing span, feet

S rotor disk area or airplane wing area, square feet

C_L airplane or helicopter lift coefficient $L/\frac{1}{2}\rho V^2S$

ω angular velocity of helicopter (pitching or rolling), radians per second

δ angular displacement of rotor cone due to angular velocity of helicopter, radians

ϵ angular displacement of rotor cone due to a control displacement or fuselage angle-of-attack change, radians

P period of oscillation, seconds

g acceleration due to gravity, feet per second per second

M pitching moment, foot-pounds

M_ω damping in pitch or roll (rate of change of pitching or rolling moment with pitching or rolling velocity), foot-pounds per radian per second

M_V stability with speed (rate of change of moment with translational velocity), foot-pounds per foot per second

M_α static stability with angle of attack (rate of change of moment with angle of attack) foot-pounds per radian

T_α rate of change of thrust with angle of attack, pounds per radian

C_m pitching-moment coefficient $M/\frac{1}{2}\rho V^2SR$ for helicopter; $M/\frac{1}{2}\rho V^2S\bar{c}$ for airplane, where $\bar{c}$ is mean aerodynamic chord

α helicopter or airplane fuselage angle of attack, degrees

l distance between rotor thrust vector and helicopter center of gravity in trimmed flight, feet

e offset of flapping hinges from the rotor center of rotation, feet

h height of rotor hub above helicopter center of gravity, feet

Helicopter nose-up moments, angular displacements, and angular velocities are assumed to be positive. For lateral motions from hovering, moments, angular displacements, and angular velocities which tend to raise the advancing side of the fuselage are positive. Changes in translational velocities in the direction of increasing velocity, as well as upward forces, are also positive.

Stability Definitions

The following stability definitions are given for terms used herein:

Trim. An aircraft is trimmed in steady flight when the resultant force and moment on the aircraft are equal to zero.

Aircraft stability. Stability is concerned with the behavior of an aircraft after it is disturbed slightly from the trimmed condition.

Static stability. An aircraft is statically stable if there is an initial tendency for it to return to its trim condition after an angular displacement or after a change in translational velocity from that condition; it is unstable if it tends to diverge from trim after being displaced. An aircraft is neutrally stable if it tends to remain in the condition to which it has been displaced.

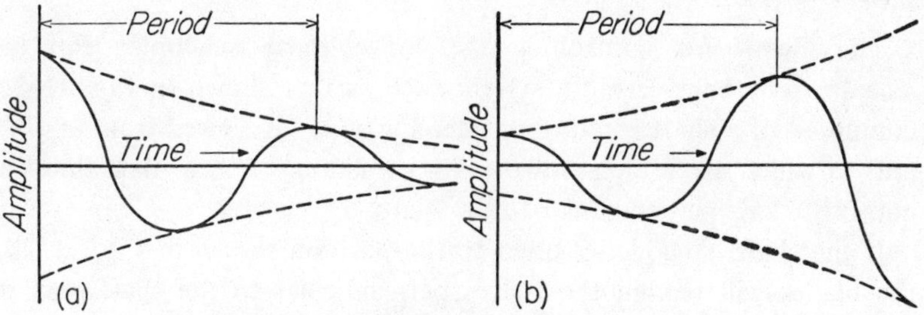

Fig. 11–1 Stable and unstable oscillations.
(a) Stable
(b) Unstable

Dynamic stability. The dynamic stability of an aircraft deals with the oscillation of the aircraft about its trim position following a disturbance from trim. Figure 11–1 illustrates a typical variation of amplitude of two oscillations with time. The period of these oscillations, which is defined as the time required for the oscillation to go through one cycle, is shown in this figure. If the envelope of the oscillation (dash line) decreases in magnitude with time, the oscillation is dynamically stable; if it increases with time, the oscillation is dynamically unstable. The time to double or half the amplitude of the oscillation is defined as the time necessary for the amplitude of the envelope to double or half. This quantity is a measure of the degree of stability or instability of the oscillation in that a small time to half the amplitude indicates a rapidly convergent or highly stable oscillation; whereas a small time to double amplitude indicates a rapidly divergent or highly unstable oscillation.

Rotor Characteristics

In order to best understand the discussion in the succeeding sections, it is worth while to review certain characteristics of rotor behavior, some of which were discussed from a different viewpoint in previous chapters. These characteristics follow.

Rotor Control

The means for controlling the conventional helicopter can be visualized by considering a system such as that shown in Fig. 11–2a composed of a shaft rotating counterclockwise (as viewed from above) and to which are attached two blades which are free to flap about a chordwise axis perpendicular to the shaft.

If the shaft is suddenly tilted to the position shown in Fig. 11–2b, the blades will realign themselves perpendicular to the shaft only if forces normal to the tip-path plane are produced as a result of the shaft tilt to force the blades to follow. Inasmuch as the blades are hinged, no mechanical forces can be transmitted. If the rotor were located in a vacuum, no aerodynamic forces would be produced and, hence, the tip-path plane would remain in its original position. Examination of the schematic, detailed views of the rotor hub in Figs. 11–2c and 11–2d shows that this condition of no tip-path plane tilt is mechanically possible. Under actual operating conditions, however, the tip-path plane will follow after a short interval of time because of the air forces that are produced as a result of the shaft tilt.

As can be seen in Fig. 11–2b, the tilt of the shaft causes the angle of attack of the blades to change cyclically. Thus, the blade moving to the left has an increased lift and moves up to a maximum positive displacement one-quarter revolution after the position of maximum lift. The blade moving to the right has a decreased lift and moves down to a maximum negative displacement one-quarter revolution after the position of maximum negative lift. Therefore, a short time later, the plane of rotation is again perpendicular to the rotor shaft as shown in Fig. 11–2e. Thus, although by tilting the shaft it was impossible to force physically the hinged blades to align themselves with the shaft, the tilt produced a cyclic change in blade angle of attack such that the air forces brought the blades into proper alignment. (See also discussion in Chapter 7.)

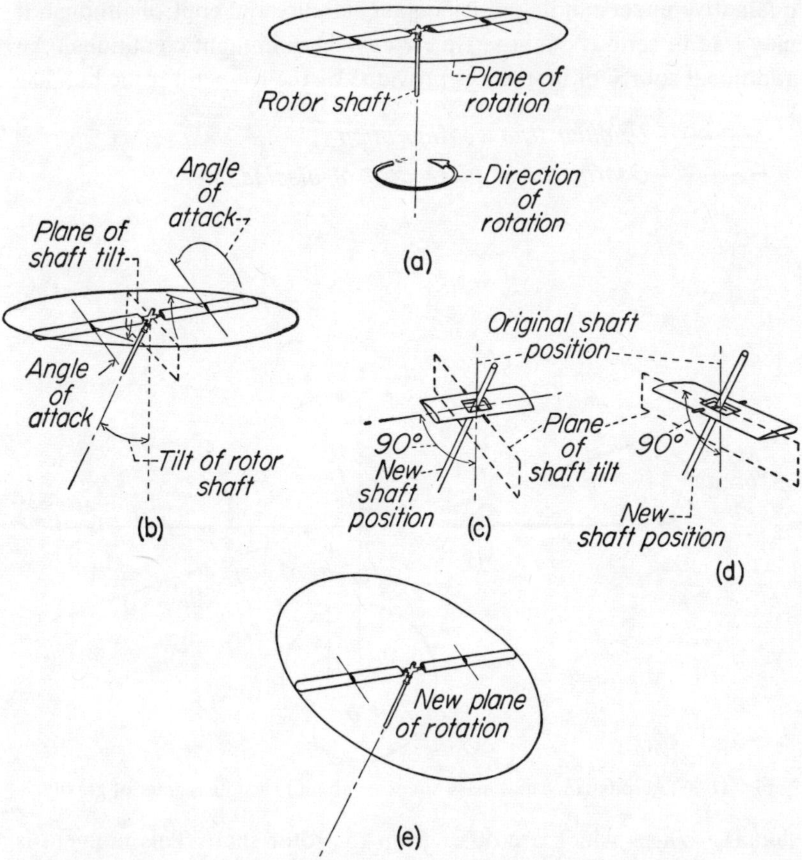

Fig. 11–2 The effect of a rotor shaft tilt on the plane of rotation.

This idea can be applied directly to a helicopter in that, if the rotor shaft is tilted, the rotor will quickly realign itself with respect to the shaft. A movement of the control stick of a conventional helicopter is equivalent to tilting the shaft with respect to the fuselage. The resulting tilt of the rotor with respect to the fuselage will produce a moment about the helicopter center of gravity because the rotor thrust, which acts approximately perpendicular to the tip-path plane, is displaced from the center of gravity. (In this discussion, rotor thrust is assumed to act perpendicular to the tip-path plane. This assumption is sufficient for a qualitative understanding of helicopter stability and control although it may lead to serious quantitative errors in certain flight conditions.) An additional source of moment is provided by the rotor tilt if the hub has

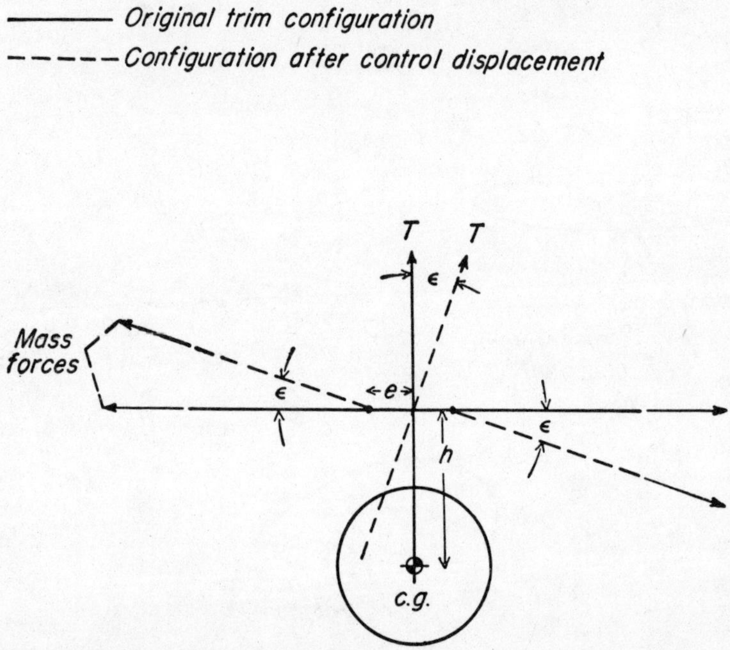

Fig. 11–3 Aerodynamic and mass moments about helicopter center of gravity.

flapping hinges which are offset from the rotor shaft. This moment is caused by the centrifugal force on the blades.

The two sources of moment are illustrated in Fig. 11–3, which shows that the moment about the helicopter center of gravity produced by a unit stick displacement (which may be defined as the control power) is increased by an increase in the vertical distance between the rotor hub and the helicopter center of gravity, or by an increase in the flapping-hinge offset. Although the vertical height of the center of gravity is not usually a design variable, control power may be adjusted to a considerable extent in the design stage through the use of offset flapping hinges. An increase in control power through the use of offset flapping hinges is desirable in order to maintain adequate control during maneuvers in which rotor thrust is reduced, inasmuch as this source of moment is independent of rotor thrust. Offset flapping hinges also permit an increase in the allowable center-of-gravity range of the single-rotor helicopter. (The moment caused by a shift in center of gravity must be compensated for by a control moment in order to maintain the helicopter in trim. Inasmuch as the amount of control displacement is limited for mechanical reasons, an increase in the amount of control power is the feasible way to allow for a greater center-of-gravity travel.)

Damping in Pitch (*or Roll*)

The foregoing discussion points out that because of blade inertia, some delay exists between a rapid shaft tilt and the realignment of the rotor with the shaft. Thus, if the shaft continues to tilt, the tip-path plane will continue to lag behind the rotor shaft and in so doing supplies the aerodynamic moment necessary to overcome continuously the flapping inertia of the rotor during steady pitching or rolling.

If the aerodynamic and inertia flapping moments are equated, the following result for the angular displacement of the rotor plane with respect to the shaft per unit tilting velocity of the shaft is obtained for the hovering case:[1]

$$\frac{\delta}{\omega} = \frac{16}{\gamma\Omega} \tag{1}$$

The dimensions of the quantities of either side of equation (1) will be noted to be the units of time. The quantity $16/\gamma\Omega$ can be interpreted physically as follows: if the rotor shaft is tilting at any constant angular velocity, the thrust vector reaches a given attitude in space $16/\gamma\Omega$ seconds after the rotor shaft has reached that attitude.

If a helicopter is tilted at an angular velocity ω, as shown in Fig. 11–4, the ensuing lag of the rotor plane displaces the thrust vector and thus produces a moment about the center of gravity. If offset hinges are present, there is an additional moment caused by the tilt, as explained

[1] A derivation of equation (1) may be found in reference 28 (Appendix IIB). The effect of tip-speed ratio, which is of second order, may be easily derived.

in the section titled "Rotor Control". Moments attributable to tilting velocity are known as *damping in pitch* or *damping in roll*, depending upon the axis about which the tilting occurs and can be expressed mathematically as $\Delta M/\Delta\omega$ or M_ω. Because this moment is opposite to the tilting velocity for the conventional rotor with the assumption that the thrust is perpendicular to the tip-path plane, M_ω is stabilizing and, according to the convention, negative in sign.

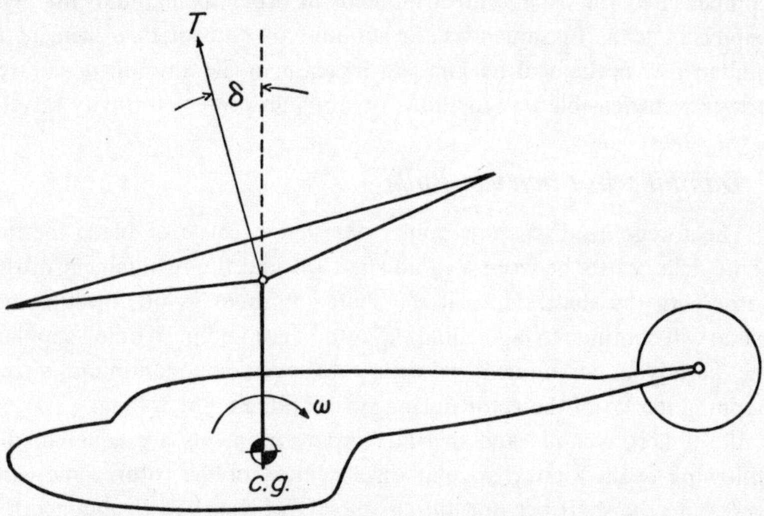

Fig. 11–4 Damping moment arising from pitching velocity.

As indicated in equation (1), the amount of rotor tilt for a given angular velocity is inversely proportional to the rotor speed and to the blade mass factor γ (and, hence, directly proportional to blade moment of inertia). It follows, therefore, that small helicopters, which operate at high rotor speed in order to maintain reasonable tip speeds, will tend to have less rotor damping than large helicopters. The value of blade mass factor γ is reduced to a considerable extent for helicopters powered by blade tip-jet units. Tip-jet powered helicopters would therefore be expected to have considerably more rotor damping than conventionally powered helicopters. In addition, γ may vary appreciably because of differences in blade construction, although this effect is less marked than the effect of adding tip units.

In addition to the effects of rotor speed and γ, rotor damping may be increased by the use of devices that act upon the control system in such a manner as to increase the displacement of the rotor from its trim position due to a given rate of roll or pitch. An example of such a device is a rate gyro that would apply opposite control by an amount proportional to the rolling or pitching velocity of the helicopter. Increased effective damping is achieved by the Bell Stabilizer Bar (Fig. 2–12) and the Hiller Control Rotor (Fig. 2–16), both of which act on the same principle as the rate gyro.

Thus far, design factors and devices that affect rotor damping by affecting the amount of rotor tilt for a given pitching or rolling velocity have been discussed. Rotor damping may also be varied by changing the moment caused by a given rotor displacement. The damping moment produced by a given rotor displacement will depend upon the rotor height and the amount of offset of the flapping hinges.

Control Sensitivity

The combination of control power and damping in roll (or pitch) together determine an important flying-qualities characteristic. This characteristic is called *control sensitivity* and may be defined as the maximum rate of roll (or pitch) achieved by a unit displacement of the controls. Control sensitivity may be defined in three alternate ways as follows:

$$\text{Control sensitivity} \begin{cases} = \dfrac{\text{control power}}{\text{rotor damping}} \\[2ex] = \dfrac{\dfrac{\text{control moment}}{\text{stick displacement}}}{\dfrac{\text{damping moment}}{\text{angular velocity}}} \\[2ex] = \dfrac{\text{angular velocity}}{\text{stick displacement}} \end{cases}$$

Physically, the manner in which control power and damping determine control sensitivity may be understood from the following argument. If the control stick is displaced (laterally, for example) and held, the helicopter will initially accelerate angularly at a constant rate that is inversely proportional to the moment of inertia of the helicopter about its longitudinal axis. [This result follows from Newton's law $M = I(d\omega/dt)$.] As the angular velocity builds up, the opposing damping-in-roll moment increases in proportion until an angular velocity is reached at which the damping moment is equal to the control moment. The helicopter is therefore stabilized at that angular velocity, because the resultant moment on the helicopter is zero. It is thus apparent that if the rotor damping is large with respect to the control power, then the maximum rate of roll reached by the helicopter by a given stick displacement would be small, inasmuch as a sufficiently large damping moment would be produced at a small rolling velocity to balance the control moment. Alternately, it is clear that if rotor damping is small with respect to the control power, then the maximum angular velocity attained by a given stick displacement would be large.

Helicopters with conventional control systems are subject to high control sensitivity. In fact, according to reference VI–6 (Appendix IIA), the maximum rate of roll achieved by a small, two-place helicopter may be as great as those of some modern fighter airplanes at the speeds for their maximum rates of roll. This is true not because of high control power, but rather because of low damping, which, for the helicopter, is a fraction of that for airplanes. This same reference goes on to state that high control sensitivity can lead to overcontrolling, which in turn results in a short-period, pilot-induced lateral oscillation.

It is worth while to point out which of the physical characteristics of the helicopter can be varied so as to reduce excessive control sensitivity. The height of the rotor and the offset of the flapping hinges do not affect control sensitivity because they change control power and rotor damping in proportion. Design factors and devices which increase rotor damping without affecting control power result in reduced control sensitivity. Thus, large helicopters operating at low rotor speed, helicopters with tip-jet units, and helicopters with devices such as the Bell Stabilizer Bar and the Hiller Control Rotor will have more desirable values of control sensitivity.

Although rotor height and flapping-hinge offset do not affect control sensitivity, the ratio of these values to the fuselage moment of inertia will determine to a large extent the time necessary to reach the maximum angular velocity.

Rotor Static Stability with Speed and with Angle of Attack

Inasmuch as an aircraft can be displaced in pitch (angle-of-attack change) or by a change in forward speed, two aspects of static stability exist because of the two sets of forces and moments produced by these two changes.

ROTOR STABILITY WITH SPEED. Consider a rotor mounted on a helicopter which is subjected to a translational velocity. The effect of

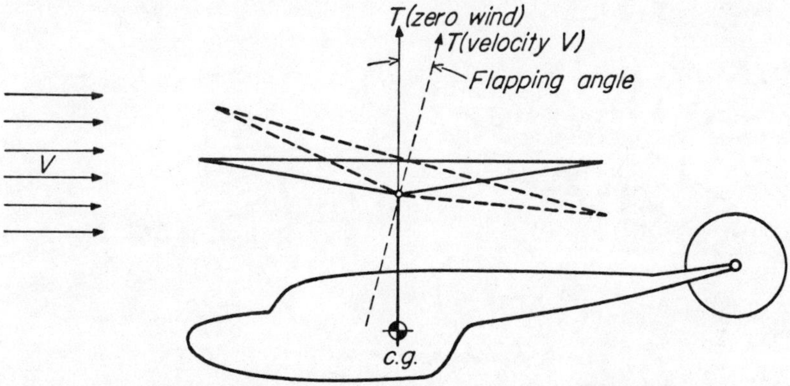

Fig. 11–5 Source of moment due to speed change.

this translational velocity is to tilt the tip-path plane in a direction away from the velocity of translation as shown in Fig. 11–5. This tilting of the rotor plane is a result of blade flapping. The rotor plane will tilt farther backwards (that is, flapping will increase) with increasing translational speeds, inasmuch as the velocity of the advancing blades becomes increasingly greater than the velocity of the retreating blades. Figure 11–5 indicates that this tilt of the rotor plane due to translational velocity will produce a moment about the helicopter center of gravity. The moment will be nose-up with increasing speed and nose-down with decreasing speed. The variation of moments due to changes in translational velocity is a measure of stability with speed, which can be expressed mathematically as $\Delta M/\Delta V$ or M_V. Inasmuch as nose-up moments are considered positive, M_V is always positive for the conventional helicopter rotor.

ROTOR STABILITY WITH ANGLE OF ATTACK. As was shown in Fig. 11–2, a change in attitude of the hovering helicopter (which is prevented from translating) results in an equal tilt of the rotor plane with the result that no rotor moment or change in thrust occurs. Thus, the hovering helicopter has neutral stability with attitude change. In forward flight, however, a change in longitudinal attitude (fuselage angle of attack)

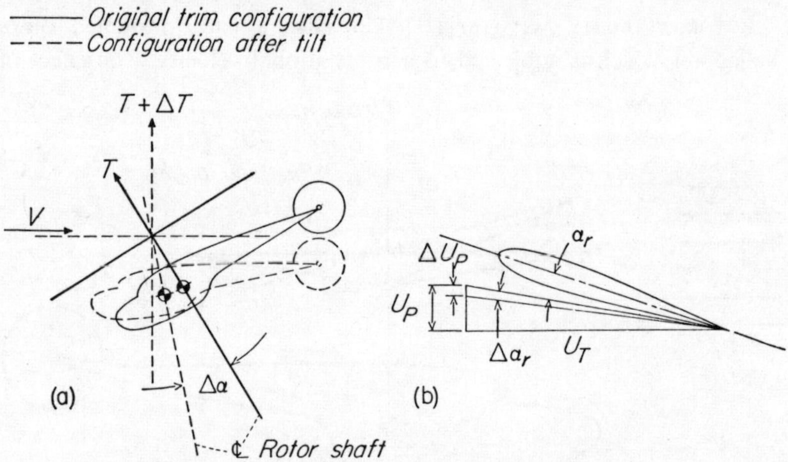

Fig. 11–6 Source of moment due to angle change.
(a) Side view of helicopter
(b) Typical blade element

will produce a rotor moment and a thrust change. This moment, which is due to a change in fuselage angle of attack at constant velocity, arises from the change in flapping (tilt of the rotor plane relative to the fuselage) and can be understood by an examination of Fig. 11–6.

Consider a nose-up change in fuselage angle of attack α from the trim value as shown in Fig. 11–6a. The changes in relative velocities and angle of attack of a typical blade element, which result from this change in fuselage angle, are shown in Fig. 11–6b where U_P, U_T, and α_r represent trimmed values. The change in blade-section angle of attack $\Delta\alpha_r$ is equal to $\Delta U_P/U_T$ (for the usual assumption of small angles included in helicopter analyses), and the change in lift at this section, which is proportional to $\Delta\alpha_r U_T^2$, is therefore proportional to $\Delta U_P U_T$. Inasmuch as ΔU_P is constant over the rotor disk (the component of flight velocity through the disk is constant over the disk), the change in lift due to the change in fuselage angle of attack is greater on the advancing blade where U_T is highest. This unequal increase in lift between the advancing and retreating blades is compensated for by increased flapping or a backward tilt of the rotor cone with respect to the fuselage. At the same time, the increased lift at all sections results in an increase

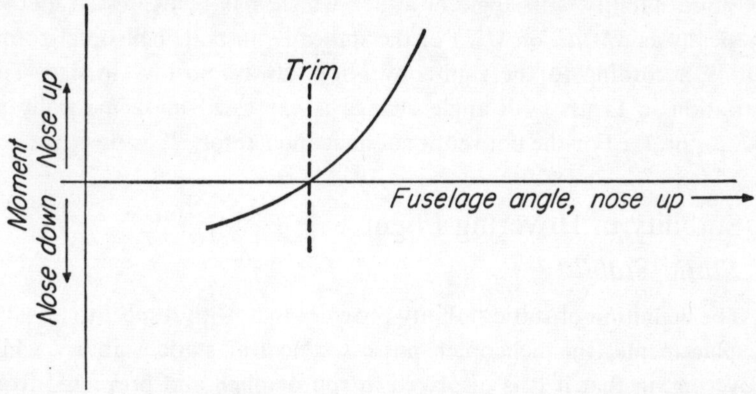

Fig. 11–7 Moment contributed by rotor about center of gravity as a function of fuselage angle of attack.

in the magnitude of the rotor thrust.

Figure 11–6a shows that this tilt of the thrust vector with respect to the fuselage, which results from the nose-up change in fuselage angle, produces a nose-up moment about the fuselage center of gravity which is accentuated by the increased magnitude of the rotor thrust. If a nose-down change in fuselage angle had been considered, the result would have been a forward tilt of the rotor cone relative to the fuselage and a reduction in thrust. Inasmuch as a change in angle results in a change in magnitude as well as a tilt of the thrust vector, doubling a nose-up change in angle more than doubles the nose-up moment. Conversely, doubling a nose-down change in angle results in less than a doubled nose-down moment but nevertheless a nose-down moment.

The preceding discussion shows that the variation of moment about the center of gravity with angle of attack at constant speed for the helicopter rotor would be as shown in Fig. 11–7. This figure shows that the rotor is unstable with fuselage angle of attack and that a given change in angle of attack from trim produces a greater moment change in the nose-up direction than in the nose-down direction. The figure also shows that the instability with angle of attack becomes greater with larger nose-up angle-of-attack changes and smaller with larger nose-down angle-of-attack changes.

The variation of moment due to changes in fuselage angle is a measure of static stability with angle of attack which may be expressed mathematically as $\Delta M/\Delta\alpha$ or M_α. For the statically unstable helicopter rotor, M_α is, according to the sign convention, always positive in sign. The variation in thrust with angle change is expressed mathematically as $\Delta T/\Delta\alpha$ or T_α. For the conventional helicopter rotor, T_α is positive.

Stability in Hovering Flight

Static Stability

The definition of static stability provides that, with respect to angular displacements, the helicopter possesses neutral static stability while hovering, in that if it is displaced in roll or pitch and prevented from translational motion, no moments will arise to tend to restore it to its original position. The concept can be understood by remembering that the resultant rotor thrust always passes through the helicopter center of gravity irrespective of the angular position of the helicopter.

It might be pointed out that the conventional fixed-wing airplane in forward flight is also neutrally stable in roll in that no restoring or upsetting moments are produced when the airplane is displaced in roll. Although no restoring moments will be produced by the angular displacement of the airplane, this displacement will result in a lateral velocity due to the unbalanced lateral component of lift force. Once the airplane is moving laterally, the dihedral of the wings, combined with the sideslip velocity, produces a moment tending to reduce its lateral velocity by tilting the airplane in a direction opposite to its initial tilt. This effect can be seen in Fig. 11–8. Thus, an airplane with sufficient wing dihedral is statically stable with regard to changes in lateral velocity.

A similar situation exists for the hovering helicopter. An angular displacement of the helicopter, while directly producing no restoring moment, will result in a translational velocity due to the unbalanced horizontal component of the thrust force. As a result of the velocity, a moment is produced which tilts the helicopter so that the horizontal component of the thrust vector acts to reduce the translational speed to

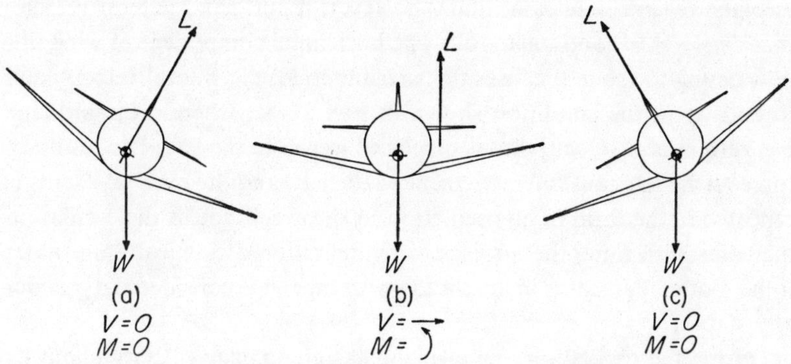

Fig. 11–8 Lateral motion of a fixed-wing airplane.

its initial zero value. Thus, because of its positive rotor stability with speed, the helicopter is statically stable with regard to changes in translational velocity. The moment produced by a translational velocity should be noted to be analogous to the moment produced by wing dihedral and sideslip velocity for the fixed-wing airplane in forward flight.

Dynamic Stability in Hovering

The dynamic behavior of the hovering helicopter when upset in roll or pitch can best be explained by first examining the elements that influence the behavior of the fixed-wing airplane in forward flight when upset in roll, inasmuch as the behavior of both aircraft in these conditions is similar in many respects.

ANALOGY WITH THE AIRPLANE. In order to study the dynamic behavior of the airplane, a more detailed discussion of its behavior when displaced in roll is desirable. Consider again the airplane displaced in roll to the right as in Fig. 11–8a. A resultant force to the right that causes the airplane to sideslip to the right can be observed. Once the airplane is moving laterally, the dihedral of the wings combined with the sideslip velocity produces a moment tending to restore the airplane to a level attitude as in Fig. 11–8b. If the airplane is assumed to be restrained from yawing about its vertical axis so that no other effects are present, this moment will succeed in leveling the airplane. When the airplane reaches a level attitude, however, it still has a lateral velocity that causes it to continue to roll. The horizontal component of wing lift, now acting to the left, causes the airplane to lose its lateral velocity and to end up in the condition shown in Fig. 11–8c, wherein the airplane has zero lateral velocity but is displaced in roll to the left. The resultant force to the left causes a movement to the left, and the cycle of events is repeated in the form of an oscillation. If the amplitude of the oscillation increases with time, the airplane is by definition dynamically unstable; if the motion decreases in amplitude with time, it is considered dynamically stable.

During the oscillation, the airplane has an angular velocity about its longitudinal axis. At the instant when the airplane is in the position shown in Fig. 11–8b, for example, it is rolling to the left. The result of the rolling velocity is to reduce the angle of attack of the right wing. (See Fig. 11–9.) Similarly, the angle of attack of the left wing is increased. Thus a clockwise moment is produced that tends to oppose the counterclockwise angular velocity of the airplane. The initial angular displacement of an airplane thus results in an oscillation during which the airplane is acted upon by two opposing moments: the first, a moment produced by the sideslip velocity; and the second, a damping moment produced by the angular velocity of the airplane.

HELICOPTER MOTION FOLLOWING A DISTURBANCE. The motion following an initial angular displacement of a helicopter, as well as the moments acting on it during the oscillation, is analogous to the motion (and moments) just described. Just as for the airplane, it is desirable in the study of the dynamic behavior of the hovering helicopter to discuss the motion of the helicopter following an angular displacement in greater detail than was done in the section entitled "Static Stability."

If the hovering helicopter is displaced in roll to the right (Fig. 11–10a), the resultant force to the right will cause the helicopter to move to the configuration shown in Fig. 11–10b. The helicopter, in moving from the position of Fig. 11–10a to that of Fig. 11–10b, is subjected to a counterclockwise moment because of its stability with speed. This moment rolls the helicopter until it reaches the configuration shown in Fig. 11–10c.

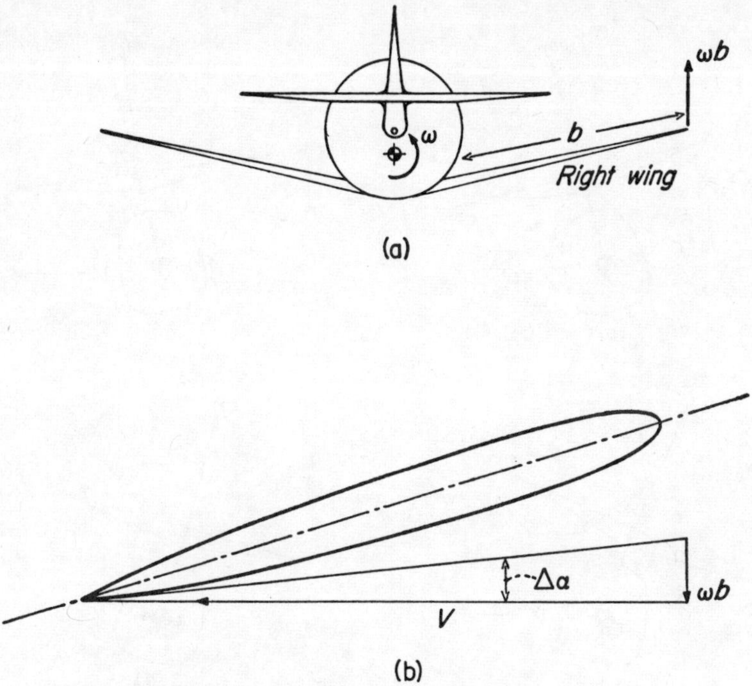

Fig. 11–9 Airplane damping due to roll.
(a) Rear view
(b) Side view of right wing

A horizontal force now tends to slow down the helicopter, so that it returns to zero horizontal velocity in the position of Fig. 11–10d. Because a horizontal force to the left is now present, the helicopter starts to move to the left. By proceeding in the manner described for the first half of the cycle, the helicopter reaches the position shown in Fig. 11–10a, at which time one cycle of the oscillation will have been completed, and the process repeats.

Just as is true of the fixed-wing airplane in a lateral oscillation, the helicopter has an angular velocity about its own axis during the oscillation which also results in a moment due to damping in roll. This moment has an important effect on the oscillation. Examine the position of the helicopter shown in Fig. 11–10c. At this instant the helicopter has a counterclockwise angular velocity which causes a small clockwise tilt

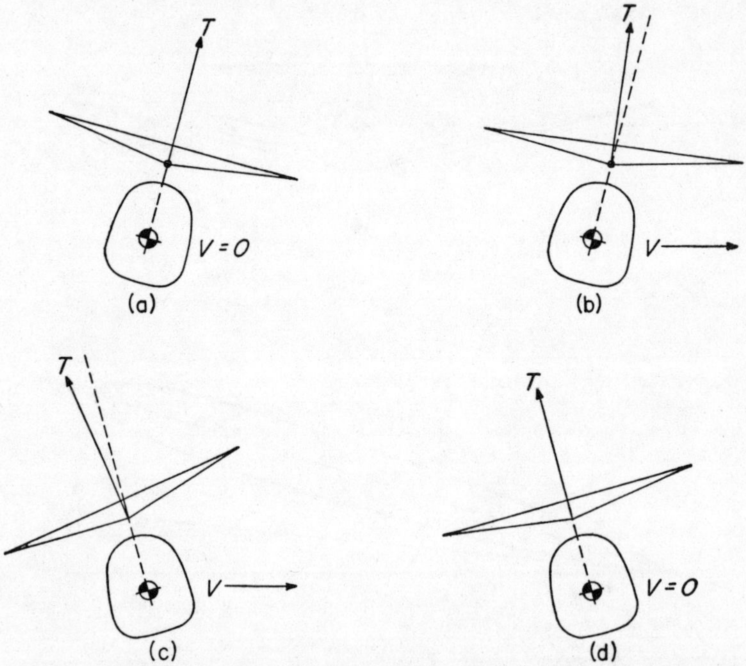

Fig. 11–10 Oscillation of a helicopter following an angular displacement in hovering (damping neglected).

of the rotor cone from that shown with damping neglected. The actual configuration of the rotor, with damping considered, is as shown in Fig. 11–11. As can be seen from this figure, the angular velocity of the helicopter causes the rotor cone to lag behind the position it would have if no damping were present.

Thus far the separate effects resulting from an angular displacement in attitude of the helicopter have been examined. It has been seen that the result of the displacement is an oscillation, and it will now be shown that stability with speed and damping in pitch (or roll) are most important in influencing the period of the oscillation. (The factors that influence the divergence or convergence of the oscillation are indicated subsequently herein.)

In order to examine the combined effects of stability with speed and damping in roll, the motion following an angular displacement of a hovering helicopter is examined in successive steps. For the sake of clarity, stability with speed and damping in roll are assumed to act

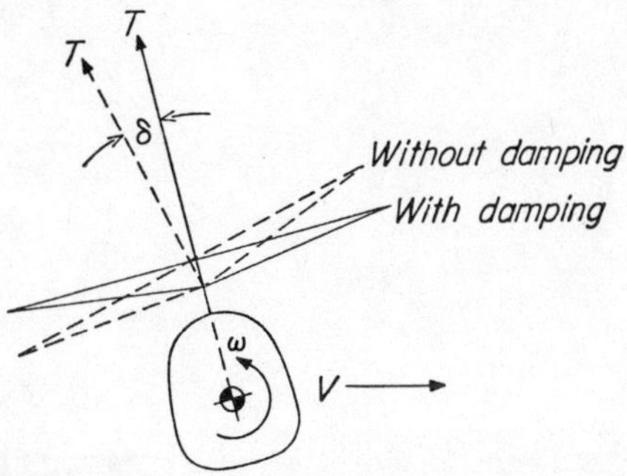

Fig. 11–11 Position of rotor cone with and without damping for helicopter of Fig. 11–10c.

successively, although their effects actually occur simultaneously. Each of the following cycles of events should, therefore, be considered as occurring over a very short interval of time. Also, the moment of inertia of the fuselage is assumed to be negligible for the immediate discussion.

Consider a hovering helicopter displaced in roll (or in pitch) to an attitude shown in Fig. 11–12a. Although no moment is produced about the center of gravity of the helicopter, a resulting force occurs to the right which will cause a velocity to the right, and the helicopter is displaced to the configuration of Fig. 11–12b. In this configuration, the thrust vector has been inclined to the left and produces a counterclockwise moment about the center of gravity as a result of stability with speed. Inasmuch as the fuselage moment of inertia was assumed to be negligible, this moment in turn quickly produces a counterclockwise angular velocity so that in a short interval of time, the helicopter is in the configuration of Fig. 11–12c. Because of damping in roll, the counterclockwise angular velocity has permitted the fuselage to overtake the rotor cone, so that, after a negligible interval of time, the rotor

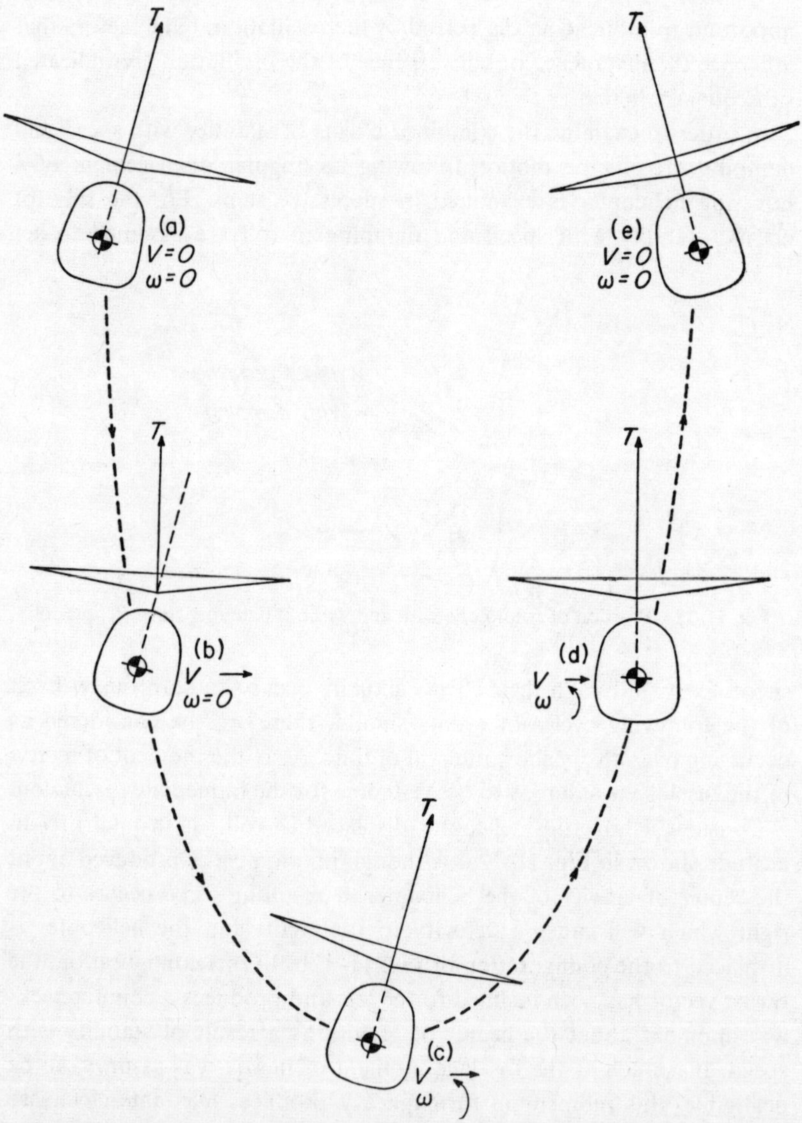

Fig. 11–12 Lateral oscillation of a helicopter following an angular displacement in hovering (damping included).

tilt originally produced by the speed stability is neutralized. Inasmuch as a horizontal component of force to the right still exists, the helicopter continues to accelerate in that direction and the process is repeated; that is, the additional translational velocity causes an additional thrust vector tilt to the left which produces a counterclockwise moment and an increase in angular velocity. Because of the damping in roll, this increased angular velocity permits the fuselage to align itself with the thrust vector so that again, after a negligible time interval, the additional tilt produced by the speed stability is neutralized.

Because each cycle has thus far rotated the helicopter toward a level attitude, the helicopter soon attains a horizontal attitude as shown in Fig. 11–12d. The previous cycles of events continue to occur in the same way except that from now on the thrust vector is tilted to the left, and the velocity of the helicopter is thus reduced until it reaches the position of Fig. 11–12e where it has zero angular and translational velocity. This position corresponds to that of Fig. 11–12a. Because a horizontal component to the left is still present, the helicopter starts to move left, the process represented by Figs. 11–12a to 11–12e is repeated, and the helicopter continues to oscillate back and forth. The time required for the helicopter to move from the position shown in Fig. 11–12a to that of Fig. 11–12e is one-half the period of the oscillation.

In reference 28 (Appendix IIB), a formula is derived for the period of the oscillation of a hovering helicopter having zero fuselage moment of inertia, which can be written as

$$P = \frac{2\pi}{\sqrt{g}} \sqrt{\frac{-M_\omega}{M_V}} \tag{2}$$

The formula for damping in roll, M_ω, is approximately $-Th(\delta/\omega)$. The speed stability M_V can be approximately calculated from an equation that represents the variation of longitudinal flapping with translational velocity.

From the preceding discussion, the effect of stability with speed and damping in pitch on the period can be physically interpreted. Consider the helicopter moving from the position shown in Fig. 11–12a to that shown in Fig. 11–12b. The larger the speed stability, the greater the thrust vector tilt in Fig. 11–12b. A larger angular velocity results and, therefore, the position of Fig. 11–12c is reached more quickly. An increase in speed stability thus reduces the period of the oscillation. Equation (2) gives the same result inasmuch as the speed stability term appears in the denominator. The effect of speed stability on period appears to explain the experimentally observed difference noted in reference IV–6 (Appendix IIA) between the period of the pitching and the rolling oscillation for the conventional single-rotor helicopter. If the tail rotor shaft, for example, is mounted above the center of gravity, the tail rotor will add to the helicopter's speed stability during lateral motion, and thus the period will be decreased. This effect arises from the change in tail rotor thrust due to the change in inflow that occurs while the tail rotor is experiencing a lateral velocity.

The effect of damping in roll can be seen by comparing Figs. 11–12b and 11–12c. The larger the damping in roll, the smaller the angular velocity necessary to neutralize the thrust vector tilt that was produced by the speed stability in Fig. 11–12b. Slower changes in attitude result and thus the position of Fig. 11–12c is reached later than if less damping were present. An increase in damping in roll thus increases the period of the oscillation. Equation (2) gives the same result inasmuch as the damping-in-roll term appears in the numerator.

According to the mathematics of reference 28 (Appendix IIB), the presence of a finite fuselage moment of inertia results in a higher period of the oscillation than that given by equation (2). The general effects, however, of speed stability and damping in roll are believed to be valid also for the case of finite moment of inertia.

Although a physical representation of the effect of the various parameters on the convergence or divergence of the hovering oscillation is difficult, their effects have been investigated theoretically. In reference 23 (Appendix IIB), it was concluded that the dynamic instability of the conventional helicopter in hovering flight can be reduced by decreasing the moment of inertia of the helicopter fuselage, by increasing the moment of inertia of the rotor blades about their flapping hinges (which increases the damping in pitch), by increasing the vertical height of the rotor above the center of gravity of the helicopter, and by offsetting the flapping hinges from the center of the rotor.

Usually, most of these factors are fixed by other design considerations and therefore cannot be easily varied. Inasmuch as single-rotor helicopters with conventional control systems have shown themselves to be dynamically unstable, means for improving the dynamic stability characteristics of helicopters by the addition of special devices which act upon the control system have been discussed in several papers, among which are references 25 and 26 (Appendix IIB).

Longitudinal Stability in Forward Flight

Static Stability

As was done in the study of stability of the helicopter in the hovering condition, some airplane stability concepts are useful in the interpretation of the physical parameters affecting helicopter static stability in forward flight.

ANALOGY WITH THE AIRPLANE. Inasmuch as an airplane can be displaced from trim in pitch (angle-of-attack change) or by a change in forward speed, in general two aspects of static stability exist because of the two sets of forces and moments produced by these two changes.

If an airplane is flying in a trimmed position and the angle of attack is increased while its speed is kept constant, the airplane is statically stable with respect to angle of attack if the resulting aerodynamic moment is a nose-down moment. (Airplane static stability with angle of attack is dependent upon center-of-gravity position, inasmuch as variations in center-of-gravity position affect the moment arm of the lift forces on the wing and tail.)

Consider now the static stability of an airplane with changes in speed but with angle of attack kept constant. If power and Mach number effects are neglected, which is justified for the present discussion, a variation in speed from trim speed while the angle of attack and flight path are kept constant (as could be done in a wind tunnel) produces no aerodynamic moment about the center of gravity. In other words, the airplane is neutrally statically stable with speed at constant angle of attack and flight path angle because no change is obtained in lift or moment coefficients with speed. A given speed change from trim merely changes all of the aerodynamic forces and moments acting on the airplane in the same proportion, and the airplane is thereby maintained in trim.

With these concepts in mind, the static stability of a given airplane with fixed center-of-gravity location can be expressed by the plots of moment coefficient against angle of attack and speed of Fig. 11–13, data

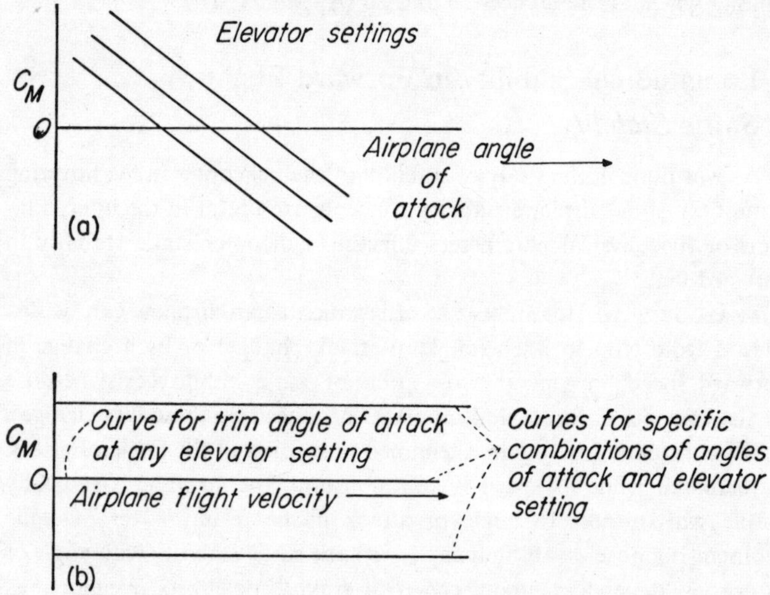

Fig. 11–13 Basic static stability curves of airplane in gliding flight.
(a) C_m against α
(b) C_m against V

for which can be obtained from wind-tunnel tests. Because the moment coefficient at constant angle of attack and control deflection is independent of speed as shown in Fig. 11–13b, the single static-stability curve of Fig. 11–14, which does not depend on speed, can be obtained from Fig. 11–13a alone. Figure 11–14 was obtained from Fig. 11–13a by picking off the elevator settings and their corresponding trim angles of attack, the trim angles of attack being readily converted to lift coefficients. This type of plot is the conventional way of representing the static stability of an airplane because it can be easily obtained from flight tests of an airplane trimmed in steady level flight (that is, $L = W$, $C_M = 0$). A positive slope to the curve of Fig. 11–14 means that the airplane is stable stick fixed, in that a forward movement of the control stick (or down elevator) is required for trim at a decreased airplane angle of attack (or C_L).

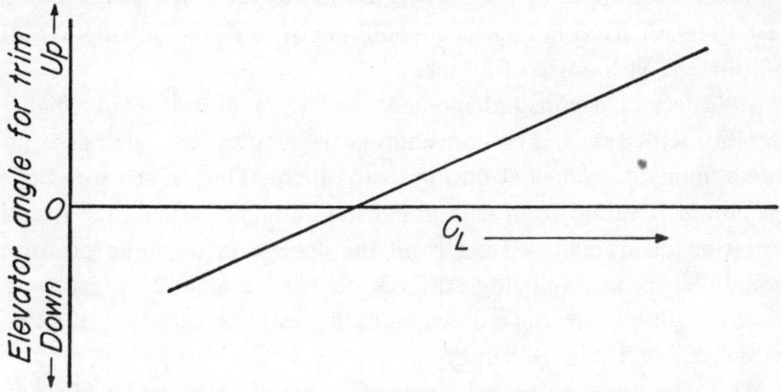

Fig. 11–14 Static stability of airplane as measured in gliding flight.

It should be emphasized that the single curve in Fig. 11–14 completely defines the static stability of an airplane (at fixed center-of-gravity position) only because the static stability of an airplane with speed at constant angle of attack and flight path angle is neutral. When the effect of propeller operation is considered, however, a single curve such as that given in Fig. 11–14 is no longer sufficient as the airplane is no longer neutrally stable with speed at constant angle of attack. Because the helicopter has positive and not neutral static stability with speed, it is therefore apparent that, like the airplane in the power-on condition, a single curve does not suffice.

STATIC STABILITY OF THE HELICOPTER. The static stability of the helicopter in forward flight depends upon the moments produced on the helicopter by a change in speed from trim during flight at a constant angle of attack, as well as moments produced by a change in angle of attack from trim at constant speed. The moment contributed by the rotor as a result of either of these changes has already been discussed in the section entitled "Rotor Characteristics."

For the actual helicopter, the fuselage and stabilizing surfaces (if any) will also contribute aerodynamic moments which vary when either the speed or angle of attack is changed. These moments are brought about in three different ways:

(1) Effect of a variation of moment coefficient with angle of attack. The conventional helicopter fuselage has an unstable variation of moment with angle of attack which adds to the rotor angle-of-attack instability. A fixed tail surface would contribute a stabilizing variation of moment with angle of attack.

(2) Effect of a constant moment coefficient during steady flight on stability with speed. The conventional helicopter fuselage has a nose-down moment coefficient during steady flight. Thus, if the speed of the helicopter is varied from trim at constant angle of attack, the resulting variation in moment arising from the change in dynamic pressure is destabilizing. If stabilizing surfaces contribute a nose-up moment in steady flight by carrying a down load, the resulting variation of moment with speed will be stabilizing.

(3) Effect on stability with angle of attack of a thrust-axis offset from the helicopter center of gravity. The thrust axis is offset from the helicopter center of gravity during steady flight in order to compensate for an aerodynamic pitching moment acting on the fuselage or stabilizing surfaces, or if the helicopter has offset flapping hinges and the center of gravity is not on the rotor shaft. The conventional helicopter fuselage has a nose-down moment in steady flight which is compensated for by the thrust vector being offset ahead of the helicopter center of gravity. This offset results in the rotor contributing an additional unstable moment variation with angle-of-attack change as can be understood by examining Fig. 11–15.

An increase in the fuselage angle of attack results in a nose-up rotor moment greater (by an amount equal to the product of the thrust increment and the initial center-of-gravity offset) than the moment produced by the rotor with no center-of-gravity offset. Thus, nose-down fuselage moments, which require the thrust axis to be offset forward of the center of gravity, add to the angle-of-attack instability of the rotor. If stabilizing surfaces contribute a nose-up moment in steady flight by carrying a down load, the resulting offset between the thrust vector and the helicopter center of gravity counteracts the rotor instability with angle of attack or, if the offset is great enough, will even make the helicopter in effect statically stable with angle of attack.

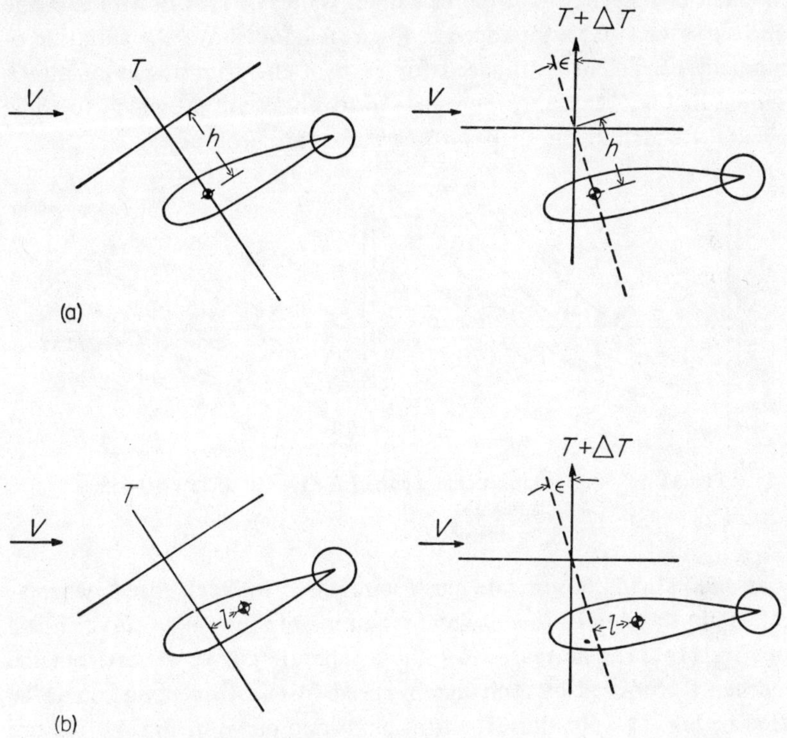

Fig. 11–15 Effect of center-of-gravity offset on pitching moment arising from fuselage angle-of-attack change.
(a) No center-of-gravity offset
(b) Center-of-gravity offset aft of thrust vector

Offset flapping hinges can make a similar contribution to rotor static stability with angle of attack. For example, if the center of gravity is forward of the rotor shaft, then, because of the offset hinges, the center of gravity will also be forward of the thrust vector. Hence, as for the case of a down load on a tail surface, the rotor instability with angle of attack is counteracted.

The two types of forward-flight static stability can be represented by the moment-coefficient curves of Figs. 11–16a and 11–16b which can be obtained from wind-tunnel tests. Figure 11–16a shows the variation of moment coefficient about the helicopter center of gravity with fuselage angle of attack at various speeds. Figure 11–16b shows the variation of moment coefficient with speed for each of the trim angles of attack shown in Fig. 11–16a. (Figures 11–16 to 11–19 are presented to show general trends but the shapes of the curves are arbitrary.)

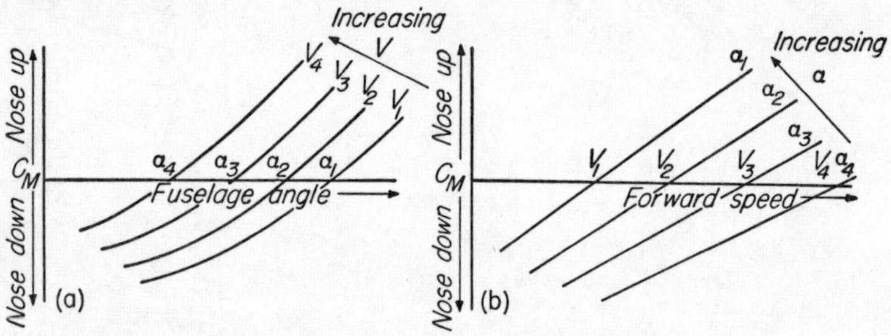

Fig. 11–16 Basic static stability curves for typical tailless helicopter.
(a) C_m against α
(b) C_m against V

In Fig. 11–16a, a separate curve is required for each speed; whereas the static stability of the airplane requires only the single curve shown in Fig. 11–13a. The reason for these separate curves arises from the moments produced by a change in speed from a trim point as can be seen in Fig. 11–16b; thus, the trim point and curve of Fig. 11–16a are shifted.

The amount of static stability or instability of the helicopter is quantitatively defined by the curves of Fig. 11–17, which represent the slopes of the curves of Fig. 11–16 at the trim conditions. Specifically, the curve of Fig. 11–17a was obtained by picking off values of airspeed and $\Delta C_m/\Delta\alpha$ at $C_m = 0$ from the curves of Fig. 11–16a. Similarly, the curve of Fig. 11–17b was obtained from the curves of Fig. 11–16b.

Although methods of obtaining curves similar to those of Fig. 11–17 from flight tests have not yet been fully explored, one technique of measuring the static stability of a helicopter in flight, which is used by the NACA, might be mentioned briefly. The method consists of measuring the variation of stick position with thrust coefficient while maintaining constant tip-speed ratio and pitch-lever position. Inasmuch as an increase in thrust coefficient obtained in this manner is accompanied by an increase in fuselage angle of attack, an aft motion of the stick with increasing thrust coefficient indicates stability with angle of attack.

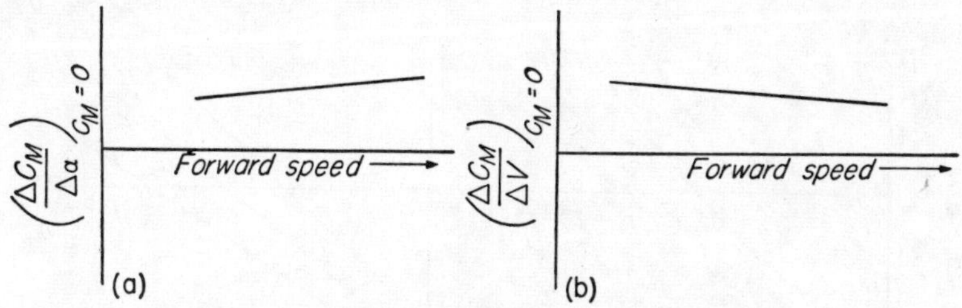

Fig. 11–17 Slopes of curves of Fig. 11–16 at $C_m = 0$.
(a) Stability with angle of attack
(b) Stability with speed

The curves of Fig. 11–17 represent a typical tailless helicopter (one with no horizontal tail surface) in power-on flight in that it is unstable with angle of attack and stable with speed. According to reference VI–6 (Appendix IIA), this instability with angle of attack is a principal stability deficiency of the conventional tailless helicopter in forward flight.

It should be emphasized that the curves of Figs. 11–16 and 11–17 represent the characteristics of a helicopter having given center-of-gravity and stick positions, gross weight, rotor speed, and collective pitch, and flying at a given altitude. The effect of variations in gross weight, rotor speed, and altitude can be accounted for by plotting the stability data in nondimensional form. One possible method of plotting is shown in Figs. 11–18 and 11–19.

In order to account for a change in stick position, the contribution of the fuselage and tail surfaces (if any) to the total value of C_m must be known. The effect of a center-of-gravity change with fixed stick position can be effectively accounted for by correcting each value of C_m in Fig. 11–18 by an amount equal to $C_L(\Delta l/R)$. For the special case of no-moment contribution by the fuselage or tail surface, a center-of-gravity change at a given flight condition results in a change in fuselage attitude which is compensated for by a change in stick position, and the stability of the aircraft is unaffected.[1] If, however, either the fuselage or a fixed tail surface do contribute moments that change with angle of attack, or if the helicopter has offset flapping hinges, a center-of-gravity change will, by tilting the fuselage, change the fuselage moments

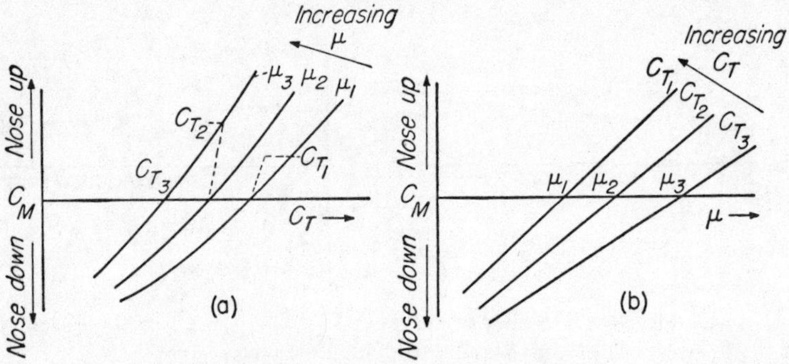

Fig. 11–18 Basic nondimensional static-stability curves for typical tailless helicopter.
(a) C_m against C_T
(b) C_m against μ

and thus change the horizontal distance between the thrust vector and the center of gravity in trimmed flight. As discussed previously, this change in center-of-gravity offset during trimmed flight does affect the stability of the helicopter.

In order to take account of variations in collective pitch, curves similar to those in Figs. 11–18 and 11–19 would be needed for several pitch values. In the practical case, it might be more advantageous to plot these curves for constant power instead of constant collective pitch.

Curves similar to those in Fig. 11–18 not only take account of variations in the trim value of rotor speed but also variations in rotor speed which will normally occur during changes in fuselage angle of attack or forward speed. This variation in rotor speed affects the static stability

[1] For a given flight condition, the attitude of the rotor plane in space is fixed. Thus, in order to maintain a given flight condition when the center of gravity is shifted and a tilt of the fuselage and rotor plane results, the control stick must be moved to a position such that the rotor plane returns to its initial attitude. (See discussion in Chapter 7.)

of the helicopter. For example, the autorotating rotor has different stability characteristics than the powered rotor. The primary reason for this difference is the fact that the rotor speed of the autorotating rotor is not controlled by the engine but is free to vary with changes in

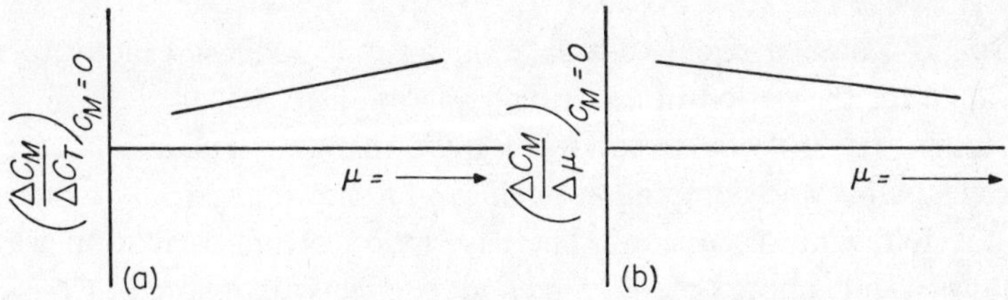

Fig. 11–19 Slopes of curves of Fig. 18 at $C_m = 0$.
(a) Stability with angle of attack.
(b) Stability with speed

forward speed or angle of attack. Reference 24 (Appendix IIB) states that the effect of these variations in rotor speed is to make the autorotating rotor neutrally stable with changes in speed at constant angle of attack and positively stable with changes in angle of attack at constant speed. Thus, the power-speed characteristics of the helicopter engine affect the stability of the helicopter.

Dynamic Stability in Forward Flight

Many of the factors that influence the dynamic stability of the helicopter in forward flight can be understood from the information already presented about the dynamic behavior of the helicopter in hovering. If the helicopter is assumed to have neutral static stability with respect to changes in angle of attack (as it has in hovering oscillations as a result of near-zero airspeeds), then the period of the longitudinal oscillation in forward flight is primarily influenced by the same quantities as the hovering oscillation; namely, static stability with speed and damping in pitch. This contention is borne out by an examination of the approximate equation in reference 24 (Appendix IIB) for the period of the longitudinal oscillation of a helicopter in forward flight. This equation, which may be written as follows, neglects, among other things, the moment of inertia of the helicopter (moment of inertia is expected to increase the period):

$$P = 2\pi \sqrt{\frac{\left(-M_\omega - \frac{WV}{g}\frac{M_a}{T_a}\right)}{M_V g}} \tag{3}$$

If M_α is assumed equal to zero, this formula reduces exactly to the formula for the period in hovering [equation (2)].

HELICOPTER MOTION FOLLOWING A DISTURBANCE. The importance of speed stability and damping in pitch can be shown physically by means of the following discussion. (The description of the oscillation which follows is only approximate, as secondary effects are ignored.) Consider a longitudinal oscillation of a helicopter having neutral stability with angle of attack. Assume the helicopter to be flying at a trimmed condition in level flight, at which time a disturbance causes it to nose down and start to descend as shown in Fig. 11–20a. The component of weight along the flight path will accelerate the helicopter and increase its speed until the helicopter reaches the position shown in Fig. 11–20b. Because of speed stability, this increased velocity produces a backward tilt of the rotor plane and a nose-up moment, which in turn causes a nose-up angular acceleration.

The angular acceleration leads to an angular velocity of such magnitude that the damping in pitch allows the fuselage to overtake the rotor thrust; thus the vector tilt due to static stability with speed is neutralized. As long as there is a component of weight along the flight path, the helicopter speed will continue to increase and the preceding steps will be repeated. The continually increasing angular velocity of the helicopter during these steps results in a continuously increasing fuselage angle of attack. In turn, the thrust will continuously increase until it levels off the glide path and the helicopter reaches the position shown in Fig. 11–20c.

In this position, the helicopter has approximately maximum forward speed, maximum nose-up angular velocity, and maximum fuselage angle of attack. Inasmuch as the thrust at this point is greater than the helicopter weight (because the angle of attack is greater than the trim value), the helicopter will start to climb. The component of weight along the flight path now opposes the forward motion and the helicopter

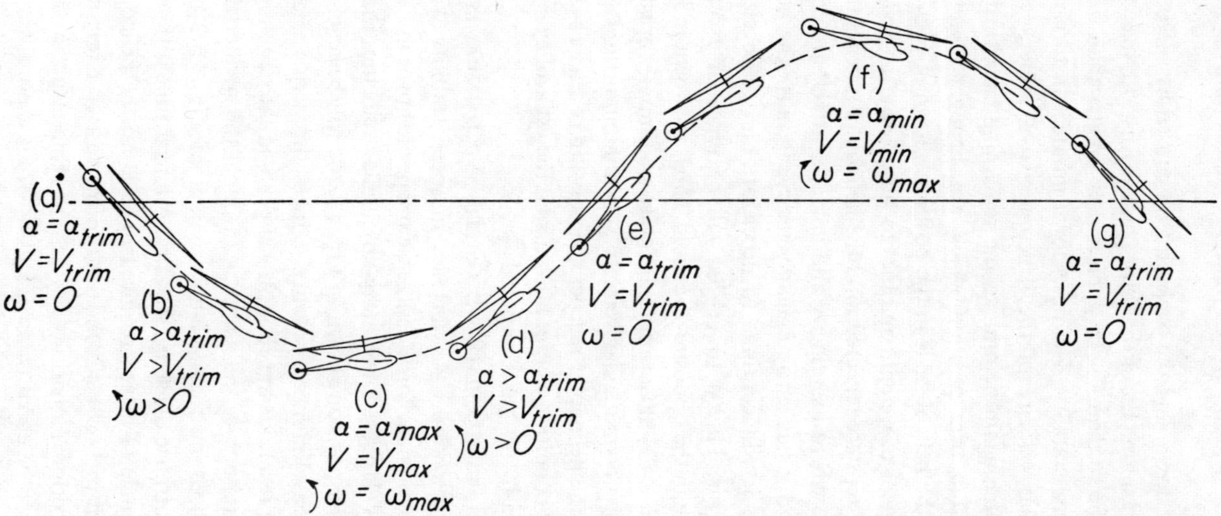

Fig. 11–20 Longitudinal oscillation of typical helicopter in forward flight.

begins to slow down, and the backward tilt caused by speed stability is reduced. The resulting tilt of the rotor plane is forward, inasmuch as the forward tilt due to damping in pitch is now greater than the rearward tilt due to speed stability. The nose-down moment in turn reduces the nose-up angular velocity of the helicopter to a value such that the damping in pitch again neutralizes the remaining backward tilt of the rotor plane from trim position which was brought about by speed stability, and the helicopter is in the position shown in Fig. 11–20d.

The component of weight continues to slow down the helicopter and the preceding steps are repeated until the helicopter reaches the position of Fig. 11–20e where its velocity and angle of attack are equal to the trim values and it has zero angular velocity. Because the helicopter is now climbing, it will continue to decelerate and the cycle of events depicted by Figs. 11–20a to 11–20e will be repeated except that all changes will be in the opposite direction. Thus, as shown in Fig. 11–20f, the helicopter will have approximately minimum forward speed, maximum nose-down angular velocity and minimum fuselage angle of attack. When the helicopter reaches the position of Fig. 11–20g, it is in the same flight condition as Fig. 11–20a, and the cycle of events depicted in Figs. 11–20a to 11–20e is repeated.

EFFECT OF STATIC STABILITY WITH SPEED AND DAMPING IN PITCH ON PERIOD OF OSCILLATION. An increase in speed stability will cause a larger nose-up moment for the increase in speed shown in Fig. 11–20b. This moment will cause larger nose-up angular velocities than hitherto attained and the position shown in Fig. 11–20c will be reached sooner. Thus, an increase in speed stability reduces the period. Equation (3) gives the same result inasmuch as M_V appears in the denominator. The larger the magnitude of the damping in pitch, the smaller the angular velocity produced by the nose-up moment of Fig. 11–20b which is required to neutralize the speed stability. A longer time is thus necessary to reach the angle of attack required to level off the helicopter in the position shown in Fig. 11–20c. Thus, an increase in damping in pitch increases the period. Equation (3) gives the same result in that M_α has been assumed to be equal to zero, and $-M_\omega$ is a positive quantity in the numerator.

EFFECT OF ANGLE-OF-ATTACK STATIC STABILITY ON PERIOD OF OSCILLATION. Equation (3) shows that the effect of static stability with angle of attack M_α is to add to, or subtract from, the effect of damping in pitch M_ω. If a helicopter is statically unstable with angle of attack, M_α is positive and inasmuch as T_α is positive, the term $(WV/g)\,(M_\alpha/T_\alpha)$ is positive. Thus, the magnitude of the numerator and, consequently, the period, is reduced. Inasmuch as moments due to changes in angle of attack and angular velocity vary during the oscillation, they must be approximately in phase in order that they may be added algebraically. Figures 11–20c and 11–20f show that α and ω reach peak values simultaneously.

Physically, the effect of angle-of-attack stability M_α on the damping in pitch and thus on the period can be seen from a study of Fig. 11–20c. When the helicopter is in this position, its nose-up angular velocity, which is a maximum, produces a maximum nose-down moment due to damping in pitch. At the same time, the angle of attack, which is also a maximum, results in a maximum nose-up moment in that the helicopter was assumed to be statically unstable with angle of attack. Thus, the effect of static instability with angle of attack is to reduce the effect of damping in pitch and, consequently, the period of the oscillation. It follows that, if a stabilizing device such as a tail surface is installed on a helicopter to make it statically stable with angle of attack, the period of the oscillation will be increased.

INFLUENCE OF WV/g AND T_α ON PERIOD OF OSCILLATION. As previously discussed, M_α, if stable, adds to, or if unstable, subtracts from the effect of M_ω. The relative contributions of these two quantities depend upon the relative magnitudes of the angle-of-attack change and the pitching velocity. The effects of WV/g and T_α are present because they determine the magnitude of the angle-of-attack change for a given pitching velocity. These two terms affect the maximum change in angle of attack for a given maximum pitching velocity as follows: at any point in the oscillation, the thrust force will differ from the weight of the helicopter by an amount equal to the centrifugal force produced by the curved flight path. A change in angle of attack is necessary to produce this change in thrust. If the change in thrust with angle of attack T_α is increased, a given increase in thrust can be obtained by a smaller change in angle of attack. Thus, the larger the value of T_α, the smaller the effect of M_α. This conclusion is substantiated by equation (3) inasmuch as M_α is divided by T_α. The magnitude of the centrifugal force acting on the helicopter per unit of pitching velocity depends upon WV/g. Therefore, the larger the value of this quantity, the greater the required change in thrust, the greater the change in angle of attack during the oscillation, and the greater the effect of M_α. Equation (3) gives the same result, inasmuch as M_α is multiplied by WV/g.

EFFECT OF STABILITY PARAMETERS ON DIVERGENCE OF OSCILLATION. An example of the influence of the stability parameters that were previously discussed on helicopter handling qualities is their effect on the rate of divergence of an oscillation in forward flight. In practice, the rate of divergence may have an important effect on handling qualities, particularly if the divergence is so great that only a fraction of one cycle can be tolerated. (See Appendix IIA, reference VI–6.) According to an approximate analysis (reference 24 of Appendix IIB), a helicopter that is statically unstable with angle of attack will also be dynamically unstable, but a large amount of damping in pitch or a sacrifice in speed stability will reduce the influence of a given amount of static instability. Thus, it appears desirable to incorporate in the helicopter some means of producing stability with angle of attack or a large amount of damping in pitch. The approximate theory of reference 24 also indicates that the effect of fuselage moment of inertia is to increase the dynamic instability; that is, the moment of inertia of the fuselage causes the oscillation to diverge more rapidly.

Summary Remarks

The discussion of single-rotor helicopters with conventional control systems that has been given in this chapter may be summarized by the following paragraphs:

(1) Rotor control is obtained by tilting the thrust vector with respect to the center of gravity. If offset hinges are present, this tilt produces an additional moment due to mass forces in the blades. Control power, which is the control moment obtained per unit stick displacement, is

affected by the height of the rotor above the center of gravity and the amount of offset of the flapping hinges.

(2) Rotor damping arises from a tilt of the thrust vector with respect to the center of gravity caused by blade inertia during pitching or rolling of the helicopter. The amount of damping moment for a given angular velocity is increased by an increase in rotor height, flapping-hinge offset, and blade moment of inertia, and by a reduction in rotor speed. It can also be increased by special devices such as the rate-gyro component of an autopilot, the Bell gyro-bar, and the Hiller control rotor, all of which apply control proportional to the pitching or rolling velocity.

(3) Control sensitivity, which is an important handling characteristic, is defined as the maximum rate of roll per unit stick deflection. It depends upon the ratio of control power to rotor damping. Control sensitivity is not affected by rotor height or use of offset flapping hinges inasmuch as those factors vary control power and rotor damping in proportion. Excessive control sensitivity can be reduced by varying those factors that increase rotor damping without increasing control power.

(4) For the helicopter, two aspects of static stability must be considered; namely, stability with speed and stability with angle of attack. Unlike the airplane, the helicopter has a moment variation with speed at constant angle of attack. This moment variation is stable in that an increase in speed results in a nose-up moment. The helicopter rotor is unstable with angle of attack at constant speed in that a nose-up change in angle of attack results in a nose-up moment. The instability with nose-up changes is greater than that with nose-down changes. Also, the instability with large nose-up changes is greater than the instability with small nose-up changes. In addition, most fuselages are also unstable and thus most conventional helicopters are unstable with angle of attack, which is a major flying-quality deficiency. The presence of instability with angle of attack will not show up in a plot of stick position against speed. Thus, measurements of stick position against angle of attack at constant speed must also be made for a complete evaluation of the static stability of the helicopter.

(5) The static stability of the helicopter in forward flight is unaffected by a center-of-gravity shift if no moments are contributed by components

other than the rotor. If there are other moment contributions, as for example, from a fixed-tail surface, the static stability is affected.

(6) When disturbed from a hovering condition, the resulting motion of a helicopter is an oscillation, the period of which depends primarily upon two factors; namely, moments due to changes in speed (stability with speed) and moments due to the angular velocity of the helicopter (damping in pitch or roll).

(7) If neutral angle-of-attack stability is assumed and if fuselage inertia effects are neglected, then the motion of a helicopter following a disturbance in forward flight is an oscillation, the period of which depends, as in the hovering condition, mainly upon stability with speed and damping in pitch. The presence of static instability of the helicopter with angle of attack causes the oscillation to decrease in period.

(8) According to an approximate theory, dynamic instability in forward flight can be reduced by the addition of positive static stability with angle of attack, by increasing the damping in pitch, or by a sacrifice in speed stability.

Means are being sought for the improvement of helicopter handling qualities by the use of devices which alter the magnitude of one or more of the pertinent stability factors. For example, several devices already in use either increase the damping in pitch or add positive static stability with angle of attack.

12

AN INTRODUCTION TO HELICOPTER VIBRATION PROBLEMS

The present chapter is intended to give the reader a physical understanding of helicopter vibration problems. The aim is to show what forces the rotor and helicopter as a whole may encounter and the nature of the response of the rotor and helicopter to these forces. No quantitative analyses are included.

Kinds of Vibrations

Vibrations may be divided into two types, ordinary and self-excited. Ordinary vibrations are those in which a system of springs, dampers, and masses is forced to vibrate by some alternating external force. The amplitude of the vibration per unit applied force may vary widely, depending on the proportions of the system and the frequency of the applied force. Ordinary vibrations will be discussed in detail in this chapter.

Self-excited vibrations are those requiring no external alternating force for sustention. When self-excited vibrations occur, the system is in an unstable state and any small disturbance will cause an oscillation of increasing amplitude. Examples of self-excited vibrations on the helicopter include "ground resonance" in which blade lag motion is coupled with a translational oscillation of the rotor hub, and blade flutter which involves the coupling of blade flexing and twisting with air forces. Self-excited vibrations will not be further discussed in this chapter, other than the following brief mention.

Essentially, "ground resonance" is a self-excited mechanical vibration that involves a coupling between the motion of the rotor blades about their lag hinges and the motion of the helicopter as a whole on its landing gear. When the frequencies of the two motions approach each other a violent shaking of the aircraft occurs which, if undamped, would result in its complete destruction. (In the past, "ground resonance" has been responsible for the destruction of several autogyros and helicopters.) This phenomenon was theoretically investigated by the NACA and other agencies and means were suggested for avoiding "ground resonance." (See Appendix IIA, references V–5, V–10, V–11, and Appendix IIB, reference 30.) In order to make the theory easy to use, it was put in the form of simple charts which predicted the range of rotor speeds in which the instability occurred and the amount of damping necessary to avoid dangerous frequencies.

Another example of a self-excited vibration problem peculiar to helicopters is sometimes encountered in the operation of two-bladed rotors. The phenomenon has been called blade *weaving* because of the appearance of the wavy path traced by the blade tips and is an aerodynamic instability or type of flutter. This problem was investigated theoretically in reference V–6 (Appendix IIA). The general result of the study was that a see-saw rotor with a coning angle is more unstable than an airplane wing with corresponding parameters. The additional destabilizing effect is associated with the difference in blade moments of inertia about the flapping axis and the axis of rotation. The difference in moments of inertia may be viewed as representing chordwise against "flapwise" (i.e., flapping plane) mass distribution. With a large built-in coning angle, the blade mass is distributed above and below the feathering axis; mass forces are such as to tend to move the blade masses farther from the centerline of rotation by lying down and therefore an unstable spring constant exists. Individually flapping blades, on the other hand, are subjected to a stable spring constant (unless bent to a severe *S* curve by the radial-load distribution), because the mass is distributed more chordwise than vertically.

With certain combinations of coning angles and blade design parameters, flutter can occur even when the chordwise center of mass of the blades is well ahead of the 25 per cent chord point. Proposed remedies for the flutter include decreasing the coning angle of the blades, designing the blades so that their mass is confined to the plane of rotation, increasing the control-system stiffness and forward position of the center of mass, and adding mechanical damping to the rotor system.

Sources of Energy for Ordinary Vibrations

The energy sources supplying alternating forces to the helicopter are threefold: air forces acting on the rotors, engine vibrations, and air forces acting on the fuselage and nonrotating parts of the machine. Of these, the first two are of most consequence. While engine vibrations are important with regard to pilot comfort and structural fatigue, their nature and means of isolation are not within the scope of the present discussion. The following paragraphs will therefore consider only vibrations caused by forces applied to the rotor.

Exciting Forces Applied to Rotor Blades

Alternating air forces acting on the rotor blade are almost entirely due to the periodic variations encountered in forward flight. Hovering is essentially a steady condition, and periodic air forces which do arise are due to secondary effects, such as the impulse felt by blades passing near booms or other structural members.

Because most rotor-force variations are periodic, repeating their cycle of events faithfully each revolution, all steady alternating inputs must be even multiples of the rotor speed. Thus, 1/rev., 2/rev., 3/rev., 4/rev., etc., inputs occur. Alternating forces of the orders mentioned above all occur in forward flight due to the variations in velocity and angle of attack encountered by the rotating blade. For example, the velocity of a blade element at any azimuth angle ψ is closely given as

$$U_T = \Omega r + V \sin \psi$$

Inasmuch as the applied forces vary with the square of the velocity, terms involving $\sin^2 \psi$ occur. From trigonometry, $\sin^2 \psi$ may be expressed as $1/2 - 1/2 \cos 2\psi$ so that velocity variations alone involve first and second harmonic force variations. When the simultaneous angle-of-attack changes are taken into account, still higher periodic forces are found.

In general, the amplitude of the applied air force in a given condition decreases as the order increases. Second harmonic variations are quite important, especially in the low-speed range where the induced flow over the disk varies considerably from front to rear. Third and fourth harmonic inputs are also important.

Periodic air forces are applied in both the thrust and torque (i.e., flapping and inplane) directions.

Response of the Nonflexible Hinged Blade to Periodic Forces

It has been shown in Chapter 7 that the natural frequency of the rotor blade is approximately 1 per revolution in the flapping direction and 1/3 to 1/4 per revolution in the lagging direction for conventional rotors. Therefore the air force inputs at 2/, 3/, and 4/rev. are considerably above the natural frequency of the blade. These forces accordingly produce amplitudes which depend primarily on the inertia of the blade, the motions of the blade being nearly 180° out of phase with the applied force. Higher harmonic motions of the blade as a whole, while small, are important insofar as they cause forces to be transmitted to the rest of the helicopter.

Periodic Blade Flexing

Rotor blades are not rigid but flex in both the flapping and in-plane directions in several modes. In fact, the normal rotor blade behaves more nearly like a rope than like a rigid rod. The blade may bend in any of its natural modes, as shown in Fig. 12–1. The frequency at which these bending modes occur depends on the mass distribution and stiffness characteristics of the blade as well as on the speed of rotation of the blade. (Because centrifugal force has a stiffening effect on the system, the natural frequency of each mode increases as rotational speed is increased.) It has been found that for blades of conventional mass and stiffness distribution, the first natural mode of flapwise bending occurs at a frequency between two and three times the rotor speed.

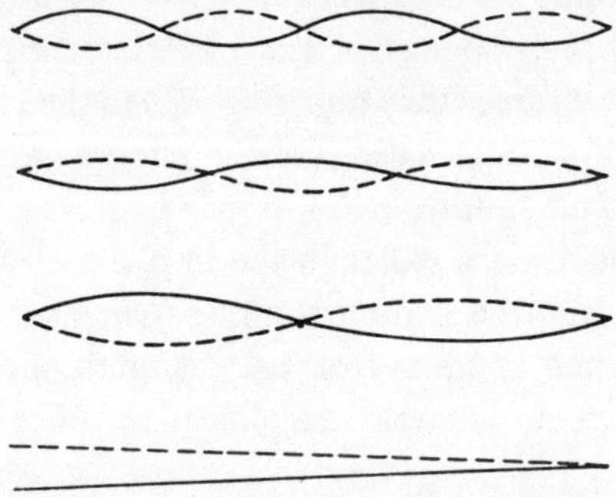

Fig. 12–1 Natural blade bending modes.

In forward flight, the spanwise distributions of lift and torque vary periodically, the load shifting toward the tip on the retreating blade. Typical load distributions are shown in Fig. 12–2. These variations in load excite the blade flexing modes and the blade responds with an

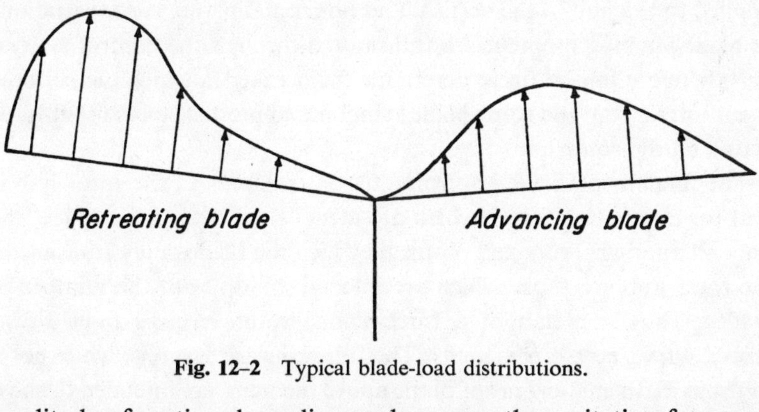

Fig. 12–2 Typical blade-load distributions.

amplitude of motion depending on how near the excitation frequency is to the blade natural frequency. When excitations occur at the natural frequency, these forces are only opposed by damping forces, because by definition the spring and mass forces in the blade are in equilibrium. Thus the forces, when applied near resonance, produce amplitudes much greater than would be expected if the forces were applied statically to the blade. This dynamic amplification is important in conventional rotor blades because the exciting forces at 2/rev. and 3/rev. are near enough to the natural bending frequency to cause large amplifications. The 2/rev. and 3/rev. flexings of the blade are important, not only because of the input forces they may transmit to the rotor hub, but also because of the high alternating blade stresses and accompanying possibilities of fatigue failure.

Conventional blades are stiffer in the in-plane direction than in the flapping direction and the natural bending frequencies are correspondingly higher. The first bending frequency is often about four times the rotational speed. Considerable amplification of 4/rev.-inputs may, therefore, occur.

Forces Transmitted to the Rotor Hub

The preceding paragraphs have considered the rotor blades as the vibrating system. It is also of interest to regard the rotor as the source of alternating forces applied to the helicopter structure. Flapping or in-plane motions, or blade flexing in either plane cause periodic forces to be applied to the hub in the vertical and horizontal plane. Also, variations in blade-pitching moment are transmitted through the control system. It is therefore important to determine the manner in which the periodic input forces from the rotor blades combine to produce a resultant force acting on the rotor hub.

The fundamental rule governing the force inputs to the rotor hub is that for alternating forces which are identical on each rotor blade, the only alternating forces and moments which the blades may transmit to the rotor hub are those which are integral multiples of the number of blades. Thus, the hub of a three-bladed rotor receives only 3/rev., 6/rev., 9/rev., etc., force inputs. The following paragraphs, while not a rigorous mathematical proof of the above theorem, are intended to show physically why it is so.

VERTICAL FORCE INPUTS. Vertical force inputs are simplest to understand. Suppose that each blade of a two-bladed rotor transmits a 2/rev. vertical force at the flapping hinge given as $F = F_o \cos 2\,\psi$ (Fig. 12–3). Each blade then exerts its maximum upward force on the hub at

$\psi = 0$ and $\psi = 180°$. Because the two blades are 180° apart, they exert their upward force at exactly the same instant, and thus the inputs add to make an alternating force on the hub of amplitude $2F_o$. A 2/rev. force on a two-bladed rotor is therefore directly transmitted. Consider, however, the resultant hub force due to a 1/rev. force acting at the blade

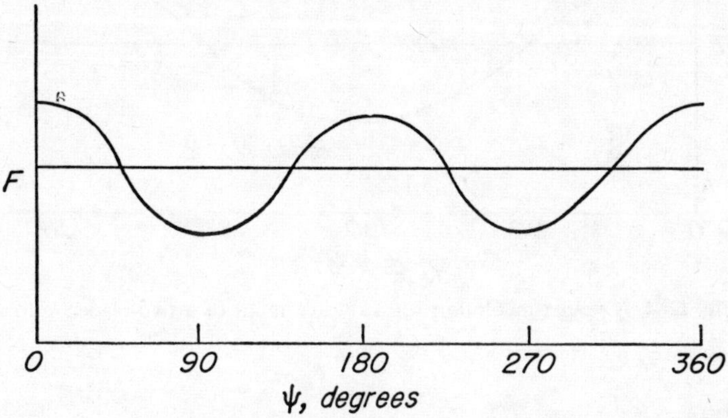

Fig. 12–3 Two-per-revolution vertical-force input of a two-bladed rotor.

hinge, $F = F_o \cos \psi$ (Fig. 12–4). Then when the blade at $\psi = 0°$ has its maximum up load, the blade at $\psi = 180°$ has its maximum down load and the net vertical force on the hub is 0. Similarly, the blade forces cancel each other at all other azimuth positions. By similar reasoning it will be found that for a two-bladed rotor, blade inputs at 1/rev., 3/rev., 5/rev., etc. cancel while integral multiples of the number of blades, 2/rev., 4/rev., 6/rev., etc., add. The same rules apply to a rotor with any number of blades, the integral multiples being additive and all other frequencies canceling out.

TORSIONAL INPUTS. Alternating forces applied by the blades to the hub in the in-plane direction may produce alternating torsional moments about the rotor shaft. Rules governing these moments are exactly the same as for vertical inputs. Only harmonics which are integral multiples of the number of blades can produce shaft moments.

IN-PLANE FORCES. Alternating forces in the longitudinal or lateral direction may be transmitted to the shaft by alternating forces acting about the lag hinge in the in-plane direction. Consider the general case where a tangential force at the lag hinge is given by $F_o \sin n\psi$, where n is any integer. (Thus, a first, second, third, etc., harmonic variation is

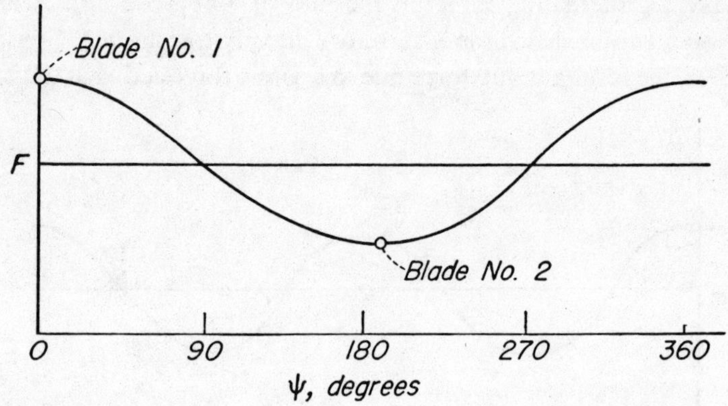

Fig. 12–4 One-per-revolution vertical-force input of a two-bladed rotor.

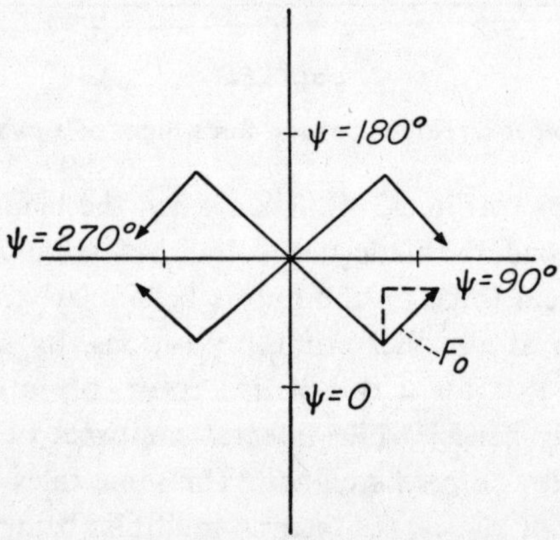

Fig. 12–5 Amplitude of a two-per-revolution tangential lag-hinge force.

considered.) At any instant, the component of F acting in, say, the longitudinal direction is

$$F_{\text{long. comp.}} = F_0 \sin n\psi \sin \psi$$

Reference to trigonometric tables will show that the product of two sine (or two cosine or sine $\times$ cosine) terms is the sum of sine or cosine terms of frequency $n + 1$ and $n - 1$. Thus, a tangential force transmitted by the rotating blade to the hub at a frequency of n per revolution produces longitudinal or lateral forces at the hub of frequencies $n + 1$ and $n - 1$. As was found for vertical inputs, only frequencies which are integral multiples of the number of blades are additive and result in a force at the hub; all other frequencies cancel out. Thus, a three-bladed

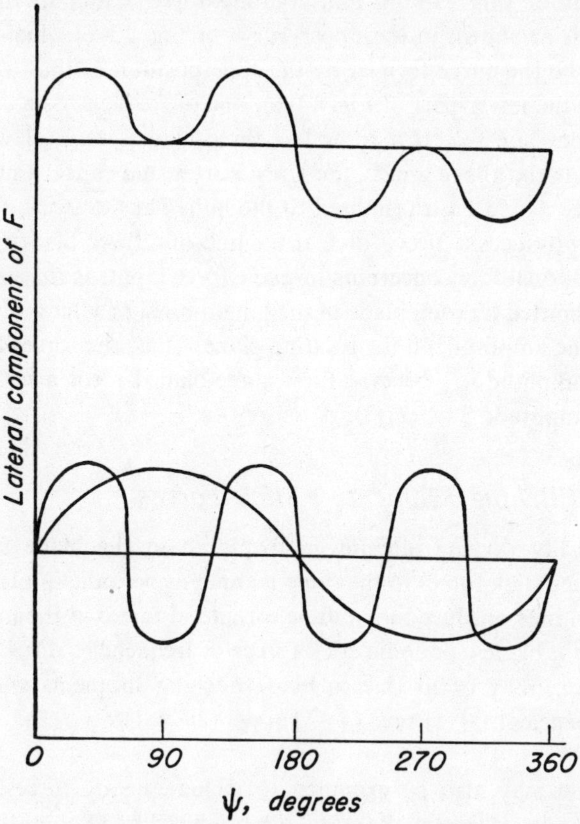

Fig. 12–6 Lateral component of force of Fig. 12–5.

rotor can experience only 3/rev., 6/rev., 9/rev., etc., inputs, but note that in the case of in-plane forces these longitudinal or lateral inputs at, say, 3/rev. are due to in-plane blade forces at 2/rev. or 4/rev. frequency.

An example may help to clarify the above discussion. Consider a tangential force acting at the lag hinge of 2/rev. frequency. Suppose, for example, that the force is given by $F_o \sin 2\psi$. This force is shown in

Fig. 12–5 at different azimuth positions. The force has a maximum positive value F_o at $\psi = 45°$ and $225°$ (positive direction is taken in the direction of rotation); it has zero amplitude at $\psi = 0$, 90, 180, and 270°, and it has a maximum negative amplitude at $\psi = 135°$ and 315°. The component of this varying and rotating force acting in the lateral direction is as shown in the upper curve of Fig. 12–6, which is easily shown to be the curve formed by the superposition of the two curves, (shown in the lower part of Fig. 12–6), one of frequency $\sin \psi$ and one of frequency $\sin 3\psi$. If a rotor has three blades, each of which are contributing the above inputs, the 1/rev. part would cancel and only the 3/rev. force would be transmitted to the hub. Thus, a 2/rev. force at a lag hinge produces a 3/rev. force at the hub of a three-bladed rotor.

An additional fact concerning in-plane force inputs is that the amplitude transmitted by each blade in the longitudinal or lateral direction is one-half the amplitude in the rotating plane. Thus, the forces at the lag hinge of amplitude F_o become for a three-bladed rotor a longitudinal force of amplitude $3 \times (F_o/2)$.

Blade Pitching Moments—Stick Forces

Periodically varying pitching moments about the blade feathering axis produce stick forces in the same manner as periodic in-plane forces at the lag hinge produce longitudinal or lateral forces at the hub. For a rotor with n blades, periodic stick forces of frequencies n/rev., $2n$/rev., $3n$/rev., etc., may occur due to blade pitching moments which vary with frequencies $(n + 1)$/rev., $(n - 1)$/rev., $(2n + 1)$/rev., $(2n - 1)$/rev., etc.

This rule may also be extended to include steady forces, because $0 \times n$/rev. (i.e., steady stick forces) are produced by $O \times n + 1$/rev. or 1/rev. blade-pitching moments. Thus for any number of blades 1/rev. pitching moments produce steady stick forces.

The above statements assume, of course, that the control system is completely reversible; that is, that a pitching motion of the blade will cause a movement of the stick. In a conventional rotor-control system, oscillating stick forces are due entirely to blade-pitching moments (having both mass and aerodynamic origins). Steady forces may be due to 1/rev. blade pitching moments or to some external source such as a spring on the control stick.

As an example of forces due to blade pitching moments, consider a three-bladed rotor in which each blade is subjected to periodic moments of 1/rev., 2/rev., 3/rev., 4/rev., and 5/rev. According to the rules stated above, and assuming a reversible control system, stick forces would be produced as follows:

ON THREE-BLADED ROTOR

Blade Pitching Moment	*Stick Forces Produced*
1/rev.	Steady—direction depending on phase of input
2/rev.	3/rev. shake
3/rev.	None in control stick—swashplate encounters 3/rev. vertical force (felt in pitch control)
4/rev.	3/rev. shake
5/rev.	6/rev. shake

The amplitude of the moment transmitted per blade, as was found for in-plane forces, is always one-half the amplitude of the moment in the rotating system.

The rules discussed in the preceding paragraphs are summarized in Tables 12–1 and 12–2 for rotors of two, three, and four blades.

Response of the Helicopter to Forces Applied at the Rotor Hub

The amplitudes of motion which are caused by a given force amplitude applied at the rotor hub depend on the spring and mass characteristics of the helicopter. In general, the amplitude of motion of a solid mass equal to the mass of the helicopter would be small under the applied forces. For example, a 3000-pound mass responds to a 300-pound force applied at 6 cycles per second (approximately 3/rev.) with an amplitude of about 0.01 inch. If, however, the several masses and "springs" which make up the structure have a natural frequency near the exciting frequency, considerable amplification may occur and uncomfortable vibration may result. In designing the structure, care must, therefore, be taken to see that natural frequencies do not occur in the operating range of rotor speeds or multiples of the speed.

TABLE 12-1

VERTICAL FORCE TRANSMITTED BY BLADES TO HUB OR SHAFT TORQUE TRANSMITTED BY BLADES TO HUB

Vertical Force at Flap Hinge Due to One Blade *or* Torque at Hub Due to One Blade	Two Blades	Three Blades	Four Blades
$F = F_o$	F_o	F_o	F_o
$F = F_o \sin \psi$	0	0	0
$F = F_o \cos \psi$	0	0	0
$F = F_o \sin 2\psi$	$2F_o \sin 2\psi$	0	0
$F = F_o \cos 2\psi$	$2F_o \cos 2\psi$	0	0
$F = F_o \sin 3\psi$	0	$3F_o \sin 3\psi$	0
$F = F_o \cos 3\psi$	0	$3F_o \cos 3\psi$	0
$F = F_o \sin 4\psi$	$2F_o \sin 4\psi$	0	$4F_o \sin 4\psi$
$F = F_o \cos 4\psi$	$2F_o \cos 4\psi$	0	$4F_o \cos 4\psi$
$F = F_o \sin 5\psi$	0	0	0
$F = F_o \cos 5\psi$	0	0	0
$F = F_o \sin 6\psi$	$2F_o \sin 6\psi$	$3F_o \sin 6\psi$	0

TABLE 12–2

FORCE TRANSMITTED BY BLADE IN-PLANE FORCES TO ROTOR HUB IN LONGITUDINAL AND LATERAL DIRECTION OR FORCE TRANSMITTED BY BLADE PITCHING MOMENTS TO PILOT'S STICK IN LONGITUDINAL AND LATERAL DIRECTIONS

In-plane Tangential Force at Lag Hinge Due to One Blade *or* Blade Pitching Moment Due to One Blade	Two Blades		Three Blades		Four Blades	
	Lateral Force at Hub or Stick	Longitudinal Force at Hub or Stick	Lateral Force at Hub or Stick	Longitudinal Force at Hub or Stick	Lateral Force at Hub or Stick	Longitudinal Force at Hub or Stick
$F = F_o$	0	0	0	0	0	0
$F = F_o \sin \psi$ $F = F_o \cos \psi$	$F_o \sin 2\psi$ $F_o + F_o \cos 2\psi$	$F_o - F_o \cos 2\psi$ $F_o \sin 2\psi$	0 $3/2\ F_o$	$3/2\ F_o$ 0	0 $2F_o$	$2F_o$ 0
$F = F_o \sin 2\psi$ $F = F_o \cos 2\psi$	0 0	0 0	$3/2\ F_o \sin 3\psi$ $3/2\ F_o \cos 3\psi$	$-3/2\ F_o \mathrm{co}\ 3\psi$ $3/2\ F_o \sin 3\psi$	0 0	0 0
$F = F_o \sin 3\psi$	$F_o \sin 2\psi$ $F_o \sin 4\psi$	$F_o \cos 2\psi$ $-F_o \cos 4\psi$	0	0	$2F_o \sin 4\psi$	$-2F_o \cos 4\psi$
$F = F_o \mathrm{co}\ 3\psi$	$F_o \cos 2\psi$ $F_o \cos 4\psi$	$F_o \sin 2\psi$ $F_o \sin 4\psi$	0	0	$2F_o \cos 4\psi$	$2F_o \sin 4\psi$
$F = F_o \sin 4\psi$ $F = F_o \cos 4\psi$	0	0 0	$3/2\ F_o \sin 3\psi$ $3/2\ F_o \cos 3\psi$	$3/2\ F_o \cos 3\psi$ $3/2\ F_o \sin 3\psi$	0 0	0 0
$F = F_o \sin 5\psi$ $F = F_o \cos 5\psi$	0 0	0 0	$3/2\ F_o \sin 6\psi$ $3/2\ F_o \cos 6\psi$	$-3/2\ F_o \cos 6\psi$ $3/2\ F_o \sin 6\psi$	$2F_o \sin 4\psi$ $2F_o \cos 4\psi$	$2F_o \cos 4\psi$ $2F_o \sin 4\psi$

Sign Convention: Forces—(1) Blade in-plane force at lag hinge is positive when tending to accelerate rotor hub. (2) Blade pitch-up moment is positive. (3) Longitudinal: forward direction is positive; lateral: toward advancing side is positive.

APPENDIX I

SYMBOLS[1]

The following list of symbols, which is subdivided for ready reference under appropriate headings, contains nomenclature that is used throughout this book. The symbols given, which are the ones most generally used in studies of helicopter aerodynamics, are those which have been adopted as standard by the NACA Subcommittee on Helicopters. The Subcommittee contains representatives of the NACA, the Civil Aeronautics Administration, the armed services, the helicopter industry, and educational institutions.

Physical Quantities

The symbols representing the physical characteristics of the helicopter and of the medium in which it operates are as follows:

W gross weight of helicopter, pounds

b number of blades per rotor

R blade radius, feet

r radial distance to blade element, feet

x ratio of blade-element radius to rotor-blade radius (r/R)

c blade-section chord, feet

c_e equivalent blade chord (on thrust basis), feet $\dfrac{\int_0^R cr^2\,dr}{\int_0^R r^2\,dr}$

σ rotor solidity $(bc_e/\pi R)$

θ blade-section pitch angle; angle between line of zero lift of blade

[1] The list of symbols presented herein was taken from reference VIII–2 (Appendix IIA).

section and plane perpendicular to axis of no feathering, radians. (See section "Air Flow Relative to Rotor" for definition of axis of no feathering.)

θ_0 blade-pitch angle at hub, radians

θ_1 difference between hub and tip pitch angles; positive when tip angle is larger, radians

e distance from drag hinge (vertical pin) to axis of rotation, feet

m mass of blade per foot of radius, slugs per foot

I_1 mass moment of inertia of blade about flapping hinge, slug-feet^2

γ mass constant of rotor blades; expresses ratio of air forces to mass forces ($c_e \rho a R^4 / I_1$). (See section "Blade-Element Aerodynamic Characteristics" for definition of a.)

ρ mass density of air, slugs per cubic foot

Air-Flow Parameters

AIR FLOW RELATIVE TO ROTOR. Before the symbols associated with air flow relative to the rotor are listed, it is advisable to point out that the axis that is used as a reference for the system is the "control axis," or the "axis of no feathering." This axis has been defined in Chapter 7 as the axis about which there is no first harmonic feathering or cyclic-pitch variation. The plane perpendicular to this axis has been termed the "rotor disk" in many papers on rotating-wing aircraft.

The symbols for the velocities, velocity parameters, and angles that are used in defining the air flow relative to the rotor follow:

V true airspeed of helicopter along flight path, feet per second

V_h horizontal component of true airspeed of helicopter, feet per second

V_v vertical component of true airspeed of helicopter, feet per second

Ω rotor angular velocity, radians per second

α rotor angle of attack; angle between axis of no feathering and plane perpendicular to flight path, positive when axis is pointing rearward, radians

v induced inflow velocity at rotor (always positive), feet per second

μ tip-speed ratio ($V \cos \alpha / \Omega R$)

λ inflow ratio $[(V \sin \alpha - v)/\Omega R]$

V' resultant velocity at rotor; vector sum of translational and induced velocities, feet per second

ψ blade azimuth angle measured from downwind position in direction of rotation, radians

AIR FLOW RELATIVE TO BLADE ELEMENT. The symbols for the velocities and angles defining the air flow relative to the rotor-blade elements are listed as follows:

U_T component at blade element of resultant velocity perpendicular to blade-span axis and to axis of no feathering, feet per second

U_P component at blade element of resultant velocity perpendicular both to blade-span axis and U_T, feet per second

U_R component at blade element of resultant velocity parallel to blade-span axis and perpendicular to U_T, feet per second

U resultant velocity perpendicular to blade-span axis at blade element, feet per second

ϕ inflow angle at blade element in plane perpendicular to blade-span axis, radians $\left(\tan^{-1} \frac{U_P}{U_T}\right)$

α_r blade-element angle of attack, measured from line of zero lift, radians $(\theta + \phi)$

$\alpha_{(x)(\psi)}$ blade-element angle of attack at any radial position and at any blade azimuth angle, degrees; for example $\alpha_{(1.0)(270°)}$ is blade-element angle of attack at tip of retreating blade at 270 degrees azimuth position

Aerodynamic Characteristics

BLADE-ELEMENT AERODYNAMIC CHARACTERISTICS. The symbols for the two-dimensional aerodynamic characteristics of the airfoil sections comprising the rotor-blade elements are listed as follows:

c_l section lift coefficient

c_{d_0} section profile-drag coefficient

$\delta_0, \delta_1, \delta_2$ coefficients in power series expressing c_{d_o} as a function of α_r ($c_{d_o} = \delta_0 + \delta_1\alpha_r + \delta_2\alpha_r^2$)

a slope of curve of section lift coefficient against section angle of attack (radian measure)

ROTOR AERODYNAMIC CHARACTERISTICS. The symbols for the quantities that define the aerodynamic characteristics of the rotor are listed as follows:

L lift, pounds
D drag, pounds
T rotor thrust, pounds
Y lateral force, pounds
Q rotor-shaft torque, pound-feet
P rotor-shaft power, pound-feet per second
L' rolling moment, pound-feet
M pitching moment, pound-feet
N yawing moment, pound-feet

C_L lift coefficient $\left(\frac{L}{\frac{1}{2}\rho V^2\pi R^2}\right)$

C_D drag coefficient $\left(\frac{D}{\frac{1}{2}\rho V^2\pi R^2}\right)$

C_T thrust coefficient $\left(\frac{T}{\pi R^2\rho(\Omega R)^2}\right)$

C_Y lateral-force coefficient $\left(\frac{Y}{\frac{1}{2}\rho V^2\pi R^2}\right)$

C_Q rotor-shaft torque coefficient $\left(\frac{Q}{\pi R^2\rho(\Omega R)^2 R}\right)$

C_P rotor-shaft power coefficient $\left(\frac{P}{\pi R^2\rho(\Omega R)^3}\right)$

C_l *rolling-moment coefficient* $\left(\frac{L'}{\frac{1}{2}\rho V^2\pi R^2 R}\right)$

C_m pitching-moment coefficient $\left(\frac{M}{\frac{1}{2}\rho V^2\pi R^2 R}\right)$

C_n yawing-moment coefficient $\left(\frac{N}{\frac{1}{2}\rho V^2 \pi R^2 R}\right)$

B tip-loss factor; blade elements outboard of radius BR are assumed to have profile drag but no lift

M rotor figure of merit $\left(0.707 \frac{C_T^{3/2}}{C_Q}\right)$

Rotor-Blade Motion

FLAPPING MOTION. Blade flapping motion may be described as the variation with azimuth angle of the blade flapping angle, the flapping angle being defined as the angle between the blade-span axis and the plane perpendicular to the axis of no feathering. This motion may be expressed as a function of the azimuth angle by the Fourier series:

$$\beta = a_0 - a_1 \cos \psi - b_1 \sin \psi - a_2 \cos 2\psi - b_2 \sin 2\psi - \ldots$$

where β blade flapping angle at particular azimuth position

a_0 constant term in Fourier series that expresses β (radians); hence, the rotor coning angle

a_n coefficient of $\cos n\psi$ in expression for β

b_n coefficient of $\sin n\psi$ in expression for β

FEATHERING MOTION. Feathering motion may be described as the variation with azimuth angle of the blade pitch angle at a representative radius, usually taken at 0.75 radius. This motion may be expressed as a function of the azimuth angle by the Fourier series:

$$\theta = A_0 - A_1 \cos \psi - B_1 \sin \psi - A_2 \cos 2\psi - B_2 \sin 2\psi - \ldots$$

where θ blade pitch angle at particular azimuth position

A_0 constant term in Fourier series that expresses θ (radians); hence, the mean blade pitch angle at the representative radius

A_n coefficient of $\cos n\psi$ in expression for θ

B_n coefficient of $\sin n\psi$ in expression for θ

IN-PLANE MOTION. The blade drag angle is defined as the angle between the blade-span axis and the line drawn through the rotor center of rotation and the drag hinge, the angle being positive in the direction of rotation. Changes in the blade drag angle are termed the *in-plane motion* of the rotor blades. This motion may be expressed as a function of the azimuth angle by the Fourier series:

$$\zeta = E_0 + E_1 \cos \psi + F_1 \sin \psi + E_2 \cos 2\psi + F_2 \sin 2\psi + \ldots$$

where ζ blade drag angle at particular azimuth position

E_0 constant term in Fourier series that expresses ζ (radians); hence, the mean blade drag angle

E_n coefficient of $\cos n\psi$ in expression for ζ

F_n coefficient of $\sin n\psi$ in expression for ζ

Performance

Parameters useful for expressing helicopter performance are as follows:

$(D/L)_o$ rotor profile drag-lift ratio

$(D/L)_i$ rotor induced drag-lift ratio

$(D/L)_p$ parasite drag of helicopter components other than lifting rotors divided by rotor lift

$(D/L)_c$ drag-lift ratio representing angle of climb, positive in climb $\left(\tan^{-1} \frac{V_v}{V_h}\right)$

$(D/L)_r$ rotor drag-lift ratio; ratio of equivalent drag of rotor to rotor lift $[(D/L)_o + (D/L)_i]$

$(D/L)_u$ component of rotor resultant force along flight path (that is, useful component of rotor resultant force) divided by rotor lift $[(D/L)_p + (D/L)_c]$

P/L shaft power parameter, where P is equal to rotor-shaft power divided by velocity along flight path and is therefore also equal to drag force that could be overcome by the shaft power at flight velocity $[(D/L)_r + (D/L)_u]$

f equivalent-flat-plate area representing parasite drag, based on unit drag coefficient, square feet (helicopter parasite drag/$\frac{1}{2}\rho V^2$)

APPENDIX IIA

BIBLIOGRAPHY OF NACA PAPERS ON ROTATING WING AIRCRAFT[1]

A key to the abbreviations for NACA papers is given as follows:

Rep	*Technical Report*
TN	*Technical Note*
TM	*Technical Memorandum*
MR	*Memorandum Report*
ARR	*Advance Restricted Report*
ACR	*Advance Confidential Report*
RM	*Research Memorandum*
RB	*Restricted Bulletin*
CB	*Confidential Bulletin*

Papers that are of limited availability are indicated in the following Bibliography by use of one, two, or three asterisks with the following significance:

- * Copies may be obtained on loan by writing to the NACA or photostats may be purchased directly from the Library of Congress.
- ** Copies available for reference in the Washington Office library of the NACA or photostats may be purchased directly from the Library of Congress.
- *** Copies available for reference in the Washington Office library of the NACA.

[1] This list was brought up to date from the papers listed in reference VIII-1, Appendix IIA).

All other papers may be obtained by writing directly to the NACA.

I. Aerodynamics of Hovering and Vertical Flight

1. Nikolsky, A. A. and Seckel, Edward, "An Analysis of the Transition of a Helicopter from Hovering to Steady Autorotational Vertical Descent," *NACA TN 1907*, 1949.
2. Nikolsky, A. A. and Seckel, Edward, "An Analytical Study of the Steady Vertical Descent in Autorotation of Single-Rotor Helicopters," *NACA TN 1906*, 1949.
3. Gessow, Alfred, "Flight Investigation of Effects of Rotor Blade Twist on Helicopter Performance in the High-Speed and Vertical-Autorotative-Descent Conditions," *NACA TN 1666*, 1948.
4. Gustafson, F. B. and Gessow, Alfred, "Analysis of Flight Performance Measurements on a Twisted, Plywood-Covered Helicopter Rotor in Various Flight Conditions," *NACA TN 1595*, 1948.
5. Gessow, Alfred, "Effect of Rotor Blade Twist and Plan-Form Taper on Helicopter Hovering Performance," *NACA TN 1542*, 1947.
6. Lipson, Stanley, "Static Thrust Investigation of Full-Scale PV–2 Helicopter Rotors Having NACA 0012.6 and 23012.6 Airfoil Sections," *NACA MR L6D24*, 1946.
7. Gustafson, F. B. and Gessow, Alfred, "Effect of Rotor-Tip Speed on Helicopter Hovering Performance and Maximum Forward Speed," *NACA ARR L6A16*, 1946.
8. Dingeldein, Richard C. and Schaefer, Raymond F., "Static-Thrust Tests of Six Rotor-Blade Designs on a Helicopter in the Langley Full-Scale Tunnel," *NACA ARR L:F25b*, 1945.

***9. Gustafson, F. B., "A Summary of the Effects of Blade Twist on Helicopter Performance—TED No. NACA 1301," *NACA MR L5H24*, Army Air Forces and Bur. Aero., 1945.

10. Gustafson, F. B. and Gessow, Alfred, "Flight Tests on the Sikorsky HNS–1 (Army YR–4B) Helicopter. II—Hovering and Vertical-Flight Performance with the Original and an Alternate Set of Main-Rotor Blades, Including a Comparison with Hovering Performance Theory," *NACA MR L5D09a*, 1945.
11. Bailey, F. J., Jr. and Voglewede, T. J., "An Estimate of the Effect of Engine Supercharging on the Take-Off Thrust of a Typical Helicopter at Different Altitudes and Temperatures," *NACA MR L5C12b*, 1945.
12. Gustafson, F. B., "Effect on Helicopter Performance of Modifications in

Profile-Drag Characteristics of Rotor-Blade Airfoil Sections," *NACA ACR L4H05*, 1944.

13. Knight, Montgomery and Hefner, Ralph A., "Analysis of Ground Effect on the Lifting Airscrew," *NACA TN 835*, 1941.

*14. Betz, A., "The Ground Effect on Lifting Propellers," *NACA TM 836*, 1937.

*15. Knight, Montgomery and Hefner, Ralph A., "Static Thrust Analysis of The Lifting Airscrew," *NACA TN 626*, 1937.

16. Wheatley, John B. and Bioletti, Carlton, "Analysis and Model Tests of Autogyro Jump Take-Off," *NACA TN 582*, 1936.

*17. Wheatley, John B., "An Aerodynamic Analysis of the Autogyro Rotor with a Comparison between Calculated and Experimental Results," *NACA Rep. 487*, 1934.

**18. Toussaint, A., "Drag or Negative Traction of Geared-Down Supporting Propellers in the Downward Vertical Glide of a Helicopter," *NACA TN 21*, 1920.

II. Aerodynamics of Forward Flight

1. Gessow, Alfred, "An Analysis of the Autorotative Performance of a Helicopter Powered by Rotor-Tip Jet Units," *NACA TN 2154*, 1950.

2. Carpenter, Paul J., "The Effect of Wind Velocity on Performance of Helicopter Rotors as Investigated with the Langley Helicopter Apparatus," *NACA TN 1698*, 1948.

3. Gessow, Alfred, "Flight Investigation of Effects of Rotor Blade Twist on Helicopter Performance in High-Speed and Vertical-Autorotative-Descent Conditions," *NACA TN 1666*, 1948.

4. Gustafson, F. B. and Gessow, Alfred, "Analysis of Flight-Performance Measurements on a Twisted, Plywood-Covered Helicopter Rotor in Various Flight Conditions," *NACA TN 1595*, 1948.

5. Dingeldein, Richard C. and Schaefer, Raymond F., "Full-Scale Investigation of the Aerodynamic Characteristics of a Typical Single-Rotor Helicopter in Forward Flight," *NACA TN 1289*, 1947.

6. Gessow, Alfred and Myers, Garry C., Jr., "Flight Tests of a Helicopter in Autorotation, Including a Comparison with Theory," *NACA TN 1267*, 1947.

7. Gustafson, F. B. and Gessow, Alfred, "Effect of Blade Stalling on the Efficiency of a Helicopter Rotor as Measured in Flight," *NACA TN 1250*, 1947.

8. Talkin, Herbert W., "Charts Showing Relations among Primary Aerodynamic Variables for Helicopter-Performance Estimation," *NACA TN 1192*, 1947.

9. Gustafson, F. B. and Myers, G. C., Jr., "Stalling of Helicopter Blades," *NACA TN 1083*, 1946.

10. Gustafson, F. B. and Gessow, Alfred, "Effect of Rotor-Tip Speed on Helicopter Hovering Performance and Maximum Forward Speed," *NACA ARR L6A16*, 1946.

***11. Gustafson, F. B., "A Summary of the Effects of Blade Twist on Helicopter Performance—TED No. NACA 1301," *NACA MR L5H24*, Army Air Forces and Bur. Aero., 1945.

12. Talkin, Herbert W., "Charts for Helicopter-Performance Estimation," *NACA ACR L5E04*, 1945.

13. Migotsky, Eugene, "Full-Scale-Tunnel Performance Tests of the PV–2 Helicopter Rotor," *NACA MR L5C29a*, 1945.

14. Coleman, Robert P., Feingold, Arnold M., and Stempin, Carl W., "Evaluation of the Induced-Velocity Field of an Idealized Helicopter Rotor," *NACA ARR L5E10*, 1945.

15. Gustafson, F. B., "Flight Tests of the Sikorsky HNS–1 (Army YR–4B) Helicopter. I—Experimental Data on Level-Flight Performance with Original Rotor Blades," *NACA MR L5C10*, 1945.

16. Bailey, F. J., Jr. and Gustafson, F. B., "Charts for Estimation of the Characteristics of a Helicopter Rotor in Forward Flight. I—Profile Drag-Lift Ratio for Untwisted Rectangular Blades," *NACA ACR L4H07*, 1944.

17. Gustafson, F. B., "Effect on Helicopter Performance of Modifications in Profile-Drag Characteristics of Rotor-Blade Airfoil Sections," *NACA ACR L4H05*, 1944.

18. Bailey, F. J., Jr., "A Simplified Theoretical Method of Determining the Characteristics of a Lifting Rotor in Forward Flight," *NACA Rep. 716*, 1941.

19. Sissingh, G., "Contribution to the Aerodynamics of Rotating-Wing Aircraft," Part II, *NACA TM 990*, 1941.

*20. Pflüger, A., "Aerodynamics of Rotating-Wing Aircraft with Blade Pitch Control," *NACA TM 929*, 1940.

*21. Sissingh, G., "Contribution to the Aerodynamics of Rotating-Wing Aircraft," *NACA TM 921*, 1939.

22. Hohenemser, K., "Performance of Rotating-Wing Aircraft," *NACA TM 871*, 1938.

23. Bailey, F. J., Jr., "A Study of the Torque Equilibrium of an Autogiro Rotor," *NACA Rep. 623*, 1938.

*24. Breguet, Louis, "The Gyroplane-Its Principles and Its Possibilities," *NACA TM 816*, 1937.

***25. Wheatley, John B. and Clay, William C., "Full-Scale Wind-Tunnel and Flight Tests of a YG–1 Autogiro," *NACA MR*, Army Air Corps, August 26, 1937.

26. Wheatley, John B., "An Analytical and Experimental Study of the Effect of Periodic Blade Twist on the Thrust, Torque, and Flapping Motion of an Autogiro Rotor," *NACA Rep. 591*, 1937.

27. Wheatley, John B. and Bioletti, C., "Wind-Tunnel Tests of 10-Foot-Diameter Autogiro Rotors," *NACA Rep. 552*, 1936.

***28. Wheatley, John B., "Control-Force and General Performance Tests of Kellett YG–1 Autogiro," *NACA MR*, Army Air Corps, March 27, 1936.

29. Wheatley, John B. and Bioletti, Carlton, "Wind-Tunnel Tests of a 10-Foot-Diameter Gyroplane Rotor," *NACA Rep. 536*, 1935.

30. Wheatley, John B and Windler, Ray, "Wind-Tunnel Tests of a Cyclogiro Rotor," *NACA TN 528*, 1935.

31. Wheatley, John B., "The Influence of Wing Setting on the Wing Load and Rotor Speed of a PCA–2 Autogiro as Determined in Flight," *NACA Rep. 523*, 1935.

**32. Schrenk, Martin, "Aerodynamic Principles of the Direct Lifting Propeller," *NACA TM 733*, 1934.

*33. Wheatley, John B., "The Aerodynamic Analysis of the Gyroplane Rotating-Wing System," *NACA TN 492*, 1934.

34. Wheatley, John B. and Hood, Manley, J., "Full-Scale Wind-Tunnel Tests of a PCA–2 Autogiro Rotor," *NACA Rep. 515*, 1935.

*35. Wheatley, John B., "An Aerodynamic Analysis of the Autogiro Rotor with a Comparison between Calculated and Experimental Results," *NACA Rep. 487*, 1934.

*36. Strandgren, C. B., "The Theory of the Strandgren Cyclogiro," *NACA TM 727*, 1933.

*37. Wheatley, John B., "Simplified Aerodynamic Analysis of the Cyclogiro Rotating-Wing System," *NACA TN 467*, 1933.

*38. Wheatley, John B., "Wing Pressure Distributions and Rotor-Blade Motion of an Autogiro as Determined in Flight," *NACA Rep. 475*, 1933.

39. Wheatley, John B., "Lift and Drag Characteristics and Gliding Performance of an Autogiro as Determined in Flight," *NACA Rep. 434*, 1932.

**40. Munk, Max M., "Model Tests on the Economy and Effectiveness of Helicopter Propellers," *NACA TN 221*, 1925.

III. Rotor Blade Motion and Stall Studies

1. Gessow, Alfred, "Flight Investigation of Effects of Rotor Blade Twist on Helicopter Performance in the High-Speed and Vertical-Autorotative-Descent Conditions," *NACA TN 1666*, 1948.

2. Gustafson, F. B. and Gessow, Alfred, "Analysis of Flight-Performance Measurements on a Twisted, Plywood-Covered Helicopter Rotor in Various Flight Conditions," *NACA TN 1595*, 1948.

3. Migotsky, Eugene, "Full-Scale Investigation of the Blade Motion of the PV–2 Helicopter Rotor," *NACA TN 1521*, 1948.

4. Myers, Garry C., Jr., "Flight Measurements of Helicopter Blade Motion with a Comparison between Theoretical and Experimental Results," *NACA TN 1266*, 1947.

5. Gustafson, F. B. and Gessow, Alfred, "Effect of Blade Stalling on the Efficiency of a Helicopter Rotor as Measured in Flight," *NACA TN 1250*, 1947.

6. Gustafson, F. B. and Myers, G. C., Jr., "Stalling of Helicopter Blades," *NACA TN 1083*, 1946.

7. Dingeldein, Richard C. and Schaefer, Raymond F., "High-Speed Photographs of a YR–4B Production Rotor Blade for Simulated Flight Conditions in the Langley Full-Scale Tunnel," *NACA MR L5C12c*, 1945.

***8. Gustafson, F. B., "Observations in Flight of the Region of Stalled Flow over the Blades of the Kellett YG–1B Tapered-Blade Autogiro Rotor," *NACA MR*, Army Air Corps, September 17, 1940.

***9. Bailey, F. J., Jr. and Boothby, W. B., "Photographic Observations of Blade Motion of the YG–1B Autogiro Equipped with Tapered Rotor Blades," *NACA MR*, Army Air Corps, May 9, 1940.

10. Bailey, F. J., Jr. and Gustafson, F. B., "Observations in Flight of the Region of Stalled Flow Over the Blades of an Autogiro Rotor," *NACA TN 741*, 1939.

***11. Wheatley, John B. and Bailey, F. J., "Flight Tests of Blade Motions and Center-of-Pressure Location on Kellett YG–1 Autogiro," *NACA MR*, July 7, 1937.

12. Wheatley, John B., "An Analysis of the Factors that Determine the Periodic Twist of an Autogiro Rotor Blade, with a Comparison of Predicted and Measured Results," *NACA Rep. 600*, 1937.

13. Wheatley, John B., "An Analytical and Experimental Study of the Effect of Periodic Blade Twist on the Thrust, Torque and Flapping Motion of an Autogiro Rotor," *NACA Rep. 591*, 1937.

14. Wheatley, John B., "A Study of Autogiro Rotor-Blade Oscillations in the Plane of the Rotor Disk," *NACA TN 581*, 1936.

***15. Wheatley, John B., "Blade Motion and Bouncing Tests of KD–1 Autogiro," *NACA MR*, June 13, 1935.

IV. Airfoils

1. Schaeffer, Raymond F. and Smith, Hamilton A., "Aerodynamic Characteristics of the NACA 8–*H*–12 Airfoil Section at Six Reynolds Numbers from 1.8×10^6 to 11.0×10^6," *NACA TN 1998*, 1949.

2. Schaeffer, Raymond F., Loftin, Laurence K., Jr., and Horton, Elmer A., "Two-Dimensional Investigation of Five Related NACA Airfoil Sections Designed for Rotating-Wing Aircraft," *NACA TN 1922*, 1949.

3. Gustafson, F. B., "The Application of Airfoil Studies to Helicopter Rotor Design," *NACA TN 1812*, 1949.

4. Stivers, Louis S., Jr. and Rice, Fred J., Jr., "Aerodynamic Characteristics of Four NACA Airfoil Sections Designed for Helicopter Rotor Blades," *NACA RB L5K02*, 1946.

5. Kemp, W. B., "Wind-Tunnel Tests of a Portion of a PV–2 Helicopter Rotor Blade," *NACA MR L5C29b*, Bur. Aero., 1945.

6. Tetervin, Neal, "Airfoil Section Data from Tests of 10 Practical Construction Sections of Helicopter Rotor Blades Submitted by the Sikorsky Aircraft Division, United Aircraft Corporation," *NACA MR*, September 6, 1944.

7. Tetervin, Neal, "Tests in the NACA Two-Dimensional Low-Turbulence Tunnel of Airfoil Sections Designed to Have Small Pitching Moments and High Lift-Drag Ratios," *NACA CB 3113*, 1943.

V. Vibration

1. Morduchow, Morris, "A Theoretical Analysis of Elastic Vibrations of Fixed-Ended and Hinged Helicopter Blades in Hovering and Vertical Flight," *NACA TN 1999*, 1950.

2. Morduchow, Morris, "On Internal Damping of Rotating Beams," *NACA TN 1996*, 1949.

3. Carpenter, Paul J. and Peitzer, Herbert E., "Response of a Helicopter Rotor to Oscillatory Pitch and Throttle Movements," *NACA TN 1888*, 1949.

4. Reissner, H. and Morduchow, M., "A Theoretical Study of the Dynamic Properties of Helicopter Blade Systems," *NACA TN 1430*, 1948.
5. Coleman, Robert P. and Feingold, Arnold M., "Theory of Ground Vibrations of a Two-Blade Helicopter Rotor on Anisotropic Flexible Supports," *NACA TN 1184*, 1947.
6. Coleman, Robert P. and Stempin, Carl W., "A Preliminary Theoretical Study of Aerodynamic Instability of a Two-Blade Helicopter Rotor," *NACA RM L6H23*, 1946.
7. Coleman, Robert P., "A Preliminary Theoretical Study of Helicopter-Blade Flutter Involving Dependence upon Coning Angle and Pitch Setting," *NACA MR L6G12*, 1946.
8. Kelley, Bartram, "Response of Helicopter Rotors to Periodic Forces," *NACA ARR 5A09*, 1945.
9. Coleman, Robert P., Feingold, A. M., and Stempin, C. W., "Evaluation of the Induced-Velocity Field of an Idealized Helicopter Rotor," *NACA ARR L5E10*, 1945.
10. Feingold, Arnold, M., "Theory of Mechanical Oscillations of Rotors with Two Hinged Blades," *NACA ARR 3113*, 1943.
11. Coleman, Robert P., "Theory of Self-Excited Mechanical Oscillations of Hinged Rotor Blades," *NACA ARR 3G29*, 1943.
***12. Bailey, F. J., Jr. and Gustafson, F. B., "Flight Measurements of Stick Vibration of the YG–1B Autogiro Equipped with Tapered Rotor Blades," *NACA MR*, Army Air Corps, February 28, 1940.
13. Bailey, F. J., Jr., "Flight Investigation of Control-Stick Vibration of the YG–1B Autogiro," *NACA TN 764*, 1940.

VI. Stability

1. Amer, Kenneth B., "Theory of Helicopter Damping in Pitch or Roll and a Comparison with Flight Measurements," *NACA TN 2136*, 1950.
2. Carpenter, Paul J. and Paulnock, Russell S., "Hovering and Low-Speed Performance and Control Characteristics of an Aerodynamic Servo-controlled Helicopter Rotor System as Determined on the Langley Helicopter Tower," *NACA TN 2086*, 1950.
3. Gustafson, F. B., Amer, Kenneth B., Haig, C. R., and Reeder, John P., "Longitudinal Flying Qualities of Several Single-Rotor Helicopters in Forward Flight," *NACA TN 1983*, 1949.
4. Gessow, Alfred and Amer, Kenneth B., "An Introduction to the Physical Aspects of Helicopter Stability," *NACA TN 1982*, 1949.
5. Reeder, John P. and Haig, Chester, "Some Tests of the Longitudinal

Stability and Control of an H–13B Helicopter in Forward Flight," *NACA RM L9E25a*, 1949.

6. Reeder, John P. and Gustafson, F. B., "Notes on the Flying Qualities of Helicopters," *NACA TN 1799*, 1949.

7. Gustafson, F. B. and Reeder, J. P., "Helicopter Stability," *NACA RM L7K04*, 1948.

8. Myers, Garry C., Jr., "Flight Measurements of Helicopter Blade Motion with a Comparison between Theoretical and Experimental Results," *NACA TN 1266*, 1947.

***9. Bailey, F. J., Jr., "A Flight Determination of the Moments of the YG–1B Tapered Blade Rotor about the Hub Trunnions," *NACA MR*, Army Air Corps, November 15, 1939.

10. Hohenemser, K., "Dynamic Stability of a Helicopter with Hinged Rotor Blades," *NACA TM 907*, 1939.

*11. Schrenk, M., "Static Longitudinal Stability and Longitudinal Control of Autogiro Rotors," *NACA TM 879*, 1938.

***12. Wheatley, John B., "Control-Force and General Performance Tests of Kellett YG–1 Autogiro," *NACA MR*, Army Air Corps, March 27, 1936.

***13. Wheatley, John B., "Control-Force Measurements in Flight of the Pitcairn YG–2 Autogiro," *NACA MR*, Army Air Corps, February 26, 1936.

**14. Crocco, G. Arturo, "Inherent Stability of Helicopters," *NACA TM 234*, 1923.

**15. Bateman, H., "Stability of the Parachute and Helicopter," *NACA Rep. 80*, 1920.

VII. Landings and Loads

1. de Guillenschmidt, P., "Calculation of the Bending Stresses in Helicopter Rotor Blades," *NACA TM 1312*, 1951.

2. Duberg, John E. and Luecker, Arthur R., "Comparisons of Methods of Computing Bending Moments in Helicopter Rotor Blades in the Plane of Flapping," *NACA ARR L5E23*, 1945.

***3. Gilruth, R. R., "Results of Landing Tests of Kellett YG–1 Autogiro," (*ACR 35–278*) *NACA MR*, Army Air Corps and Bur. Air Commerce, November 15, 1937.

4. Peck, William C., "Landing Characteristics of an Autogiro," *NACA TN 508*, 1934.

VIII. General

1. Gessow, Alfred, "Bibliography of NACA Papers on Rotating-Wing Aircraft, July 1948." *NACA RM L7J30*, 1948.

2. Gessow, Alfred, "Standard Symbols for Helicopters," *NACA TN 1604*, 1948.

*3. Focke, H., "The Focke Helicopter," *NACA TM 858*, 1938.

*4. Küssner, Hans Georg, "Helicopter Problems," *NACA TM 827*, 1937.

**5. Seiferth, R., "Testing a Windmill Airplane ('Autogiro')," *NACA TM 394*, 1927.

**6. Klemin, Alexander, "An Introduction to the Helicopter," *NACA TM 340*, 1925.

**7. Moreno-Caracciolo, M., "The 'Autogiro'," *NACA TM 218*, 1923.

**8. Oemichen, Étienne, "My Experiments with Helicopters," *NACA TM 199*, 1923.

**9. Balaban, Karl, "Evolution of the Helicopter," *NACA TM 196*, 1923.

**10. Warner, Edward P., "The Prospects of the Helicopter," *NACA TM 107*, 1922.

**11. Von Kármán, Theodor, "Recent European Developments in Helicopters," *NACA TN 47*, 1921. (Prepared by NACA Paris Office.)

**12. Lamé, M., "Abstracts from the French Technical Press. Study of the Resistance Offered by Propellers Rotating in an Airstream," *NACA TM 21*, 1921.

**13. Anon., "The Oemichen Peugeot Helicopter," *NACA TM 13*, 1921. (Translated by NACA Paris Office.)

**14. Warner, Edward P., "The Problem of the Helicopter," *NACA TN 4*, 1920.

APPENDIX IIB

REPRESENTATIVE REFERENCE MATERIAL FROM SOURCES OTHER THAN NACA

The rotating-wing aircraft papers represented by the following list of titles is a sample of the large amount of material available for additional reference.

A key to abbreviations used in this appendix is as follows:

Br. ARC R & M	*British Aeronautical Research Committee Reports and Memoranda*
Jour. Aero. Sci.	*Journal of the Aeronautical Sciences*
Br. RAE (*TN* or *Rep.*)	*British Royal Aircraft Establishment* (*Technical Note* or *Report*)
AFEE	*Airborne Forces Experimental Establishment* (British)
Royal Soc. Proc.	*British Royal Society Proceedings* (British)
Proc. Third Annual Forum, Amer. Helic. Soc.	*Proceedings Third Annual Forum, American Helicopter Society*
Aero. Eng. Rev.	*Aeronautical Engineering Review*
Jour. Royal Aero. Soc.	*Journal Royal Aeronautical Society* (British)

1. Klemin, Alexander, "Principles of Rotating-Wing Aircraft," *Journal of the Franklin Institute*, Vol. 227, Nos. 2 and 3, 1939.
2. Glauert, H., "A General Theory of the Autogiro," *Br. ARC R & M No. 1111*, 1926.
3. Glauert, H., "Lift and Torque of an Autogiro on the Ground," *Br. ARC R & M No. 1131*, 1927.
4. Lock, C. N. H., "Further Development of Autogiro Theory," *Br. ARC R & M No. 1127*, 1927.

5. Glauert, H. and Lock, C. N. H., "A Summary of the Experimental and Theoretical Investigations of the Characteristics of an Autogiro," *Br. ARC R & M No. 1162*, 1928.

6. Lock, C. N. H. and Townend, H., "Wind Tunnel Experiments on a Model Autogiro at Small Angles of Incidence," *Br. ARC R & M No. 1154*, 1928.

7. Beaven, J. A. and Lock, C. N. H., "The Effect of Blade Twist on the Characteristics of the C.30 Autogiro," *Br. ARC R & M No. 1727*, 1936.

8. Hufton, P. A., Woodward Nutt, A. E., Bigg, F. J., and Beavan, J. A., "General Investigation into the Characteristics of a C.30 Autogiro," *Br. ARC R & M No. 1859*, 1939.

9. Lock, C. N. H., Bateman, H., and Townend, H., "An Extension of the Vortex Theory of Airscrews with Applications to Airscrews of Small Pitch and Including Experimental Results," *Br. ARC R & M No. 1014*, 1925.

10. Glauert, H., "On the Vertical Ascent of a Helicopter," *Br. ARC R & M No. 1132*, 1927.

11. Lock, C. N. H., "The Application of Goldstein's Theory to the Practical Design of Airscrews," *Br. ARC R & M No. 1377*, 1931.

12. Kaman, C. H., "Aerodynamic Considerations of Rotors in Hovering and Vertical Climb Conditions," *Jour. Aero. Sci.*, Vol. 10, No. 7, July 1943, pp. 201–208.

13. Brotherhood, P., "An Investigation in Flight of the Induced Velocity Distribution Under a Helicopter Rotor When Hovering," *Br. ARC RAE Rep. No. Aero. 2212*, June 1947.

14. Zbrozek, J., "Ground Effect on the Lifting Rotor," *Br. ARC RAE TN No. Aero. 1903*, July 1947.

15. Fitzwilliams, O. L. L. and Mather, H. A., "The Vertical Landing of a Helicopter When the Kinetic Energy of the Rotors is Used as a Temporary Source of Power," *AFEE No. Res. 23*, 1947.

16. Squire, H. B., "The Flight of a Helicopter," *Br. ARC R & M No. 1730*, 1935.

17. Goldstein, S., "On the Vortex Theory of Screw Propellers," *Royal Soc. Proc. (A) 123, 440*. 1929.

18. Glauert, H., "Airplane Propellers," Vol. IV, Div. L of "Aerodynamic Theory," W. F. Durand, ed., Julius Springer, Berlin, 1935.

19. Stewart, William, "Flight Testing of Helicopters," *Jour. Royal Aero. Soc.*, Vol. 52, No. 449, May 1948.

20. Gessow, Alfred, "A Comparison of Rotor Theory with Experimental

Data Obtained from NACA Flight Tests of a Helicopter in Various Flight Conditions," *Proc. Third Annual Forum, Amer. Helic. Soc.*, Phila., Pa., 1947.

21. Zbrozek, J. K., "Stability and Control of Single Rotor Helicopter with Hinged Blades," *Aircraft Engineering*, Vol. 21, No. 240, February 1949.

22. Gessow, Alfred and Amer, Kenneth B., "An Explanation of Some Important Stability Parameters That Influence Helicopter Flying Qualities," *Aero. Eng. Rev.*, August 1950, Vol. 9, No. 8, pp. 28–35.

23. Sissingh, G., "Contributions to the Dynamic Stability of Rotary-Wing Aircraft with Articulated Blades," *Translation No. F–TS–690–RE*, Air Materiel Command, August 1946.

24. Hohenemser, K., "Longitudinal Stability of the Helicopter in Forward Flight," *Translation No. F–TS–688–RE*, Air Materiel Command, August 2, 1946.

25. Miller, R. H., "Helicopter Control and Stability in Hovering Flight," *Jour. Aero. Sci.*, Vol. 15, No. 8, August 1948, pp. 453–472.

26. Stuart, J., III, "The Helicopter Control Rotor," *Aero. Eng. Rev.*, August, 1948, Vol. 7, No. 8, pp. 33–37.

27. Prewitt, R. H. and Wagner, R. A., "Frequency and Vibration Problems of Rotors," *Jour. Aero. Sci.*, Vol. 7, No. 10, August 1940, pp. 444–450.

28. Hohenemser, K., "Stability in Hovering of the Helicopter with Central Rotor Location," *Translation No. F–TS–687–RE*, Air Materiel Command, August 1, 1946.

29. Seibel, C., "Periodic Aerodynamic Forces on Rotors in Forward Flight," *Jour. Aero. Sci.*, Vol. 11, No. 10, October 1944, pp. 339–342.

30. Deutsch, M. L., "Ground Vibrations of Helicopters," *Jour. Aero. Sci.*, Vol. 13, No. 5, May 1946, pp. 223–228.

31. Prewitt, R. H., "The Design of Rotor Blades," *Jour. Aero. Sci.*, Vol. 9, No. 7, May 1942, pp. 255–260.

32. Johnson, W. C. and Mayne, R., "Stress Analysis of Rotor Blades by a Tabular Dynamic Method," *Goodyear Aircraft Corporation Report R–107–4*, 1946.

33. Flax, Alexander H., "The Bending of Rotor Blades," *Jour. Aero. Sci.*, Vol. 14, No. 1, January 1947, pp. 42–50.

34. Yuan, Shao Wen, "Bending of Rotor Blades in the Plane of Rotation," *Jour. Aero. Sci.*, Vol. 14, No. 5, May 1947, pp. 285–293.

35. Owen, J. B. B., "The Stressing of Gyroplane Blades in Steady Flight," *R & M No. 1875*, 1939.

36. Owen, J. B. B., "The Stressing of Gyroplane Blades in Accelerated Flight," *R & M No. 1878*, 1939.

37. Miller, R. H., "Jet Propulsion Applied to Helicopter Rotors," *Jour. Aero. Sci.*, Vol. 13, No. 12, December 1946, pp. 639–645.

38. Kurkjian, H., "Test Instrumentation Problems of the Helicopter," *Proc. Third Annual Forum*, Amer. Helic. Soc., Phila., Pa., 1947.

39. Francis, Devon, "The Story of the Helicopter," Coward-McCann, Inc., New York, 1946.

40. Hafner, Raoul, "Rotor Systems and Control Problems in the Helicopter," Paper read before Royal Aero. Soc., September 1947.

INDEX